Terracotta Reader

The Editors :

Parth J Shah is founder president of Centre for Civil Society, New Delhi, a think tank for public policy solutions within the framework of rule of law, limited government, and competitive markets.

Parth's research and advocacy work centres on the themes of economic freedom (law, liberty, and livelihood campaign), choice and competition in education (fund students, not schools), property right approach for the environment (terracotta vision of stewardship), good governance (new public management), and rule of law (tort reforms for consumer health and safety).

The Livelihood Freedom Test, education vouchers, food stamps, community ownership of natural resources, ward-based urban management, Duty to Publish Act are some of the radical ideas he has introduced into public policy debate. He has conceptualised and organised pioneering liberal educational programmes for the Indian youth : Liberty and Society Seminar, Liberty, Art and Culture Seminar, Researching Reality Internship and Fellowship.

He has edited *The Morality of Markets, Profiles in Courage: Dissent on Indian Socialism, Friedman on India, Do Corporations have Social Responsibility?* and co-edited *Terracotta Reader, BR Shenoy: Economic Prophecies, BR Shenoy: Theoretical Vision, and Agenda for Change.* He has addressed a number of local, national and international workshops and conferences, and writes for newspapers and magazines. He is the youngest Indian member of the Mont Pelerin Society, the premier international association of classical liberals. He is in the vanguard of India's Second Freedom Movement.

Vidisha Maitra is a Research Associate at the Centre for Civil Society. Her experience at the Centre includes research on community forestry, ecosystem services of forests, decentralised water delivery systems, urban governance issues for the Centre's publication *State of Governance: Delhi Citizen Handbook 2003,* and co-coordination of the Research Internship Programme 2004. She holds a Masters degree in Economics from the Delhi School of Economics.

TERRACOTTA READER

A Market Approach to the Environment

EDITORS

PARTH J SHAH

VIDISHA MAITRA

PUBLISHED BY ACADEMIC FOUNDATION IN ASSOCIATION WITH
CENTRE FOR CIVIL SOCIETY, NEW DELHI

Centre for Civil Society
New Delhi

Academic Foundation
New Delhi

 First published in 2005 by

Academic Foundation
4772-73 / 23 Bharat Ram Road, (23 Ansari Road), Darya Ganj, New Delhi - 110 002 (India).
Phones : 23245001 / 02 / 03 / 04. Fax : +91-11-23245005. E-mail : academic@vsnl.com
www : academicfoundation.com

 Published in association with :

Centre for Civil Society, New Delhi
www.ccsindia.org

Copyright : Academic Foundation, New Delhi .

© 2005.

Terracotta Reader : A Market Approach to the Environment
edited by Parth J Shah and Vidisha Maitra.

ISBN 81-7188-426-1

Designed and typeset by Italics India, New Delhi.
Printed and bound in India.

CONTENTS

The State of Humanity

Property Rights, Markets
and Sustained Development

Forests and Wildlife

Water and Fisheries

Living with Risk: The Precautionary Principle

The New Threats: Climate Change and Biotechnology

Energy and Waste Management

Free Trade and Green Trade

Environmentalism: The New Imperialism

Epilogue

Gloom and doom environmentalism is a growth industry in India. For most of our environmentalists the sky is always falling. The print and electronic media is replete with accounts of environmental disasters that have happened or are in the making. In line with the adage that bad news sells, environmental success stories are few and far between and buried on inside pages of newspapers. For example, every year in June as the World Environment Day rolls around there is much hand wringing about the terrible state of India's environment accompanied by the usual refrain of government apathy, as well as calls for action. The irony of such statements is inescapable—on one hand while we blame the government for mismanaging the environment, on the other it is to the same state that we look for succour. But then if we were not schizoid we would not be Indians!

In the new millennium it is imperative we break out of this "gloom and doom" mindset and of depending on the government to solve our problems. Sure enough, the country faces serious environmental challenges. But at the same time, our awareness and understanding of these issues has increased. The experience of many countries rich and poor alike shows there are robust alternatives to government intervention in the field of environment. Institutions of civil society (community organisations, think-and-do tanks, markets and such like) can and do play an important role in environmental management. It is these institutions that we must seek to strengthen and not the state. Our hope for salvation lies in a vibrant and active civil society and not in a moribund and corrupt state. A new and constructive environmentalism should include this and other features.

Who better to espouse this than the Centre for Civil Society led by the indefatigable Parth Shah and his dedicated band of 'liberal warriors'? It is a pleasure to watch how over the years since it was founded on 15 August 1997 (the golden jubilee of our independence),

CCS has injected reason and logic into emotive public debates on issues ranging from education and intellectual property to public sector divestment. This anthology of readings on the environment compiled by CCS comes not a day too soon. Amidst the cacophony of gloom and doom and shrill cries for greater control by the state here are voices that call for a more reasoned approach, one that is informed by a liberal perspective—limit the role of the state and enhance the space for civil society and markets.

With respect to civil society there is a growing body of evidence that for centuries local communities have self-organised to manage natural resources such as village pastures and forests and irrigation networks. In fact, it is often the case that government agencies such as state forest departments can hinder local self-organisations. For instance, the Delhi-based environmental NGO, Centre for Science and Environment (CSE) has convincingly argued that rules imposed by governments without consulting local participants and excluding communities who have been using forest resources for generations, are bound to create conflict and are not conducive to sustainable use of these resources.

On the role of markets, as two leading environmental economists Maureen Cropper and Wallace Oates wrote in their landmark survey of the subject in the *Journal of Economic Literature* a dozen years ago "(W)hen the environmental revolution arrived in the late 1960s, the economics profession was ready and waiting. Economists had what they saw as a coherent and compelling view of the nature of pollution with a straightforward set of policy implications. ... Economists saw pollution as the consequence of an absence of prices for certain scarce environmental resources (such as clean air and water), and they prescribed the introduction of surrogate prices in the form of unit taxes or "effluent fees" to provide the needed signals to economize on the use of these resources." In the intervening years we have seen a range of countries implementing these ideas from those in Europe and North America to developing economies such as China, Malaysia, Philippines, Colombia, and a host of others. India, of course, is conspicuous by its absence from this growing list of countries that are using market-based instruments for environmental management.

These then are the ideas that permeate this collection of writings on environment. You may not agree with all of them. Sometimes neither do the contributors among themselves. The idea is not to manufacture consent but to provoke you to think 'outside the box' with a refreshingly different set of views from what you typically get in 'green' writing. And for this we must thank CCS.

Shreekant Gupta

Associate Professor of Environmental and
Natural Resource Economics,
Delhi School of Economics,
University of Delhi

Acknowledgements

As in any collection of previously published articles, this volume has benefited tremendously from the help and cooperation of numerous authors, journal editors, and publishers. We would specially like to thank David Boaz of the Cato Institute, Jane Shaw of the Property and Environment Research Centre, Rita Simon of the University of Maryland, Fred L Smith of Competitive Enterprise Institute, Richard Ebeling of the Foundation for Economic Education, and Philip Booth of the Institute of Economic Affairs for allowing us to reprint some of the articles in this collection.

We received helpful advice and inputs from Indur Goklany of the US Environmental Protection Agency, Nirmal Sengupta of the Indira Gandhi Institute for Development Research, Michael De Alessi of Reason Foundation, Julian Morris and Kendra Okonski of International Policy Network, Prof Jagdish Bhagwati of Columbia University, Sushil Saigal of Winrock India, Ashish Kothari of Kalpavriksh, and HB Soumya at the Centre for Civil Society.

Shreekant Gupta of the Delhi School of Economics provided many valuable suggestions and also penned the preface to the book. Somnath Bandyopadhyay of the Aga Khan Foundation took time off a busy work schedule to write an extremely thought provoking epilogue.

The John Templeton Foundation provided financial support and Rituraj Kapila of the Academic Foundation took over the task of producing this attractive volume.

Contributors to this Volume

John Baden is the founder and chairman of the Foundation for Research on Economics and the Environment (FREE). Dr. Baden received his Ph.D from Indiana University in 1969 and was awarded a National Science Foundation Postdoctoral Fellowship in environmental policy.

Robert C. Balling, Jr. Robert Balling is the director of the U.S Office of Climatology and an associate professor of geography at Arizona State University. He received his Ph.D in geography from the University of Oklahoma in 1979.

Somnath Bandyopadhyay is Programme Officer for rural development programmes with the Aga Khan Foundation in New Delhi. He has also served as Senior Ecologist with the Gujarat Ecology Commission. He holds a Ph.D in Environmental Sciences from the Jawaharlal Nehru University in Delhi.

Roger Bate is a visiting fellow at the American Enterprise Institute and adjunct fellow at the Competitive Enterprise Institute. He holds a Ph.D from Cambridge University and is a board member of Africa Fighting Malaria (AFM), a not-for-profit health advocacy group based in South Africa and in the United States.

Jagdish Bhagwati is University Professor at Columbia University and Senior Fellow in International Economics at the Council on Foreign Relations. He has served as Special Adviser to the UN on Globalisation and External Adviser to the Director General, WTO. Prof Bhagwati's latest book *In Defense of Globalisation* was published in 2004 to worldwide acclaim.

Gregory Conko is a policy analyst and Director of Food Safety Policy with the Competitive Enterprise Institute. He is also the Vice President and a member of the Board of Directors of the AgBioWorld Foundation. He graduated from the American University in 1992 with a B.A in Political Science and History.

Michael De Alessi is Director of Natural Resource Policy for the Reason Public Policy Institute in Los Angeles and the author of *Fishing for Solutions*. He holds an M.S in Engineering Economic Systems from Stanford University and an M.A in Marine Policy from the Rosenstiel School of Marine and Atmospheric Science at the University of Miami.

Hannes H. Gissurarson is professor of political science at the University of Iceland. He holds a Masters degree in history from the University of Iceland and a Ph.D in political science from Oxford University.

Indur M. Goklany has worked with the US federal and state governments and the private sector on global warming, biotechnology, biodiversity, and other environmental issues for more than 25 years. He has served as chief of the Technical Assessment Division of the National Commission on Air Quality and a consultant to the Environmental Protection Agency's Office of Policy, Planning, and Evaluation.

Shreekant Gupta is Associate Professor of environmental and natural resource economics at the Delhi School of Economics. He has over 15 years of experience in applied economic research and policy analysis, mainly on environmental issues such market-based instruments (MBIs), environmental valuation, climate change and poverty and environment linkages. He received his Ph.D in Economics from the University of Maryland.

Deepak Lal is James S Coleman Professor of International Development Studies, University of California at Los Angeles and Professor Emeritus of Political Economy, University College London. He has served as a consultant to the Indian Planning Commission, ILO, UNCTAD, OECD, UNIDO, the World Bank, and the ministries of planning in Korea and Sri Lanka.

Natasha Landell-Mills is a Research Associate with the International Institute for Environment and Development, London and is involved in their Instruments for Private Sector Forestry Project.

Vidisha Maitra is a Research Associate at the Centre for Civil Society, an independent public policy think-tank in Delhi. She holds a Masters degree in Economics from the Delhi School of Economics.

Ambrish Mehta is a trustee and activist for Action Research in Community Health and Development (ARCH), Mangrol, Gujarat, and the Initiative for Open Society. He has also served in the Gujarat Ecology Commission as Nodal officer for the implementation of NGO Environment Action Fund (NEAF).

Roger E. Meiners is professor of economics and law at the University of Texas at Arlington, and Senior Associate at the Property and Environment Research Centre (PERC). He received his law degree from the University of Miami in 1978 and his Ph.D in Economics from Virginia Polytechnic Institute in 1977.

Julian Morris is Director of the International Policy Network and editor of *Sustainable Development: Promoting Progress or Perpetuating Poverty?* He completed his M.Phil in Land Economics from Cambridge University in

1995 and has served as visiting professor of Economics at Buckingham University.

Kendra Okonski is Director of Sustainable Development for the International Policy Network, London. She holds a degree in Economics from Hillsdale College, USA.

Elinor Ostrom is Arthur F. Bentley Professor of Political Science Indiana University, Bloomington and the author of *Governing the Commons: The Evolution of Institutions for Collective Action*. Professor Ostrom received a Ph.D in political science from UCLA in 1965.

Trupti Parekh is a trustee and full-time activist for Action Research in Community Health and Development (ARCH), a private non-profit organisation based in Mangrol in eastern Gujarat, and the Initiative for Open Society.

Ian Powell works for Shell International and has written extensively on markets for ecosystem services, in association with Forest Trends, Washington D.C.

C S Prakash is a professor of Plant Molecular Genetics and Director of the Center for Plant Biotechnology Research at Tuskegee University, and the founder of the AgBioWorld Foundation. He obtained his Ph.D in forestry and genetics from the Australian National University, Canberra.

Nirmal Sengupta teaches at the Indira Gandhi Institute of Development Research in Mumbai. His research interests include law and economics, and management of water resources. He received his Ph.D in Economics from the Indian Statistical Institute.

Parth J. Shah is the founder and President of the Centre for Civil Society (CCS), an independent research and educational think-tank based in New Delhi. He received his Ph.D in Economics from Auburn University in the USA, and taught economics at the University of Michigan, Dearborn before starting CCS. In India his research focuses on private initiatives and reforms in the education system, and property rights approach to environmental problems and natural resource management.

James Shikwati is the founder and Executive Director of the Inter Region Economic Network (IREN Kenya), a non-profit public policy research and educational organisation that promotes market-based responses to contemporary socio-economic and environmental issues. His areas of expertise include Economics and African Development.

Randy T. Simmons is professor of political science and Chairman of the political science department at Utah State University, Senior Associate at the Property and Environment Research Centre (PERC), and a Senior Scholar with the Competitive Enterprise Institute (CEI). Simmons received his Ph.D. in political science from the University of Oregon in 1980.

The late Julian Simon was professor of business administration at the University of Maryland and a Senior Fellow at the Cato Institute. He wrote numerous books and articles, and is best known for his work on population, natural resources, and immigration.

Fred L. Smith, Jr. is the founder and President of the Competitive Enterprise Institute, Washington D.C. He is co-editor (with Michael Greve) of the book *Environmental Politics: Public Costs, Private Rewards*. Mr. Smith has a degree in mathematics and political science from Tulane University.

Robert J. Smith is the President of the Center for Private Conservation in Washington, D.C, where he focuses on wildlife, endangered species, property rights and private stewardship. He is the author of *Earth Resources: Private Ownership vs. Public Waste*.

Andy White serves as Senior Director of Policy and Market Analysis at Forest Trends, Washington D.C. He holds a B.S in Forest Science from Humboldt State University, an M.A in Anthropology, an M.S in Forestry and a Ph.D in Forest Economics from the University of Minnesota.

Bruce Yandle is professor of economics emeritus at Clemson University in South Carolina and a faculty member with George Mason University's Capitol Hill Campus. He has also served as executive director of the Federal Trade Commission in Washington D.C.

Editors' Introduction

The Terracotta vs. Green Vision:
Restructuring Incentives vs. Reforming Human Nature

> What is common to many is taken least care of, for all men have greater regard for what is their own than what they possess in common with others.
>
> — *Aristotle*

'The tragedy of the commons' is a modern statement of the insight captured by Aristotle. Any arrangement where benefits accrue to one or a few but the costs are borne by many or all is a recipe for disaster. That is exactly what has happened with our environmental resources of *jal* (water), *sthal* (land), *van* (forests), and *pavan* (air). Fundamentally there are two ways to deal with this tragedy: change attitudes or change incentives.

The green movement aims at changing the attitude of humanity to preserve the environment. This book builds a terracotta movement that focuses on changing incentives to manage the environment. Terracotta means 'burnt earth,' and refers to earthenware made from this material. It is the creation of human action *on* a natural resource: terracotta products are *consumption* items, and a means of livelihood for those who *produce* them. So terracotta symbolises the philosophy that values natural resources not for their mere existence, but recognises the relationship between human beings and environment around them. The greens consider only the biosphere, the green part of the planet, and overlook the terra that supports life on the planet. They worship ecology without humans; we cherish all life, including human life.[1] Theirs is a heroic mission to change human nature, ours is a human endeavour to create a better world by restructuring incentives.

When industrialists pollute, poachers kill endangered species, fishermen grab the smallest fish, forest officers allow illegal felling, when you shower instead of bathing with a bucket, the greens suggest that the only sustainable solution is to change the attitude of industrialists, poachers, fishermen, forest officers and you. How does one change attitude? Increase awareness, raise consciousness, educate, ostracise, cajole. But humans are stubborn and relish the affluence that apparently harms the environment. A stick must go with the carrot: regulate, proscribe, prosecute, penalise. The battle to save the environment has to be perpetual because the enemy is us. It must continue until we completely change our attitude, our nature, or cease to exist.

The enemy is us, not because our nature is to pollute, poach, grab, consume, overuse, but because the incentives are perverse. Change the incentives and the outcomes will change. How does one change incentives? Reconfigure the structure of accountability—institutions and rules—so that those who benefit also pay the costs. The primary means of injecting accountability is to revise the pattern of ownership of natural resources. When that is not feasible, design use rules so that users pay the price for the use of the resource. These use rules are generally referred to as market-based instruments.

Structures of Resource Ownership

Three basic structures of resource ownerships exist: Individual or family; community; and collective or national or international. In short, individual, community, and collective ownership. The nature and workings of individual or private property rights are generally known. Garret Hardin who coined the phrase 'the tragedy of the commons,' used the example of a common village pasture for illustration. He reasoned that each cattle owner in the village has incentive to allow his cattle to overgraze since the benefits of overgrazing accrue to him while the costs are borne by all cattle owners in the village. Each one thinks that if he limits grazing of his cattle, there is no assurance that others will also do so. This cost-benefit calculus leads to a situation where all cattle overgraze the common pasture, the grass runs out before the next monsoon, and some cattle die—the tragedy of the commons.

To mend the tragedy, two methods are suggested: one, divide the common pasture into individual plots and give one to each cattle owner, or vest the ownership in an organised collective body, generally the government. The solution is to either privatise or nationalise the pasture. Each private owner would then have an incentive to limit grazing so that he does not run out of grass. With collective ownership, the government would design and enforce rules for the use of the pasture so that the grass would last until the next monsoon.

The Tragedy of the Collective

By all accounts, state ownership and management, commonly referred to as the command and control approach, has singularly failed in managing the use or in the preservation of natural resources. Instances of government failures in management of natural resources abound: in a famous study of grassland degradation in Central Asia, satellite images showed marked degradation in the grasslands of southern Siberia where the former Soviet Union had imposed state-owned agricultural collectives, while grasslands in Mongolia, which had allowed pastoralists to continue their traditional, self-organised group property regime, were in much better condition.

The reason for this state failure is precisely the one that Hardin gave for the tragedy of the commons. The benefits of the resource accrue largely to the functionaries of the state but the costs are borne by all citizens. The nationalisation solution had assumed that the interests of government officers and those of the people are the same. That government officers would behave as if the costs and benefits were born equally by all citizens, including themselves. That the forest guard would protect tigers as if they were his. Alas, that is not the case—"all men have greater regard for what is their own than what they possess in common with others."

Given the dismal failure of state ownership, it seems that the only option left is privatisation of the commons. General apprehensions about private property and markets prevent the greens from advocating privatisation as a solution. They instead demand more elaborate and detailed regulations, stricter enforcement with more money and machinery, and novel ways to shame people for their materialism and consumerism. These, they seem to assume, would ultimately change the attitude.

However, the greens are mistaken on both counts: one, a far more forceful state apparatus of the former Soviet Union or of current Cuba has failed to change human nature; and two, privatisation is not the *only* solution, community commons are equally workable. As Elinor Ostrom (Chapter 7) has theorised and documented, the commons function efficiently in a framework of institutions and rules that are incentive-compatible.

Before the advent of the modern, Weberian governments, most of the natural resources were well managed by local communities. Access to a common resource—village pastures for cattle grazing, forests for fruits and fuelwood, wild animals for hunting, river water for agricultural use—was controlled by norms and customs, either articulate or inarticulate. Ever increasing demand for these resources due to growing population, accelerating economic development, and improving technologies began to put pressure on the informal norms and customs that managed the use of these resources. Unfortunately instead of building on the informal arrangements that had worked well, a completely new method was adopted. The state took over the ownership and management of common resources. The genuine tragedy is the nationalisation of the commons.

For incentive-compatibility, the community should be exactly identified, its area of ownership should be clearly demarcated, and it should have a legally enforceable long-term if not perpetual right. Mutual understanding between user communities and governments is not enough: the transfer must be statutory. This is illustrated well by the degradation of the forests in the northeastern parts of India that are apparently community managed (Chapter 9). Many have used this example to argue against the possibility of community management of the commons. Elected District Councils of various tribes in the region control access to forests and not the government forest department. How can one explain severe degradation of forests that are managed by District Councils? It is the difference between vesting rights in *political representatives* of users (District Councils) and in the resource *users themselves*. Political representative bodies suffer from the tragedy of the collective, even at the lower level of administration. They are not incentive-compatible institutions. The rights must be vested with user groups.

The facts of successful community management of the commons and of the failure of state ownership suggest that the famous phrase of Hardin be revised—from 'the tragedy of the commons' to 'the tragedy of the collective.' It is indeed the tragedy of collective ownership, a situation where resources belong to everyone but are cared for by none or a few. In this situation, the accountability structure, the cost-benefit calculus fails to reconcile the interests of state functionaries with those of the people. The interests of the managers and owners are not harmonised.[2] On the other hand, flourishing commons have existed in the past and also exist today. It is in fact the tragedy of the 'collective' and not of the 'commons.'

Towards Solutions through Incentives

One of the most effective ways to right the incentives is to change the structure of resource ownership. The basic direction of this change should be away from collective ownership towards individual and community ownership. Private ownership refers to both the individual and community ownership. But for sake of clarity we use privatisation to describe the shift towards individual or family ownership and communitsation for the correction towards community ownership.

Land exemplifies the gradual evolution from collective to community and finally to individual ownership. Land as land, not as forests or mountains, is largely privately owned. China, which had *de facto* private ownership of land since 1978, has recently amended its constitution to recognise that legally. This transition of land ownership, according to new resource-economics (Chapter 5), has been in the right direction. And our experience of generally superior private management of land suggests that the change of ownership structure has been beneficial.

In case of forests and water, however, the shift has been in the opposite direction—from community to collective ownership. The almost universal nationalisation of these resources has led to the tragedy of the collective, as predicted by new resource economics. People predict that water would be the cause of the next world war. The best way to avoid such tragedies is to put forests and water back in the hands of communities (Chapters 9 and 13).

The rural and tribal communities whose historical claim on these resources has been expropriated through nationalisation have also lost the most remunerative resources for their livelihood. In every country, they are the poorest communities. Governments, national and international, are doing everything to help them except to give them back their water and forests.

As the CAMPFIRE programme of Zimbabwe and Nepal's Community Forestry experiment indicate, community ownership solves two problems simultaneously: better management of natural resources and wildlife and provision of dignified livelihood to the poorest communities. With the ownership natural resources, they will be able build their own future according to their values, customs, and traditions.

Changing Resource Ownership Structures

Resource	Ownership Structure		
	Collective	Community	Individual
Land			
Forests			
Water			
Fisheries			

➡ Actual Change   ⬎ Desired Change

Air, however, has always been a collective resource. At this level of knowledge and technology it is hard to imagine any change in its ownership. Nonetheless it is important that the air is not left as an open access commons. Market-based instruments (MBIs) are an attempt to put a price on the use of air. Many experiments in MBIs in as diverse environments as of the European Union, United States, and China have been relatively successful.

Experiences the world over have made it amply clear that resources in the hands of private parties—be that of individuals, communities, or corporations—are better managed than in the hands of governments. Who should be entrusted with these resources depends upon the type of resources, circumstances, and local customs and traditions. For resources that have generally been in the commons, like water and forests, the best stewards are the local communities who have been managing those commons historically. Entrusting these resources to any other entity would mean keeping the communities out forcefully—by guns and guards. Whether these guns and guards are employed by governments or by private corporations, they would not able to withstand the battles for survival by the communities. Neither corporatisation, nor collectivisation is a solution.

In addition to all the utilitarian or efficiency arguments, it must be remembered that local communities have a prior claim—a moral claim—on these resources. They have been using the resource for generations and centuries. It is on the premise of prior use that all resources have been settled in any civilised society. The privately owned land today was at some point in time a forest. Some cleared the forests for agricultural, residential or commercial use and they received property title to the cleared land. But some people did not clear the forests and lived in them. These forest dwellers are now refused the same process of land titling that we enjoyed. The people who kept the forests intact are being penalised for not clear-cutting them in the past as we did! It is gross injustice not to recognise the rights of forest dwellers. The most efficient as well as moral resolution is to take our forests from the foresters and put them in the hands of forest dwellers.

For these new ownership arrangements to work properly—for them to be incentive-compatible—reliance on the common law of torts and negligence and an efficient judiciary is inescapable (Chapter 4). State regulators simply cannot foresee all contingencies and craft regulations to deal with them. The common law of torts is better suited to accommodate uncertainties of real life. The Bhopal Gas tragedy clearly shows the superiority of the law of torts approach in dealing with aftermaths of calamities. The day after the Bhopal carnage, many lawyers flew in from the US to sign up victims for lawsuits against Union Carbide in American courts. The lawyers came because of two characteristics of the American judicial system: contingency fee and the tort laws of strict liability (far stricter than what exists in India). To protect Bhopal victims from 'exploitation' by American lawyers, the government of India decided to file the case itself on behalf of all the victims by invoking the *parens patriae* principle, that is, seeking to represent the victims as a 'parent.' Judge John Keenan (District Judge of the Southern District of New York) used the *forum non-conveniens* and asked the Indian government to be the 'parent' in Indian court. The rest is history.

Community Ownership to Private Ownership?

Many suspect that community ownership of water and forests will be just a first step towards their complete privatisation. We don't think so, for three reasons. First, the distinction between ownership and use of the resource is important. Community owned water could be delivered to households, farms, or factories through private for-profit companies or non-profit user associations. We argue that competition in delivery is necessary for the resource owners to earn a fair price as well as for efficient use of the resource (Chapter 13).

Second, the nature of the resource determines what type of ownership structure is efficient. In case of water, riparian and prior appropriation rights are generally individual rights. This option of individual ownership is however unlikely to be available for forests, if for no other reason than the route we advocate for their transfer from the hands of the state into the hands of user communities. The government should identify user communities and transfer, not sell, the forests to them. Once a community of users owns forests then it's hard to imagine their wholesale transfer into individual or corporate hands. Dividing the forests into separate parcels would destroy their essential character. Thus the nature and characteristics of each resource decides the appropriate structure of ownership and its evolution.

We rule out the option of auctioning the forests or water bodies to highest bidders. Auctioning is completely unjust. The example of the government of the state of Chattisgarh in central India is often cited to point to the future state of natural resources. The government leased out a stretch of a river to a private company, denying people living on the banks of the river the use of the water without permission and payment. This is theft, not privatisation. Natural resources such as forests and water bodies belong to the communities who have been using and managing them historically. The state does not own them. The government has no right to auction them. The resources should be given back to user communities.

Third, the institutional framework and constraints and evolutionary processes are different in the two situations: one, where natural resources has never left the hands of the communities, and two, where the resources are first nationalised and later transferred to communities. It would surely be instructive to construct a conjectural history of how ownership and management would have evolved if the resources were never nationalised. We may be tempted to mimic that trajectory as historically correct, but it is unrealistic to expect to replicate that process now. The shift from collective to community ownership today would not be without some extra restrictions on the rights of new owners. It would most likely require that large parts of forests couldn't be converted to altogether a different use, including stipulations about some minimum levels of diversity and quality of flora and fauna that must be maintained.[3] The new owner communities are unlikely to have much option but to accept the limitations. These restrictions would surely result in a different evolutionary trajectory. One cannot undo history.

Nonetheless, the future of natural resources under community ownership would be far more preferable to the one that would emerge under continued collective ownership. This would be true even if tighter rules and stricter enforcements were brought to bear. The best way to manage the environment is to communitise natural resources, to avoid the tragedy of the collective. The change from collective to community ownership restructures the incentives for proper management.

Plan *versus* Process

The incentive model discussed here could be understood from a different perspective: Plan versus Process. It is humanly impossible to plan an ecology or an economy—the amount of information necessary, the system of accountability and incentives for the enforcers of the Plan with corrective feedback mechanisms, and the managerial prowess required to make all people of the country accept and carry out the Plan simply can't be mustered. What should be done is to put institutions and processes in place that enable the people on the spot to marshal the information and to have the incentives to make right decisions to deal with the problems as they encounter or foresee them. It is not the Final Decision, even if decided upon by the best minds of the time, but the process of decision-making that is critical to solving problems consistently and comprehensively. The most fundamental question is not what decision to make but who is to make it—through what processes and under what incentives and constraints, and most importantly, with what feedback mechanisms to correct the decision if it were to turn out to be ineffective.

As Ludwig von Mises and F A Hayek have demonstrated, centralised structures are comparatively inefficient in gathering and processing information and in generating right incentives for implementation and improvisation. Moreover, the order that is observed in the universe, or ecology, or economy is not the result of any human plan. It is generated through characteristic behaviour of the individual constituents—galaxies and stars, flora and fauna, and consumers and producers—within a framework of laws, rules, norms, and customs. The order emerges through the process; it is not ordained from outside.

The Nature Conservancy exemplifies the distinction between plan and the process approach in the arena of conservation. Instead of one conservation master plan for the country, the Nature Conservancy, a private association of concerned people purchases lands that have significant ecological value. They manage the land so as to protect the endangered species or provide a nurturing habitat to several sensitive species. The Nature Conservancy preserves over 92 million acres of land, both within and outside the United States, runs the largest system of private nature sanctuaries in the world, has over 20,000 wildlife species under its watch, and is running a $1 billion campaign to save 200 of the world's Last Great Places. Several such voluntary organisations would achieve far better results in a cost efficient manner, instead of one Master Plan and a giant bureaucracy.

Iceland and New Zealand's successful system of Individual Transferable Quotas in fisheries (Chapters 15 and 16) are also an example of individual property rights that have proved to be superior to all government regulations in replenishing and conserving fish stocks. Designing a right incentive and ownership structure for thousands or millions of fishermen is a more sustainable solution than relying on a single bureaucracy to manage diverse water bodies.

With the right institutions and rules, we get thousands or millions looking after the fish than a few government officers.

The success in all these areas depends on getting the process rather than the plan right.

Wilderness *versus* Wise Use: The Conflict of Two Visions

Many view the well being of forests and that of forest dwellers as two different and mutually exclusive options. This is based on a premise that the forests can be well protected only if forest using communities are excluded, and that the needs of the forest-dependent communities can be met only if the society is ready to suffer the loss of forests. One must choose between these two alternatives. This mindset is shared not only by the forest administration and the 'greens' but also by many who have the interests of native communities foremost on their minds. The champions have come to believe that under the pressures of modern culture and corporations, local communities would ultimately degrade and destroy the forests and the forests have to be 'protected' from them and the best protection can be ensured by the right and tight control of the state.

This is the Western vision of wilderness. It is in conflict with the vision of wise use. One views humans as outsiders in the natural ecosystem and the other as integral to the ecosystem. It is the green vision versus the terracotta vision.

I=PAT or 1/PAT?

The green credo has been formalised into a mathematical identity, which Garrett Hardin has called the 'third law of ecology': The IPAT equation. I = P × A × T, where I denotes man's impact on the environment, P is population, A is affluence, and T is technology. All else equal, any increase in population, production/consumption, or improvement in technology, the law suggests, *must result* in greater environmental degradation and greater pressure on finite natural resources. Sustainability, therefore, requires that all three factors, population, economic growth and technological change, should either be slowed down or are altogether halted.

Not surprisingly, the terracotta vision is exactly the opposite. As Julian Simon has demonstrated with a tremendous amount of historical data, population is the 'ultimate resource' (Chapter 2). Each child is born with a stomach that needs to be fed, but the same child has two hands and most importantly a mind. If the economic and political system allows the child to use them fully, she will produce more than her needs. So the problem is not population but the system of liberty or lack of it. He shows that energy and resources are infinite—the only limit is our imagination and ingenuity. As we face shortage of a resource, the price of the resource rises, and this signals to all entrepreneurs that profits await for those who discover a substitute or a better method of production (Chapters 20 and 22). At the time of Industrial Revolution, charcoal was the main source of energy. The forests of England were clear cut and burnt for charcoal. As fewer and fewer trees were left, the price of charcoal rose. 'Greedy businessmen' began the search and found coal under the ground. The coal replaced charcoal; the forests of England came back up. There is a lot of coal left in England but no one mines it since petroleum is even better. Fuel cells are being developed, which run on hydrogen that comes from water! The Sheikhs of Arabia would one day be sitting on oil that no one would want.

Put the (inflation-adjusted, real) price of natural resources against time for last two hundred years or so for which we have the data, the unmistakable trend is downward. Despite Malthusian geometric increase in population and Galbraithian 'conspicuous consumption,' prices of natural resources are falling. The price that has been continuously rising is that of human labour. Human labour is scarcer than any other natural resource. Our numbers are still not large enough to result in excess supply.

Human ingenuity is exercised through technology. Watch 'Modern Marvels' on the Discovery channel to get a glimpse of what technology has accomplished to improve human life as well as the environment. Despite this consistent historical track record, the greens view technology as an accident but calamity as certainty. They use the precautionary principle as an incantation to stymie technological progress (Chapter 17). No one can guarantee that we would not face any resource problems in the future, but all the historical evidence suggests optimism.

Growth in population has not increased scarcity of natural resources and technology's track record leads to optimism about the future, even if one wants to be cautious. The third variable in the IPAT equation, 'affluence,' can be measured by gross domestic product or better by the Human Development Index of the UN. By any account, an average person in the world today lives longer, is better fed, clothed, and sheltered than at any other time in human history (Chapter 1). Interestingly, those with affluence and those without affluence have a rather different conception of 'the environment.' Those without affluence think of environment as sanitation, potable, may be running water, food storage in hot and humid weather, a smoke-free kitchen. Those with affluence think of global warming, endangered species, Amazon forests, organic food. The level of affluence affects environmental priorities. That is to be expected. What is unacceptable is to raise the concerns of the affluent so high that those of the non-affluent cannot even be discussed; they get dismissed as materialism and consumerism. International agencies and NGOs take pride in dealing with the environmental concerns of the affluent but become prudish in promoting development to address the concerns of the poor. This is the worst form of imperialism the poor have ever suffered (Chapter 25).

Imperialism of the Affluent: Green versus Brown Oustees

Internationally, environmental concerns of the developed North supersede those of the developing South. The same phenomenon plays out nationally where demands of the urban educated class dominate those of the poor rural inhabitants. For a city dweller, the endangered specie is the tiger, for a villager it is he who is endangered. Urbanites campaign for forest and species protection but the costs of protection are imposed on forest dwellers. They are forced to vacate the area that is declared as a national park or a sanctuary. Their displacement however is rarely highlighted. We hardly ever see NGOs and celebrities standing up against their evacuation. The green oustees get little sympathy or support. Brown oustees—those displaced by developmental projects like dams and roads—get all of it. This is another form of imperialism.

The terracotta vision turns the IPAT equation upside down: $I = 1/P \times A \times T$. The impact is *inversely* related to P, A, and T. Another way to consider this vision is to read 'I' not as 'impact' but as 'improvement.' Environmental improvement depends directly on P, A, and T.

In the incentive-restructure approach, as opposed to the attitude-reform approach, the impact on the environment comes from the institutions and rules about ownership and use of resources that create the tragedy of the collective. The impact depends on the proportion of resources that are collectively owned and managed (C), the ratio of resources whose use is not priced (P), the extent to which tort laws are under-utilised to determine liability and negligence (T), and the level of anti-science, anti-reason attitude (A). I=CPAT.

The principles recommended here are old wisdom: user rights and user ownership, and user responsibility for managing common services and common resources. This wisdom somehow died in recent years—historians may identify and debate the reasons—but now is the right time to recognise it. The modern disciplines of new public management, new resource economics, public choice, and new institutional economics provide further support to the old wisdom. New technologies have made it possible for people to acquire necessary information and take prudent decisions. All the ingredients for sustained development and wise use of natural resources are present; we need only the courage and foresight to bring them together.

PARTH J. SHAH
VIDISHA MAITRA

Notes

1. According to Hinduism and Buddhism, the cosmos consists of five elements: earth, water, air, fire and ether. Earthenware items that have been fired are called terracotta. Thus terracotta has a significant meaning in Hindu philosophy because it is made of earth, with the element of water and air and is burnt in fire.

2. A similar principal-agent problem occurs in the management of a corporation. The interests of the managers would not always mesh with those of the stockholders. Various market mechanisms exist, the threat of take-overs and mergers being one, to ensure that managers do not stray too far away from maximising shareholder value.

3. These restrictions would vary from country to country and from place to place within a country. As discussed in detail later (Chapter 9), the stipulations in the CAMPFIRE programme of Zimbabwe are different from those in the Community Forestry Programme of Nepal, which are distinct from those on the land purchased by Nature Conservancy.

The State of Humanity

1

The Globalisation of Human Well-Being

INDUR M. GOKLANY

Much of the debate over globalisation and its merits has revolved around the issue of income inequality, and whether in the past few decades globalisation has made the rich richer and the poor poorer.[1] It has been claimed that "as globalisation has intensified, the gap between per capita incomes in rich and poor countries has widened."[2] Dollar and Kraay, pro-globalisation economists at the World Bank, have challenged such statements countering that "the best evidence available shows the exact opposite to be true...[and that]... the current wave of globalisation, which started around 1980, has actually promoted economic equality and reduced poverty."[3] Regardless of where the truth of these statements might lie, these arguments miss the point. The central issue with respect to globalisation is neither income inequality nor whether it is getting larger; rather it is whether globalisation advances human well-being, and if inequalities in well-being have, indeed, expanded, whether that is because the rich have advanced at the *expense* of the poor.

But as opponents of globalisation frequently note, human well-being is not synonymous with wealth[4] nor — to echo a catchy anti-globalisation slogan — can you eat GDP.[5] To conflate the two is to confuse ends with means. While wealth or per capita income (as measured by the gross domestic product, GDP, per capita) is probably the best indicator of material well-being, its greater importance stems from the fact that it either helps provide societies (and individuals) the means to improve other and, I would assert, more important measures of human well-being (e.g., freedom from hunger, level of health, mortality rates, child labour, educational levels, access to safe water and sanitation, and life expectancy)[6] or it is associated with other desirable indicators (e.g., adherence to the rule of law, government transparency, economic freedom and, to some extent, political freedom).[7] In fact, as shown in Figure 1, which will be discussed in greater depth below, analyses of cross country data show that although these other indicators generally improve as per capita income rises, their relationships are not linear.[8]

Figure 1

Well-Being versus Wealth in the 1990s

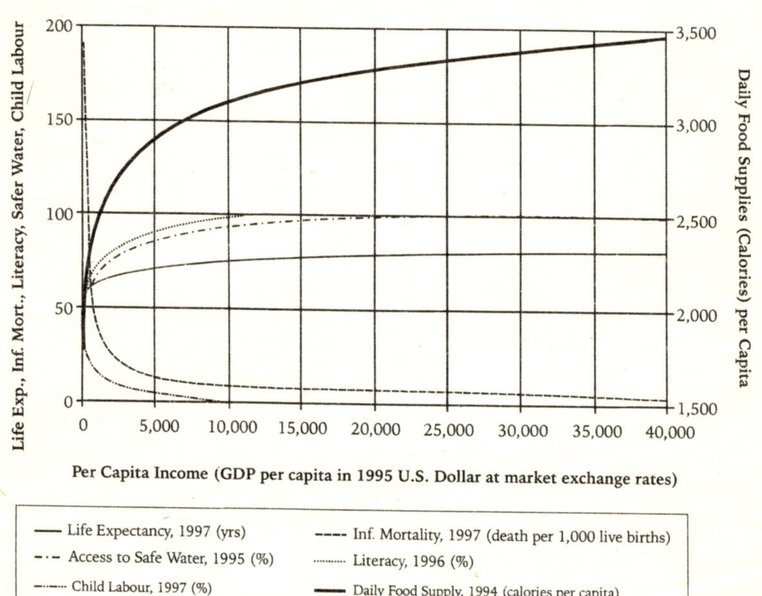

The improvements are usually rapid at low levels of economic development, but slow down or, in some cases halt altogether as they reach their practical or theoretical limits, e.g., 100 per cent for literacy and access to safe water, and 0 per cent for child labour (measured as the per cent of children aged 10-14 years in the labour force). Therefore, per capita income would not, by itself, be a good measure of human well-being, and any determination of whether globalisation has benefited humanity in general, or favours the rich at the expense of the poor, should be based on an examination of how these more relevant measures of human well-being have evolved as globalisation has advanced.

Indicators of Human Well-being

Specifically, I will examine trends in five indicators that measure distinct, though related, aspects of well-being. Three of these are measures of misery and deprivation and reflect "negative" well-being, one is a "positive" measure of well-being, and the last is the United Nations Development Programme's so-called "human development index" (HDI) which combines per capita income with two of the positive indicators of well-being.[9]

The "negative" indicators that will be examined are available food supplies per capita (low levels of which are surrogates for hunger and malnourishment), infant mortality and the prevalence of child labour. The first two — indicative at the extremes of Famine and Death, two of the Four Horsemen of the Apocalypse — have through the ages been synonymous with fear and misery. Less than half a century ago, famine, natural or man-made, still seemed to have mankind within its awful reach. This once-chronic condition claimed more than 30 million Chinese in 1959-61 alone.[10] An increase in the quantity of food is, perhaps, the first step to a healthy society. Having an adequate amount also enables the average person to focus on matters beyond mere sustenance, and to live a more fulfilling and productive life. Hunger and undernourishment, moreover, retard education and the development of human capital which, in turn, could slow down both technological change in every human endeavour and growth in every economic sector.[11] Thus, inadequate food supplies could not only add to misery but also slow progress in the positive indicators of well-being.

The second negative measure, infant mortality, also broadly tracks child and maternal mortality. Perhaps nothing has sown more sorrow and grief for womankind through the ages than the untimely death of her children. For most of mankind's tenure on Earth, infant mortality has been one of nature's cruel mechanisms for keeping human populations in check.

The third negative measure is the prevalence of child labour. That children ought to enjoy a childhood free from labour, spent in acquiring an education, and generally enjoying childhood was a luxury that for centuries could only be afforded by the upper classes and the wealthy. But in most households in most cultures, children were traditionally viewed as additional hands contributing to the family's economic security working on the farm, in handicrafts, menial tasks and, in the initial phases of industrialisation, in factories. Increases in productivity due to new technologies, however, made it possible to dispense with their labour. This trend was accelerated as families became wealthier, real prices of food dropped, the children's economic contribution became less critical for the family's survival and security in old age, and as the intrinsic and economic value of education to their children's and, possibly, the family's future economic and social security began to be recognised.

The positive measure that will be analysed is life expectancy at birth, probably the single most important indicator of human well-being. Longer life expectancy is also generally accompanied by an increase in disability-free life years. According to the World Health Organisation, the disability-adjusted life expectancies for the U.S., China, and India, for instance were 70.0, 62.3 and 53.2 years, respectively, in 1997/99.[12] Contrast that with the total (unadjusted) life expectancies of these three countries in 1950/55 which were 69.0, 40.8 and 38.7 years, respectively.[13] Moreover, studies from various developed nations (the U.S., Canada and France, for example) indicate that disability in their older populations has been declining.[14] In the U.S., for instance, the disability rate dropped 1.3 per cent per year between 1982 and 1994 for persons aged 65 and over, which resulted in 1.2 million fewer disabled persons in that age group in 1994.[15] So we are living longer—and healthier. Thus both the quantity and quality of life go hand-in-hand. It might be argued with some legitimacy that because higher levels of hunger and mortality reduce life expectancy, these measures overlap. However, life expectancy does not fully capture the fear and dread associated with Famine and Death.

The last indicator that will be examined is the United Nations Development Programme's aggregate "human development index" (HDI). This indicator was developed in recognition of the fact that there is more to development than a higher income. The HDI is estimated as the average of three measures: life expectancy at birth, educational attainment, and the logarithm of per capita income — the logarithm, because each additional dollar of income adds less to the quality of life than the previous dollar. The composition of the HDI can be justified on the grounds that life expectancy, as noted, is perhaps the most significant indicator of human well-being, per capita income reflects material well-being, and educational attainment — in addition to being an end in itself — is essential for conserving and creating new human capital. With the appropriate set of institutions, i.e., in the right setting, education can accelerate the creation and diffusion of technology.[16] Moreover, education (particularly of women) seems to be a key factor in spreading knowledge regarding safe drinking water, sanitation, proper hygiene, nutrition, and other public health practices which help societies improve health, reduce mortality and increase life expectancy.[17]

Trends in Measures of Human Well-being

So are the trends in the various measures of human well-being improving as globalisation marches on? Have gaps in these measures between the rich and the poor widened and, if they have, is globalisation responsible?

Trends with Respect to Economic Development

Figure 1, based on cross-country data, shows that various indicators of human well-being improve as countries become wealthier, with improvements coming most rapidly at the lowest levels of wealth. There are several possible explanations for this association. First, economic development indeed improves these indicators. Greater wealth translates into greater resources for researching and developing new technologies which directly or indirectly advance human well-being.[18] It also means increased resources for advancing literacy and education which, too, is generally conducive to greater technological innovation and diffusion.[19] Equally importantly, wealthier societies are better able to afford new and existing-but-underused technologies.[20] For instance, with respect to health — captured in Figure 1 by both infant mortality and life expectancy — these include "old" technologies such as water treatment to produce safe water, sanitation, basic hygiene, vaccinations, antibiotics and pasteurisation,[21] as well as newer science-based technologies such as AIDS and oral rehydration therapies, organ transplants, mammograms and other diagnostic tests. They also include agricultural technologies that increase crop yields thereby increasing available food supplies, and reducing hunger and malnourishment which then reduces the toll of infectious and parasitic diseases.[22]

Historically, reductions in hunger and undernourishment have been among the first practical steps nations have taken to improve public health, reducing infant mortality and increasing life expectancy.[23] And if despite increased food production a country is still short of food, greater wealth makes it possible, through trade, to purchase food security.[24] Greater wealth also makes it more likely that a society will establish and sustain food programmes for those on the lower rungs of the economic ladder.[25] Therefore, while "you can't eat GDP,"[26] if GDP is larger you are less likely to go hungry or be undernourished (except by choice). Thus, as Figure 1 illustrates, greater wealth, through a multiplicity of mechanisms, e.g., higher literacy, higher food supplies and greater access to safe water, leads to better health.[27]

Second, the causation might be in the reverse direction. Perhaps it is advances in human well-being that stimulate economic development, rather than vice versa. Healthier people are more energetic, less prone to absenteeism and, therefore, more productive in whatever economic activity they undertake.[28] When malaria was eradicated in Mymensingh (in Bangladesh), crop yields increased 15 per cent because farmers had more time and energy for cultivation.[29] In other areas, elimination of seasonal malaria enabled farmers to plant a second crop. A study done jointly by Harvard University Centre for International Development and the London School of Hygiene and Tropical Medicine study estimates that had malaria been eradicated in 1965, Africa's GDP would have been 32 per cent higher in 2000.[30]

Moreover, healthier people can also devote more time and energy to education and their intellectual development.[31] Good health is particularly important during children's formative years. Also, the incentives for investing in developing human capital increase if individual beneficiaries expect to live to sixty rather than, say, a mere forty. Not surprisingly, educational levels increase with life expectancy.[32] Today it is not unusual to encounter aspiring doctors and researchers in their mid-thirties, in effect, spending what once was literally a lifetime to learning their trade. And having acquired this expertise, these doctors and researchers are poised to contribute to technological innovation and diffusion in their chosen fields and to guide yet others along the same path. Thus better health helps raise human capital which aids the creation and diffusion of technology, further advancing health and accelerating economic growth.

But probably both explanations are valid with causes and effects reinforcing each other in a set of interlinked cycles. One such cycle is the health-wealth cycle in which — as we have seen — wealth begets health and health, wealth. Another cycle consists of food production, food access, education and human capital, and which also helps turn the health-wealth cycle. These cycles are embedded in a larger Cycle of Progress in which economic growth and technological change reinforce each other.[33]

Yet another explanation for the association between human well-being and wealth is that what improves one also improves the other. These would be the institutions and processes that fuel the Cycle of Progress by stimulating both economic growth and technological change. They include legal and economic systems that — through free markets; secure property rights; honest, predictable and fiscally responsible governments and bureaucracies; and adherence to the rule of law — encourage competition not only in the commercial sphere but also in the scientific and intellectual spheres, and allow those who venture their labour, intellectual capital and financial resources to profit from the risks they incur.[34] These institutions are also the foundations of civil societies and democratic systems.

Trade is an integral part of the Cycle of Progress. Freer trade directly stimulates economic growth,[35] helps disseminate new technologies, and creates pressures to invent and innovate.[36] For instance, competition from foreign car makers accelerated the introduction of several automobile safety and emission control systems to the United States improving both environmental and human well-being.[37] Competitive trade also helps contain the costs of basic infrastructure, including water supply, sanitation and power generation (although the full benefits are often squandered because of corrupt, inefficient and opaque bureaucracies and governments). A vivid example of the importance of trade in improving human well-being comes from Iraq whose inability, because of trade sanctions, to fully operate and maintain its water, sanitation and electrical system or to obtain sufficient food for its population contributed to a deterioration of public health and lowered life expectancies since the Gulf War. The need to alleviate these problems was the basis for various Security Council resolutions to extend its "Oil-for Food" programme.[38] And, as will be discussed below, trade has globalised food security.[39]

Trends with Respect to Time or Technological Change

Figure 2 shows that not only has life expectancy increased with the level of economic development but that the entire life expectancy-wealth curve has risen over time.[40] This curve's upward displacement is consistent with the creation and diffusion over time of new and existing-but-underused technologies. In effect, in Figure 2 the change in time (depicted by going from the 1962 life expectancy curve to the 1997 life expectancy curve) serves as a surrogate for technological improvement.[41]

Figure 2

Life Expectancy and Infant Mortality versus Wealth 1962 and 1997

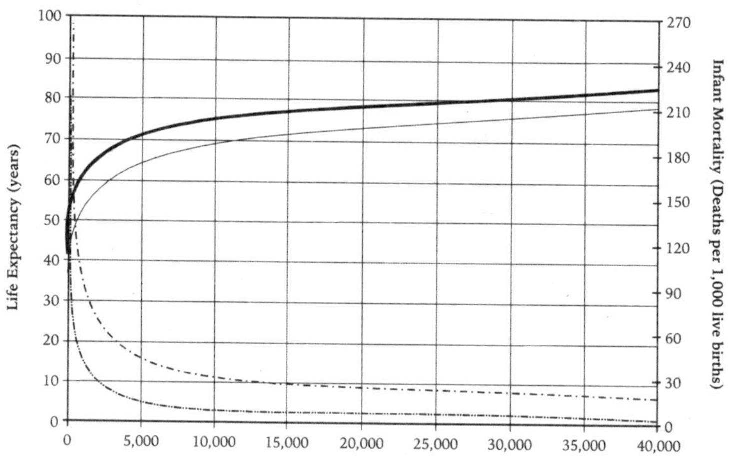

Per Capital Income (GDP per capita in 1995 U.S. Dollar at market exchange rates)

This figure also shows that infant mortality too improves with economic development and technological change (i.e., the entire curve drops with time). [42] I have shown elsewhere that these features — improvements with wealth and technology (for which time serves as a surrogate) — are common to other indicators of well-being including those shown in Figure 1. [43] Cumulatively, they indicate that for any specific level of real income, human well-being ought to be more advanced today than it was a few decades ago.

Hunger and Undernourishment

Concerns for the world's ability to feed its burgeoning population have been around at least since Malthus's *Essay on Population* two hundred years ago. Initially the concern was global. But by the 1950s and 1960s, despite the privations of the Second World War and the Great Depression, it seemed that the problem, if any, would be restricted to developing countries. Several Neo-Malthusians, such as Paul Ehrlich of *Population Bomb* fame[44] and the Paddock Brothers,[45] confidently predicted apocalyptic famines in the latter part of the 20th Century in the developing world. But remarkably, despite an unprecedented increase in the demand for food fuelled by equally unprecedented population and economic growth, its average inhabitant has never been better fed and less prone to hunger and undernourishment.

Between 1950 and 2000, world population increased by 140 per cent, and per capita income by over 170 per cent. Yet the real price of food has never been lower because of the enormous increase in agricultural productivity and trade. The latter ensures that the benefits of increased production are distributed broadly and food surpluses flow voluntarily to deficit areas. As a result, worldwide food supplies per capita have improved steadily during the past half century. Between 1961 and 1999, the average daily food supplies per person increased 24 per cent globally, from 2,257 Calories to 2,808 Calories. The increase was even more rapid in developing countries; it increased 39 per cent, from 1,932 to 2,684 Calories.[46]

The improvements for India and China — 40 per cent of humanity — are especially remarkable. By 1999, China's average daily food supplies had gone up 82 per cent to 3,044 Calories from a barely subsistence level of 1,636 Calories in 1961 (a famine year). India's went up 48 per cent to 2,417 Calories from 1,635 Calories in 1950-51.[47]

However, consistent with Figure 1, which shows per capita daily food supplies rising with wealth, improvements in per capita food supplies have been slower where for whatever reason — war, political instability or failed policies and institutions — economic development has lagged. For instance, between 1961 and 1999 average daily food supplies per capita in Sub-Saharan Africa, for instance, increased a paltry 6 per cent from 2,059 to 2,195 Calories.[48] The decline in food supplies in Eastern Europe and the Former Soviet Union (EEFSU) following the collapse of their Communist regimes only underscores the importance of economic development. [I eschew the label "economies in transition" for these countries since nowadays all economies, especially successful ones, are in perpetual transition.]

To put the improvements in per capita food supplies into context, the United Nations' Food and Agricultural Organisation (FAO) estimates that an adult in developing countries needs a minimum of 1,300 to 1,700 Calories/day merely to keep basic metabolic activities functioning at rest in a supine position! Food intake below these levels results in poor health, declining body weight, and physical and mental impairment. If one allows for moderate activity then the national daily average requirement increases to between 2,000 and 2,310 Calories per person.[49]

Therefore, since 1961, developing countries' available food supply has, on average, gone from inadequate to above adequate. But these averages mask the fact that hunger still persists today since, unlike the children of Lake Woebegone, many people unfortunately are below average. Nevertheless, between 1969-71 and 1997-99 the number of people suffering from chronic undernourishment in developing countries declined from 920 million to 790 million despite a 76 per cent growth in their population, that is, from 35 per cent to 17 per cent of their population.[50] Thus gaps between developing and developed countries in hunger and malnourishment have, in the aggregate, declined in absolute and relative terms. But the trends for Sub-Saharan Africa tell a somewhat more nuanced tale. Between 1979-81 and 1997-99, the share of population that was undernourished declined from 38 to 34 per cent but the absolute numbers increased from 168 million to 194 million.[51]

Why does economic development reduce the level of undernourishment? Cross country data show that both crop yield and per capita food supply follow the pattern indicated in Figure 2, i.e., both increase with income.[52] Crop yields increase because richer countries (or farmers) are better able to afford yield- and productivity-enhancing technologies, such as fertilisers, pesticides, better seeds and tractors.[53] But even if a country has poor yields or insufficient production, if it is rich it can import its food needs.[54] Hence, as Figure 1 shows, the richer the country, the higher its available food supplies.

Because it is always possible to have local food shortages in the midst of a worldwide glut, the importance of trade should not be underestimated. Currently, grain imports amount to 10 percent of the production in developing countries and 20 percent in Sub-Saharan Africa.[55] Without such imports, food prices in those countries would no doubt be higher and more of their people would be priced out of the market. In essence, globalisation, through trade, has enhanced food security. And in doing so it has reduced the severe health burdens that accompany hunger and undernourishment.[56]

To summarise, the developing countries where hunger and undernourishment were reduced the most are those which also experienced the most economic development. Certainly, for this indicator, globalisation leading to faster economic development and greater trade would seem to be the solution rather than the problem.

Infant Mortality

Before industrialisation, infant mortality, measured as the number of children dying before reaching their first birthday, typically exceeded 200 per 1,000 live births.[57] Starting in the nineteenth century, infant mortality started to drop in several of the currently developed countries due to advances in agriculture, nutrition, medicine, and public health. By the early 1950s, a gap had opened up between developed and developing countries as infant mortality dropped to 59 in the former and 178 in the latter.[58] By 1998 further medical advances reduced infant mortality in developed countries to 9 but because existing health care technology (including knowledge) diffused even faster from developed to developing countries, it had declined to 64 in the latter.[59] Thus, during the past half century the gap between developed and developing countries has been halved.[60]

The drop in infant mortality has been broad and deep. Since at least 1960 infant mortality has dropped more or less continuously for each of the country groups shown in Figure 3.[61] It also illustrates that in any given year, consistent with Figure 1, higher per capita income is generally associated with lower infant mortality. Between 1960 and 1999, the gaps in this indicator between high-income OECD countries and the other income groups shrank rather than increased. These gaps closed the fastest for medium income countries and the slowest for Sub-Saharan Africa. This is counterintuitive since the larger the initial gap, the faster it ought to shrink because the closer infant mortality is to zero, the harder should it be to reduce it further.

Figure 3

Infant Mortality, 1960-99

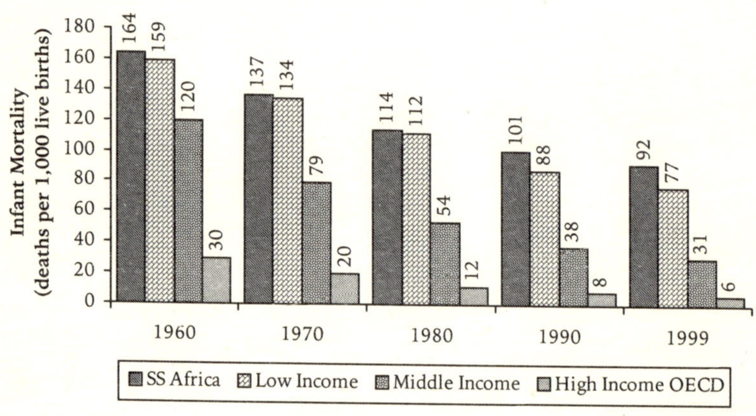

Source: World Bank, *World Development Indicators 2001* (Washington: World Bank, 2001).

Consistent with Figure 2 and the rapid technological diffusion from developed to developing countries in the past few decades, Table 1 indicates that many developing countries are far better off today than the currently developed countries were at equivalent levels of economic development.[62] In 1913 when the U.S. had a per capita income of $5,301 (in 1990 International $), its infant mortality was about 100. By contrast, in 1998 China's and India's, for example, were 31 and 71, respectively, despite having per capita incomes that were 41 to 67 per cent lower!

Table 1

Technological Progress and Infant Mortality and Life Expectancy

Country	Year	Per Capita Income (1990 International $)	Infant Mortality (deaths per 1,000 live births)	Life Expectancy at Birth (years)
U.S.	1913	5,301	~ 100	52
Ghana	1998	1,244	57	59
India	**1998**	**1,745**	**71**	**63**
China	1998	3,117	31	70
Peru	1998	3,666	40	68

Sources: Angus Maddison, *The World Economy: A Millenial Perspective* (Paris: OECD); U.S. Bureau of the Census, *Historical Statistics of the United States: Colonial Times to 1970* (Washington, DC: Government Printing Office, 1975); *World Development Indicators 2001*.

Thus, just as for hunger and undernourishment, the areas where infant mortality has improved the least are those with insufficient economic development or who, for whatever reason, have been unable to fully capitalise on existing knowledge and technology. Once again, globalisation seems to be part of the solution rather than the problem.

Life Expectancy

Because historically the decline in infant mortality was a major factor in the improvement in life expectancy, there are certain parallels between the progress in these two indicators, especially in the earlier years.

For much of human history average life expectancy used to be between 20-30 years.[63] Life expectancies in the currently developed countries increased slightly in the early part of the nineteenth century, followed by (small) declines in the middle two quarters of the 1800s (probably because of urbanisation) before commencing, with a few notable exceptions and some minor fluctuations, a sustained improvement which continues to this day.[64]

Contributing to these improvements were increases in food supplies per capita, the ascendancy of the germ theory, and the discovery of basic public health measures such as access to clean water, sanitation, pasteurisation and vaccination, antibiotics and the use of pesticides such as DDT to control malaria and other vector-borne diseases.

Because these public health and medical advances were discovered, developed and adopted first by the developed countries, a substantial gap opened up in average life expectancy between them and developing countries. By the early 1950s, the gap stood at 25.7 years in favour of the former.[65] But by the late 1990s, with the diffusion and transfer of technology (including knowledge) to developing countries, this gap had closed to 11.6 years.

A closer look at trends for different country groupings, however, reveals a more complex situation. Figure 4 compares life expectancies between high income OECD, middle income, and low income countries and Sub-Saharan Africa.

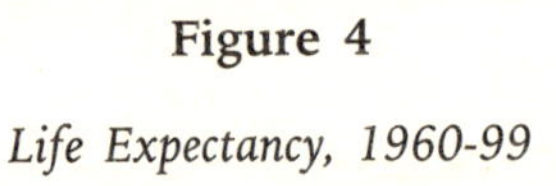

Figure 4

Life Expectancy, 1960-99

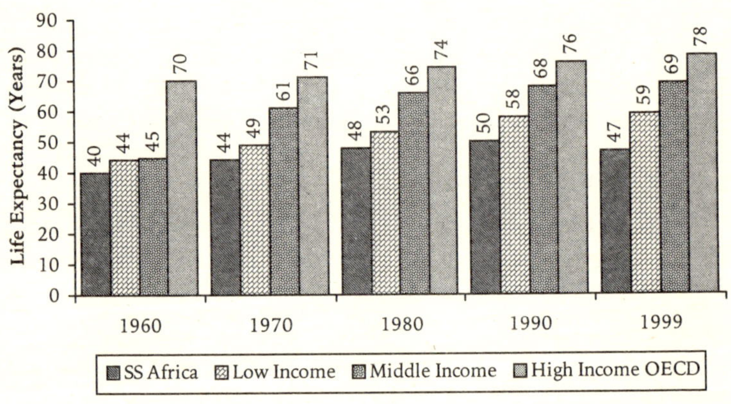

Source: World Bank, *World Development Indicators 2001* (Washington: World Bank, 2001).

Consistent with Figures 1 and 2, in any given year, life expectancy increases with per capita income. Between 1960 and 1999, life expectancy improved for high income OECD and middle income countries. However, the gap between these two sets of countries, which had shrunk from 24.5 in 1960 to 7.9 by the late 1980s, increased slightly to 8.6 by 1999, mainly because the middle income countries include many EEFSU nations whose life expectancies declined as their economies contracted during that period.[66]

The gap between high income OECD and low income countries also declined for most of the post-World War II period. But it expanded slightly from 1997 to 1999 because while life expectancy in the former continued to increase due to medical advances, it dropped slightly in the latter.[67] This drop was particularly severe in Sub-Saharan Africa where, as shown in Figure 4, life expectancy declined by 3 years in the 1990s, due to the HIV/AIDS epidemic and—in some case, even more important— the resurgence of malaria in Sub-Saharan Africa,[68] aggravated by civil unrest and cross-border conflicts in several areas. Consequently, the gap between rich and poor expanded in the 1990s reversing the direction of the trend from previous decades. But it didn't expand because the rich increased their life expectancy at the expense of the poor; rather it was because, when faced with new diseases (such as AIDS) or new forms of ancient ones (e.g., drug resistant tuberculosis) the poor lacked the economic and human resources not only to develop effective treatments but also to import and adapt treatments invented and developed in the rich countries. Notably, both economic and human resources are more likely to be augmented with globalisation than without it.

Sub-Saharan Africa's experience with AIDS is in stark contrast with that of the richer nations'. When this disease first appeared, it resulted in almost certain death everywhere — in developing as well as developed countries. The latter, particularly the U.S., launched a massive assault on the disease which led to the development of several technologies to reduce its toll. As a consequence, between 1995 and 1999 estimated U.S. deaths due to AIDS dropped by over two-thirds (from 50,610 to 16,273) although the number of cases increased by almost half (from 216,796 to 320,282). In 1996, it was the eighth leading cause of death in the U.S. By 1998 it had dropped out of the worst fifteen list.[69]

The U.S. was able to do this because it was both wealthy and had the human capital to address this disease. But despite the fact that the necessary technology now exists and, in theory, is available worldwide, similar improvements have yet to occur in Sub-Saharan countries because they cannot afford the cost of treatment, unless it is subsidised by the governments, charities or even industry from the *richer* nations. And indeed that is exactly what the worldwide effort to contain HIV/AIDS hopes to mobilise. This is as clear an illustration as any that the greater the economic resources, the greater the likelihood not only of creating new technologies but, equally importantly, of actually putting those technologies to use. And unless technologies are used, they will sit as curios on a shelf providing no benefit to humanity.

But it might be argued that the rapid spread of AIDS and other diseases was, in fact, one of the unintended consequences of globalisation. Without the transportation network that enables goods and people to move great distances, AIDS, for instance, might have been an isolated phenomenon rather than a pandemic. And indeed that much is true. But the same network also helped reduce public health problems in numerous ways. It helped reduce hunger and malnourishment by moving agricultural inputs and outputs between farms and markets. This was critical to increasing global food supplies in the past half century which, as noted, was one of the first steps to improved public health. Second, the transportation network is crucial to the worldwide diffusion of medical and public health technologies through, for instance, the distribution of medicines, vaccines, medical equipment, insecticides for vector control, and equipment for water treatment plants. But globalisation is more than the movement of goods; it also involves the movement of people and the diffusion of their ideas, knowledge and expertise. These, too, were enabled by the transportation network as doctors, nurses, agronomists, engineers, scientists — and those aspiring to those professions — moved back and forth between the developing and developed worlds.

But there is one area where globalisation of ideas and attitudes has retarded further progress toward improvements in human well-being. One reason for the resurgence of malaria in many developing countries in the 1980s and 1990s was that starting in the early 1960s, DDT, which had been instrumental in the post-World War II conquest of malaria in Europe and North America, began to be demonised in the rich countries.[70] Eventually, many of these rich countries banned DDT use and curtailed, if not eliminated, its production. Although that had virtually no effect on their public's health — the rich countries had already conquered malaria and could, moreover, afford substitutes in case they were needed to combat any recurrence — the consequences for much of the developing world were tragic. The global translocation of rich countries' attitudes toward DDT, coupled with its unavailability or higher price due to reduced production and the paternalistic insistence of western aid agencies that DDT's environmental consequences justified suspending its use for public health purposes, reduced the developing world's access to its most cost-effective weapon in its long standing war against malaria.[71] This contributed to a rebound in the malaria mortality rate in Sub-Saharan Africa.[72] That rate, which had dropped from 216 per 100,000 in 1930 to 107 in 1970, had climbed back to 165 per 100,000 in 1997.[73] Between 1990 and 1997, according to the WHO's World Health Report 1999, the malaria mortality rate in Sub-Saharan Africa increased by 17 deaths per 100,000 (from 148 to 165 per 100,000).[74] Notably, this exceeded the increase in its (total) crude death rate of 11 per 100,000 during the same period (which increased from 1,541 to 1,552 per 100,000 between 1990 and 1997).[75] That is, despite the AIDS epidemic, but for the increase in malaria deaths, Sub-Saharan Africa's mortality rate (and life expectancy) might have held their own during that period.

Nevertheless, the fact that life expectancy in the Sub-Saharan countries still exceeds the 20-30 years that was typical prior to globalisation indicates that despite the AIDS epidemic and the resurgence in malaria, the net effect of globalisation has been positive as far as life expectancy is concerned.

This conclusion—hinted at in Figure 2 by the upward displacement in the life expectancy curve as we move from 1962 to 1997— is reinforced by Table 1 which shows that life expectancies are much higher in many developing countries than they were in today's developed countries (such as the United States) at equivalent levels of economic development.

Child Labour

Figure 5 shows that the proportion of children in the work force has also been declining steadily for each of the income groups, and the richer the group, the lower that percentage. Gaps in child labour between Sub-Saharan Africa, the low and middle income countries, and the high income OECD countries have been shrinking at least since 1960. For this indicator also, the gap between high income OECD and middle income countries has diminished the most, and the least for the gap between the former and Sub-Saharan Africa.[76]

Figure 5

Child Labour, 1960-99

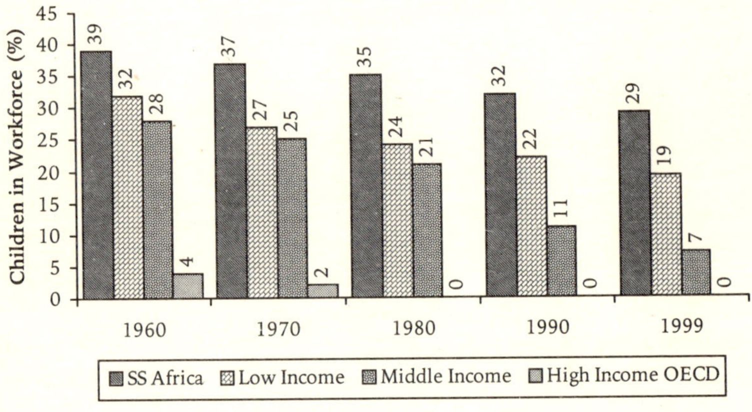

Source: World Bank, *World Development Indicators 2001* (Washington: World Bank, 2001).

Human Development Index (HDI)

Broad improvements in life expectancy, literacy and economic growth have combined to increase the HDI for most countries. Figure 6, based on the UNDP's *Human Development Report 2001*, shows that since 1975 — the first year for which that report provides data — the population-weighted HDI has improved for the so-called high, middle and low development tiers of countries, as well as for Sub-Saharan Africa (two-thirds of which are also included in the low development tier). Note that in this figure, the HDI scale tops out at one unit.

Nevertheless, despite the broad improvement for the various groups of countries, some countries' HDIs have deteriorated in the past decade or so. According to the UNDP's *Human Development Report 2001*, of the 97 countries for which data are available for 1975 and 1999, Zambia has the unique, but dubious, distinction of having a lower HDI in 1999 than in 1975 because both GDP per capita and life expectancy declined over this period. The presence of refugees from conflicts in neighbouring countries may have added to these declines. Curiously, in terms of aid as a fraction of GDP, at 22.8 percent, Zambia is also among the world's largest recipients of foreign aid.[77] Thus, its downward spiral can hardly be attributed to globalisation or to rich countries having enriched or improved themselves at the expense of the poor. Moreover, of the 128 countries for which data were available, eighteen countries (i.e., 15 per cent) — ten Sub-Saharan and eight EEFSU countries — had lower HDIs in 1999 than in 1990. Life expectancy in each of the ten Sub-Saharan countries declined during this period mainly because of HIV/AIDS and/or malaria; in all but two cases, per-capita GDP also declined. During this period, most also were directly or indirectly affected by civil unrest or spillovers from conflicts in neighbouring countries which further strained their resources. Per capita income declined in all eight EEFSU countries while life expectancies dropped in six of the eight.

Figure 6

Human Development Index, 1975-99

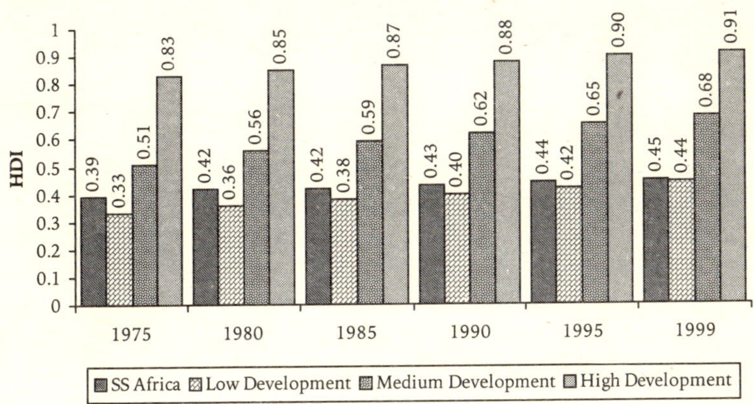

Source: United Nations Development Programme, *Human Development Report 2001* (New York: UNDP, 2001): and World Bank, *World Development Indicators 2001* (Washington: World Bank, 2001).

All else being equal, one would have expected that HDI improvements would generally be largest for the lowest tier and least for the highest tier of countries because the latter are closer to the top of the HDI scale and because with each improvement in HDI, it becomes harder to improve it further (just as each additional dollar adds less to the quality of life than the previous dollar). But in fact, as Figure 6 shows, between 1975 and 1999, the middle tier countries saw the most progress, followed, in order, by the progress for low tier, high tier, and Sub-Saharan countries. As a result, the HDI gap between the high and medium tier countries diminished the most. The gap between high and low tier countries also declined slightly but for the reasons discussed above, it expanded between the high tier and Sub-Saharan countries.

Summarising the Trends

In summary, human well-being has improved and continues to improve for the vast majority of the world's population. Because of a combination of economic growth and technological change, compared to a half century ago, today's average person lives longer and is less hungry, healthier, more educated, and more likely to have children in a schoolroom than in the workplace. During that period, indicators of well-being have improved for every country group, although life expectancies have declined in many Sub-Saharan and EEFSU countries since the late 1980s because of HIV/AIDS, malaria, or problems related to their economic deterioration.

For every indicator examined, regardless of whether the rich are richer and the poor poorer, gaps in human well-being between the rich countries and other income groups have for the most part shrunk over the past four decades. However, comparing rich countries and Sub-Saharan Africa, although the gap in infant mortality between the two has continued to close, the gap in life expectancy has expanded in the past decade or so (but not enough to erase the large reductions made previously). Despite this, in aggregate, the corresponding gap in HDI has decreased.

Globalisation and Inequality

Conventional wisdom decries income inequality, but there may be situations where some inequality would benefit humanity. Consider, for instance, that since most of the easy improvements in public health have been largely captured (except where globalisation has lagged), the search and implementation of cures and treatments for today's unconquered diseases (e.g., strokes, heart disease, and cancers) could progressively become more expensive. Richer societies are in a better position to invest in the research and development of new or improved technologies in general, and for detecting, treating or eliminating these diseases, in particular. AIDS is a case in point.

Moreover, new technologies often cost more initially. The rich, therefore, are usually the first to obtain new or innovative technology. As the rich purchase this technology, the supplier can increase production and its price drops because of economies of scale and learning-by-doing, if nothing else. Such declines allow the less wealthy to also afford that technology which then paves the way for further price drops and induces people of more modest means to enter the market. Thus, arguably, wealth inequality spurs the invention, innovation and diffusion of new technologies. This pattern has been repeated time and again for goods and services (such as telephones, VCRs, personal computers, and even vacations to exotic places) as well as health technologies (such as antibiotics, organ transplants and, now, AIDS treatments) where innovations started expensive but ended up cheaper. Therefore, some inequality in wealth probably benefits humanity. Presumably, for a given set of supply and demand characteristics for a particular technology, there is an optimum level of inequality which would maximise the rate of adoption of that technology, as well as the rate at which that improves human well-being. In other words, even if one were to ignore trends in inequalities in other, more significant indicators of human well-being, income inequality is a poor lens for viewing the merits of globalisation.

Without restricting himself only to income inequality, Amartya Sen claims that inequality is the central issue with respect to globalisation and that a "crucial question concerns the sharing of the potential gains from globalisation, between rich and poor countries, and between different groups within countries."[78]

If one accepts Sen's contention regarding the centrality of inequality, the above data indicate that whether or not income inequalities have been exacerbated, in terms of the truly critical measures of well-being — hunger, infant mortality, life expectancy, child labour — the world is much more equal now than it was a few decades ago.

But in the past dozen years the life expectancy gap between the richest and some of the poorest has expanded. Therefore, it might be argued, that with respect to this most significant of all indicators at least, globalisation might yet fail Sen's "crucial question." But it is no more reasonable to expect that globalisation would lead to equal gains among countries than, say, a course in economics would lead to equal gains in knowledge among its students. Sen, for instance, benefited much more from his education than his erstwhile classmates, not because someone else gained less but perhaps because of better preparation, harder work or, dare I say it, greater natural ability. Just as unequal sharing of benefits or outcomes does not indict education, unequal progress in human well-being does not damn globalisation.

In fact, Figures 1 through 6 suggest that where gaps in well-being have expanded, it is not because of too much globalisation, but too little. The rich are not better-off because they have taken something away from the poor, rather the poor are better-off because they have benefited from the technologies developed by the rich and their situation would have been further improved had they been better prepared to capture the benefits of globalisation. To the extent the rich can be faulted, it is that, first, their deionisation of DDT — and here globalisation is culpable — also affected attitudes in the developing countries.[79] That contributed to the resurgence in malaria during the 1980s and 1990s, because of which mortality rates are higher in Sub-Saharan Africa — and life expectancy lower — than they would otherwise have been. Second, and perhaps more importantly, by protecting favoured economic sectors through subsidies and import barriers — activities which have not necessarily improved their own economic welfare — the rich have retarded the pace of globalisation and made it harder for many developing countries to capture its benefits.

Notes

1. See, e.g., the following:

    Kevin Watkins and Aart Kraay and David Dollar, "Point/Counterpoint: Making Globalisation Work for the Poor," *Finance and Development* 39, No.1 (March 2002),

    Martin Khor, "Backlash Grows Against Globalisation," (1996), http// www.globalpolicy.org/globaliz/bcklash1.htm;

    W. Bowman Cutter, Joan Spero, and Laura D'Andrea, "New World, New Deal," *Foreign Affairs* (March/April 2000), http://www.foreignpolicy2000.org/library/issuebriefs/ readingnotes/fa_tyson.html;

    Bernard Wasow, "New World, Bum Deal?" *Foreign Affairs* (July/August 2000), http:// www.tcf.org/Opinions/In_the_News/Wasow-NewWorld_BumDeal.html;

    Jay Mazur, "Labor's New Internationalism", *Foreign Affairs* (January/February 2000);

    "The FP Interview: Lori's War," Interview originally published in *Foreign Policy* (Spring 2000), http://www.foreignpolicy.com/best_of_fp/articles/wallach.html;

    United Nations Development Program (UNDP), *Human Development Report 1999* (New York: UNDP, 1998), pp. 3, 11.

2. Cutter, Spero and D'Andrea, "New World, New Deal."

3. David Dollar and Aart Kraay, "Spreading the Wealth," *Foreign Affairs* (January-February 2000), http://www.foreignaffairs.org/articles/Dollar0102.html, visited May 26, 2002.

4. For instance, Stephen Lewis, a leading Canadian New Democrat Party politician, former Canadian ambassador to the United Nations, and erstwhile deputy executive director of UNICEF, is quoted as having said "there is something profoundly wrong with globalisation...There is more to the world than creating bigger markets. We can't ignore the human dimension": in Ryan Smith, "Lewis Flays Globalisation," January 29, 2001, on file with author. Similarly Lori Wallach, an anti-globalisation organiser who came to prominence during the Seattle protests, notes "the question is, what is going on in real measures of well-being? So, while the volume, the flow of goods may be up, and in some countries gross national product may be up, those macroeconomic indicators don't represent what's happening for the day-to-day standard of living for an enormous number of people in the world. That gets to one of the biggest critiques of the WTO in its first five years, which is that while the overall global flow of trade continues to grow, the share of trade flows held by developing countries has declined steadily. Similarly, over that five-year period, while the macroeconomic indicators have often looked good, real wages in many countries have declined, and wage inequality has increased both within and between countries"; see "The FP Interview: Lori's War."

5. Zach Dubinsky, "Amid the Tears: Protesters, Police, Politics and the People of Quebec," (April 25 - May 1, 2001), on file with author. This slogan is reminiscent of the title of a book by Eric A. Davidson, *You Can't Eat GNP: Economics as if Ecology Mattered* (Cambridge, MA: Perseus, 2000).

6. Indur M. Goklany, *Economic Growth and the State of Humanity* (Bozeman, MT: Political Economy Research Center, 2001); Indur M. Goklany, "The Future of the Industrial System," Invited Paper, *International Conference on Industrial Ecology and Sustainability*, University of Technology of Troyes, Troyes, France, September 22-25, 1999.

7. James Gwartney and Robert Lawson with Walter Park and Charles Skipton, *Economic Freedom of the World: Annual Report 2001* (Vancouver, BC: Fraser Insitute, 2001); David Dollar and Aart Kraay, "Growth Is Good for the Poor," (Washington, DC: Development Research Group, World Bank, 2000), http://www.worldbank.org/research/growth/absdollakray.htm.); James Gwartney, Randall Holcombe, and Robert Lawson, "The Scope of Government and the Wealth of Nations," *Cato Journal* 18 (1998): 163-90; Seth W. Norton, "Poverty, Property Rights, and Human Well-Being: A Cross-National Study," *Cato Journal* 18 (1998 No. 2): 233-245; Robert J. Barro, *The Determinants of Economic Growth: A Cross-Country Empirical Study* (Cambridge, MA: MIT Press, 1997).

    With respect to democracy and economic growth, Barro's *The Determinants of Economic Growth* suggests that increased economic growth tends to increase democracy (the so-called Lipset hypothesis) but democracy's effect on economic growth is mixed; apparently growth increases with democracy at low levels of democracy but declines at high levels, perhaps because redistribution impulses are harder to contain in democracies. This is echoed in William Easterly, *The Elusive Quest for Growth: Economists' Adventures and Misadventures in the Tropics* (Cambridge, MA: MIT Press, 1991), pp. 265-267. See also, Dani Rodrik, *Democracy and Economic Performance*, Kennedy School of Government, Harvard University, December 14, 1997, http://ksghome.harvard.edu/~.drodrik.academic.ksg/demoecon.pdf; Francisco L. Rivera-Batiz, *Democracy, Governance and Economic Growth: Theory and Evidence*, undated, http://www.columbia.edu/cu/economics/discpapr/DP0102 57.pdf

8. Figure 1 is based on cross country analyses reported in Indur M. Goklany, *Economic Growth and the State of Humanity*, and Indur M. Goklany, *The Precautionary Principle: A Critical Appraisal of Environmental Risk Assessment* (Washington, DC: Cato Institute, 2001). The data used to generate this figure are from *World Development Indicators 1999* (Washington, DC: World Bank, 1999) except for daily food supplies capita, which are from *World Resources 1998-1999* (Washington, DC: World Resources Institute, 1998). Each of the curves in Figure 1 is based on a best-fit equation generated using log-linear regression of the indicator on the (log of) per capita income (estimated as gross domestic product per capita), with the exception of the infant mortality curve which

was generated using a log-log regression. The curves representing access to safe water and literacy were truncated at 100 percent, while the child labour curve was truncated at 0 per cent. The slopes of each of the regression lines were significant at the 0.1 per cent, or better, level. The number of data points (N) and R^2 for the indicators were as follows: 148 and 0.645 for life expectancy, 147 and 0.745 for infant mortality, 150 and 0.629 for daily food supplies per capita, 96 and 0.520 for literacy, 51 and 0.549 for access to safe water, and 140 and 0.534 for child labour.

9. See, e.g., United Nations Development Program (UNDP), *Human Development Report 2001* (New York: UNDP, 2001).

10. Jasper Becker, *Hungry Ghosts: Mao's Secret Famine* (New York: Free Press, 1996).

11. Robert W. Fogel, "The Contribution of Improved Nutrition to the Decline of Mortality Rates in Europe and America," in: Julian L. Simon, *The State of Humanity* (Cambridge, MA: Blackwell, 1995), pp. 61-71; Robert W. Fogel, *The Fourth Great Awakening and the Future of Egalitarianism* (Chicago: University of Chicago Press, 2000); World Health Report 1999 (Geneva: World Health Organisation, 1999); Easterlin, *Growth Triumphant;* Indur M. Goklany, "Saving Habitat and Conserving Bio-diversity on a Crowded Planet," *BioScience* 48 (November 1998): 941-53.

12. *World Health Report 2000* (Geneva: World Health Organisation, 2000), pp. 176-183.

13. *World Resources 1998-1999* (Washington, DC: World Resources Institute, 1998).

14. Department of Health and Human Services, *Active Aging: A Shift in the Paradigm.* (Washington, DC: Office of Disability, Aging and Long Term Care, HHS,1997), http://aspe.hhs.gov/daltcp/reports/actaging.htm; see also Eileen M. Crimmins, Yasuhiko Saito, and Dominique Ingegneri, "Trends in Disability-free Life Expectancy in the United States, 1970-90," *Population and Development Review* 23(1997 no.3): 555-72, 689-90.

15. *Ibid.*

16. Joel Mokyr, *The Lever of Riches: Technological Creativity and Economic Progress* (New York: Oxford University Press), pp. 174-176; Gwartney *et al.*, "The Scope of Government and the Wealth of Nations"; Robert J. Barro, *The Determinants of Economic Growth*; Robert J. Barro, *Education and Economic Growth,* http://www.hrdc-drhc.gc.ca/stratpol/arb/conferences/oecd/education.pdf; William Easterly, *The Elusive Quest for Growth*, pp. 71-84.

17. Richard A. Easterlin, *Growth Triumphant: The Twenty-First Century in Historical Perspective* (Ann Arbor, MI: University of Michigan Press, 1996), pp. 9, 79. Barro, *Education and Economic Growth,* suggests that education of women at the primary level might increase economic growth through the reduction in the total fertility rate, but his analysis didn't show any significant effect on economic growth due to secondary education for women which, he opined, might be due to gender discrimination. Dean Filmer and Lant Pritchett, *Child Mortality and Public Spending on Health: How Much Does Money Matter?,* October 17, 1997, http://www.worldbank.org/html/dec/Publications/Workpapers/WPS1800series/wps1864/wps1864.pdf, show that infant and child mortality rates — indicators of public health — decline with women's education. This might be a mechanism through which women's education helps spur economic growth.

18. Not surprisingly, expenditures on research and development increase with per capita GDP. Using data for 1994, linear regression analysis of cross-country data for 1994 from *World Development Indicators 1999* shows that the slope is significant at the 5 per cent level (N = 53, R^2 = 0.506). This analysis used GDP per capita for 1994 adjusted for purchasing power parity. See also, Indur M. Goklany, "Strategies to Enhance Adaptability: Technological Change, Economic Growth and Free Trade," *Climatic Change* 30 (1995): 427-449.

20. Goklany, "Strategies to Enhance Adaptability"; Goklany, "Saving Habitat and Conserving Biodiversity."

21. Easterlin, *Growth Triumphant*, p. 161.

22. Robert W. Fogel, "The Contribution of Improved Nutrition," pp. 61-71; Robert W. Fogel, *The Fourth Great Awakening*; World Health Report 1999; Easterlin, *Growth Triumphant;* Indur M. Goklany, "Saving Habitat and Conserving Bio-diversity."

23. Fogel, *The Fourth Great Awakening*; Easterlin, *Growth Triumphant*.

24. Goklany, "Saving Habitat and Conserving Biodiversity"; Goklany, "Strategies to Enhance Adaptability"; Indur M. Goklany, "Potential Consequences of Increasing Atmospheric CO_2 Concentration Compared to Other Environmental Problems." *Technology* 7S (2000): 189-213.

25. Goklany, "Saving Habitat and Conserving Biodiversity"; Goklany, "Strategies to Enhance Adaptability."

26. Dubinsky, "Amid the Tears"; Davidson, *You Can't Eat GNP*.

27. Goklany, "Saving Habitat and Conserving Biodiversity"; see also Lant Pritchett and Lawrence H. Summers, "Wealthier is Healthier," *Journal of Human Resources* 31 (1996): 841-68.

28. World Bank, *World Development Report: Investing in Health* (New York: Oxford University Press, 1993); Fogel, "The Contribution of Increased Nutrition"; Barro, *The Determinants of Economic Growth*; *World Health Report 1999* (Geneva: World Health Organisation, 1999); Barry Bloom, "The Future of Public Health," *Nature* 402 (Supplement 1999). C63-64.

29. Easterlin, *Growth Triumphant*.

30. Harvard University Center for International Development and the London School of Hygiene and Tropical Medicine, *Economics of Malaria, Executive Summary*, 2000, http://www.malaria.org/jdsachseconomic.html.

31. *World Health Report 1999* (Geneva: World Health Organisation, 1999); Fogel, "The Contribution of Increased Nutrition."

32. Goklany, *Economic Growth and the State of Humanity*, see Fig. 7.

33. The cycle of progress is briefly described in Goklany, *Economic Growth and the State of Humanity*, pp. 26-31. See also Goklany, "The Future of the Industrial System."

34. See, e.g., Barro, *The Determinants of Economic Growth*; David Dollar and Aart Kraay, *Growth Is Good for the Poor* (Washington, DC: Development Research Group, World Bank), http://www.worldbank.org/research/growth/absdollakray.htm.; Gwartney *et al.*, *Economic Freedom of the World...2001*, Gwartney *et al.*, "The Scope of Government and the Wealth of Nations."

36. Goklany, "Strategies to Enhance Adaptability."

37. Goklany, "Strategies to Enhance Adaptability."

38. United Nations, "Security Council Extends Iraq 'Oil-for-Food' Programme for Further 186 Days," U. N. Press Release SC/6872, June 8, 2000, http://www.un.org/News/Press/docs/2000/20000608.sc6872.doc.html.

39. Goklany, "Strategies to Enhance Adaptability."

40. Figure 2 is based on cross country analyses reported in Indur M. Goklany, *Economic Growth and the State of Humanity* using data from *World Development Indicators 1999* (Washington, DC: World Bank, 1999). In this figure GDP per capita is based on constant (1995) dollars at market exchange rates. As in Figure 1, the life expectancy curves are based on best-fit equations generated using log-linear regressions. The slopes of both of these regression lines are significant, i.e., economic development leads to a statistically significant improvement in life expectancy. Equally importantly, the change in the intercepts going from 1962 to 1997 is positive and statistically significant at the 0.1 percent level. That is, the overall improvement in life expectancy between 1962 and 1997 (which can be attributed to technological change over that period) is statistically significant.

41. Goklany, *Economic Growth and the State of Humanity*.

43. Goklany, *Economic Growth and the State of Humanity*.

44. Paul Ehrlich, *The Population Bomb* (New York: Ballantine Books, 1968).

45. W. Paddock and P. Paddock, *Famine 1975! America's Decision: Who Will Survive?* (Boston, MA: Little, Brown, 1967).

46. Based on *FAOSTAT Database*, http://apps.fao.org, from Indur M. Goklany, "Agricultural Technology and the Precautionary Principle," in Roger Meiners and Bruce Yandle, eds., *Environmental Policy and and Agriculture: Conflicts, Prospects, and Implications* (Lanham, MD: Rowman and Littlefield, 2002), forthcoming.

47. *Ibid.*

48. *Ibid.*

49. Food and Agricultural Organisation (FAO), "Assessment of Feasible Progress," in *Food Security. Technical Background Documents 12–15. Volume 3* (Rome, Italy: FAO, 1996).

50. FAO, *The State of Agriculture 1996* (Rome, Italy: FAO, 1996); FAO, *The State of Food Insecurity in the World 2001*, http://www.fao.org/docrep/003/y1500e/y1500e00.htm..

51. *Ibid.*

52. Goklany, "Saving Habitat and Conserving Biodiversity"; Goklany, "The Future of the Industrial System"; Goklany, "Potential Consequences of Increasing Atmospheric CO_2 Concentration."

53. *Ibid.*

54. *Ibid.*

55. *FAOSTAT Database 2001*, htp//apps.fao.org.

56. Goklany, "Strategies to Enhance Adaptabiliy".

57. K. Hill, "The Decline in Childhood Mortality", in Simon, *The State of Humanity*, pp. 37-50.

58. *World Resources 1998-1999.*

59. Goklany, *Economic Growth and the State of Humanity*, p. 14.

60. Goklany, "The Future of the Industrial System."

61. The country groupings in this—and the following two—figures are taken from the classifications used in the World Bank's *World Development Indicators 2001.*

62. Economic data are from Angus Maddison, *The World Economy: A Millenial Perspective* (Paris: OECD); data on life expectancy (LE) and infant mortality (IM) for the U.S. in 1913 are from U.S. Bureau of the Census, *Historical Statistics of the United States: Colonial Times to 1970* (Washington, DC: Government Printing Office, 1975); the LE and IM data for 1998 are from *World Development Indicators 2001.*

63. Samuel H. Preston, "Human Mortality Throughout History and Prehistory," in Simon, *The State of Humanity*, pp. 30-36.

64. Goklany, *Economic Growth and the State of Humanity.*

65. *World Resources 2000-2001* (Washington, DC: World Resources Institute, 2000).

66. *World Development Indicators 2001.*

67. *Ibid.*

68. For example, in 1998 Zambia lost more than twice as many disability adjusted life years to malaria than to HIV/AIDS (Richard Tren, personal communication, May 14, 2002, based on stataistics from its Central Board of Health). The malaria mortality rate in Sub-Saharan Africa which stood at 197 per 100,000 in 1970 declined until the 1980s but by 1997 it had rebounded to 165; by contrast in the rest of the world it declined from 7 per 100,000 to 1 per 100,000 over that period (World Health Report 1999, p.50).

69. Goklany, *The Precautionary Principle*, pp. 9-10, and references therein.

70. Rachel Carson, *Silent Spring* (Cambridge, MA: Houghton Mifflin, 1962).

71. Roger N. Bate, "How Precaution Kills: The Demise of DDT and the Resurgence of Malaria," in *Perilous Precaution: The Folly of Disregarding Science*, ed. Roger N. Bate (Cambridge, UK: European Science and Environment Forum, 2002), pp. 70-82; Wallace Chuma, "A Renewed Role Sought for DDT in Malaria War," *Pittsburgh Post Gazette*, July 21, 2002, http://www.post-gazette.com/healthscience/20020721malaria3.asp; Goklany, *The Precautionary Principle*, pp. 13-18.

72. Bate; Chuma; Goklany, *The Precautionary Principle*, pp. 13-18.

73. World Health Organisation, *World Health Report 1999*, p. 50. It's unclear whether the mortality rate was age-adjusted for a standard population distribution. However, the change in this distribution between 1990 and 1997 is unlikely to have modified the increase in mortality rate by much.

74. *Ibid*. This increase in the mortality rate alone translates into an increase of more than 100,000 additional malaria deaths in 1997.

75. World Bank, *World Development Indicators 2002*.

76. *World Development Indicators 2001*.

77. *The Economist*, March 23-30, 2002, p. 102.

78. Amartya Sen, "A World of Extremes: Ten Theses on Globalisation," *Los Angeles Times*, July 17, 2001, http://www.globalpolicy.org/globaliz/define/0717amrt.htm.

79. Deepak Lal has warned against rich countries imposing their values on poor countries. "If the West tries to tie its moral crusade too closely to the emerging process of globalisation," he writes, "there is a danger that there will be a backlash against [that] process." Deepak Lal, "The Challenge of Globalisation: There is No Third Way," *Global Fortune: The Stumble and Rise of World Capitalism*, ed., Ian Vasquez (Washington, DC: Cato Institute, 2000), p.40.

2

Population Growth, Natural Resources, and Future Generations

JULIAN SIMON

How will the supplies of natural resources be affected by different rates of population growth? We investigate the effects of different rates of population growth, concentrating on mineral resources for simplicity.

If there is only Alpha Crusoe and a single copper mine on an island, it will be harder to get raw copper next year if Alpha makes a lot of copper pots and bronze tools this year, because copper will be harder to find and dig. And if he continues to use his mine, his son Beta Crusoe will have a tougher time getting copper than did his daddy, because he will have to dig deeper. If suddenly there are not one but two people on the island, Alpha Crusoe and Gamma Defoe, copper will be more scarce for each of them this year than if Alpha lived there alone, unless by cooperative efforts they can devise a more complex but more efficient mining operation - say, one man getting the surface work and one getting the shaft. Or, if there are two fellows this year instead of one, and if copper is therefore harder to get and more scarce, both Alpha and Gamma may spend considerable time looking for new lodes of copper.

Alpha and Gamma may follow still other courses of action. Perhaps they will invent better ways of obtaining copper from a given lode, say a better digging tool, or they may develop new materials to substitute for copper, perhaps iron.

How are the situations different if both Alpha and Gamma Crusoe are on the island, compared to Alpha being alone? In the short run, the cost of copper to Alpha will probably be higher if Gamma is there too, unless or until one of them discovers an improved production method (perhaps a method that requires two workers) or a product that can substitute for copper. And Alpha's offspring, Beta, also probably would be better off if Gamma had never shown up. But we of later generations are almost surely better off if Gamma does appear on the scene and hence (a) increases the population size, (b) increases the demand for copper, (c) increases the cost of getting it, and then (d) invents improved methods of getting it and using it, and discovers substitute products.

A larger population due to Gamma and other persons influences later costs in two beneficial ways. First, the increased demand for copper leads to increased pressure for new discoveries. Second, and perhaps even more important, a larger population implies more people to think and imagine, be ingenious, and finally make these discoveries.

The Family Analogy

The analogy of the family is sometimes (though not always) a satisfactory intuitive shortcut toward understanding the effects of population growth. For example, if a family decides to have an additional child, there is less income available to be spent on each of the original family members, as it is with a country as a whole. The family may respond to the additional "need" with the parents working more hours for additional pay, and so it is with a nation. The family may choose to save less, in order to pay for the additional expenses, or to save more in order to pay for later expenses such as education; so it is for a nation as a whole. The additional child has no immediate economic advantages to the family, but later it may contribute to the parents and other relatives; so it is for society as a whole. And like a nation, the family must balance off the immediate non-economic psychic benefits plus the later economic benefits against the immediate cost of the child. The main way that the family analogy diverges from the situation of a nation as a whole is that an additional person in the nation contributes to the stock of knowledge and to the scale of the market for the society as a whole, which benefits the entire economy, whereas an individual family is not likely to benefit much from its own discoveries.

The family model goes wrong, however, when it directs attention away from the possibility of creating new resources. If one thinks of a family on a desert island with a limited supply of pencils and paper, then more people on the island will lead to a pencil-and-paper scarcity sooner than otherwise. But for a society as a whole, there is practically no resource that is not either growable (such as trees for paper) or replaceable (except energy).

If the family starts with a given plot of land and an additional child is born, it would seem as if the result would be less land per child to be inherited. But the family can increase its "effective" land by irrigation and multiple cropping and even hydroponics, and some families respond by opening up whole new tracts of previously uncultivated land. Hence an additional child need not increase the scarcity of land and other natural resources, as appears to be inevitable when one looks at the earth as a closed resource system; instead, there is an increase in total resources.

But, you ask, how long can this go on? Surely not forever? In fact there is no logical or physical reason why the process cannot indeed go on forever. Let's return to copper as an example. Given substitute materials, development of improved methods of extraction, and discoveries of new lodes in the U.S. and in other countries and in the sea and perhaps on other planets, there is no logical reason why additional people should not increase the availability of copper or copper equivalents indefinitely.

To make the logical case more binding, the possibility of recycling copper at a faster rate due to population growth also improves the supply of the services we now get from it. To illustrate, consider a copper jug that one rubs to obtain the services of a genie. If only the single jug exists, and there are two families at opposite ends of the earth, each of them can obtain the genie very infrequently. But if the earth is populated densely, the jug can be passed rapidly from hand to hand, and all families might then have a chance to obtain the recycled jug and its genie more often than with a less dense population. So it could be with copper pots, or whatever. The apparent reason that this process cannot continue - the seeming finitude of copper in the solid earth – is invalid.

Of course, it is logically possible that the cost of the services we get now from copper and other minerals will be relatively higher in the future than now if there are more people in the future. But all past history suggests that the better guess is that cost and price will fall, just as scarcity historically has diminished along with the increase in population. Either way, however, the concept of mineral resources as "finite" is unnecessary, confusing, and misleading. And the notion of our planet as "spaceship earth," launched with a countable amount of each resource and hence having less minerals per passenger as the number of passengers is greater, is dramatic but irrelevant.

To repeat, we cannot know for certain whether the cost of the services we get from copper and other minerals will be cheaper in either year t + 50 or year t + 500 if population becomes 1 'thillion' rather than 2 'thillion' in year t. The historical data, however, show that the costs of minerals have declined faster in recent centuries, when population was larger, than in earlier centuries. This is not conclusive evidence that a bigger population implies lower costs. And higher income and a larger base of existing knowledge contribute to the cost-reduction process. But there is much less evidence - in fact, none at all - that a higher population in year t means a higher cost and greater scarcity of minerals in year t + 50 or t + 500.

Do you still doubt that the cost of mineral resources will be lower in the future than now? Do you still doubt that higher population growth now will eventually mean lower mineral costs? If so, I suggest as a mental exercise that you ask yourself: Would we be better off if people in the past had used less copper or coal? How great would our technological capacities to extract, process, and use these materials now be if we were just discovering these materials today?[1]

Jokes don't always go over well in serious books. But a joke that I have long been fond of seems appropriate here. Seventy-year-old Zeke's girlfriend has just left him after thirty years, and Zeke is in despair. His friends try hard to console him, and they especially point out again and again that in time he'll get over it, and might well meet another woman. Finally Zeke turns his tear-stained face to them and says, "But you don't understand. What am I going to do tonight?"

Similarly, you may well ask about the near-term effect of population growth on resources, after all this talk about the long run. There is more comfort for you than for Zeke, however. True, within a very short time there is little chance for the natural-resource supply to accommodate to a sudden increase in demand. But population growth is a very slow-acting phenomenon, not changing radically in any short period. And it is not until many years after the birth of a child that the additional person uses much natural resources. For both these reasons, modern industry has plenty of time to respond to changes in actual demand, and we need not fear short-run price run-ups due to increased population growth.

A Model of the Increase in Natural Resources

All natural resources - minerals, food, and energy - have become less rather than more scarce throughout human history. But it is counter-intuitive, against all common sense, for more people to result in more rather than less natural resources. So here is the theory again: More people, and increased income, cause problems of increased scarcity of resources in the short run. Heightened scarcity causes prices to rise. The higher prices present opportunity, and prompt inventors and entrepreneurs to search for solutions. Many fail, at cost to themselves. But in a free society, solutions are eventually found. And in the long run the new developments leave us better off than if the problems had not arisen. That is, prices end up lower than before the increased scarcity occurred.

The "Energy Crisis" and Population Policy

It is standard wisdom that population growth worsens the energy situation. Here is the typical view of the predecessor agency of the U.S. Department of Energy, in a brochure written for the public at large.

"In other parts of the world, particularly in the developing areas, populations are growing rapidly and each new baby further strains already inadequate energy resources. Thus, if developing areas are to grow economically, it seems clear that they must first deal with the population problem. But the rich nations, too, must control population growth. If not, there simply will not be enough energy to go around, unless per capita energy consumption is held steady or reduced - and that seems unlikely.... We must learn to conserve energy and use it more wisely or we're going to be in serious trouble"[2]

The prevalence of this unsound thinking demands that we inquire into the effects of population upon the supply of energy. We want to know: What will be the effect of more or fewer people upon the future scarcity and prices of energy?

This much we can say with some certainty: (1) With respect to the short-run future - within say thirty years - this year's population growth rate can have almost no effect on the demand for energy or on its supply. (2) In the intermediate run, energy demand is likely to be proportional to population, all else equal; hence additional people require additional energy. (3) For the longer run, whether additional population will increase scarcity, reduce scarcity, or have no effect on scarcity is theoretically indeterminate.

The outcome will depend on the net effect of increased demand on the current supplies energy as of a given moment, together with increases in potential supplies through discoveries and technological advances that will be induced by the increase in demand. In the past, increased demand for energy has been associated with reduced scarcity and cost.

There is no statistical reason to doubt the continuation of this trend. More particularly, there seems to be no reason to believe that we are now at a turning point in energy history, and no such turning point is visible in the future. This implies a trend toward lower energy prices and increased supplies.

It is important to recognise that in the context of population policy, who is "right" about the present state of energy supplies really does not matter. Yes, we will care in the year 2010 whether there will be large or small supplies of oil and gas and coal at prices relatively high or low compared to now, and even more so if government intervention in the market worsens the situation (as it usually does) and forces us to wait in line at the service station. And it matters to the State Department and the Department of Defence whether our national policies about energy pricing and development lead to large or small proportions of our energy supply being imported from abroad. But from the standpoint of our national standard of living it will matter very little even if energy prices are at the highest end of the range of possibilities as a result of relatively unfruitful technological progress and of maximum increases in demand due to maximum rises in GNP and population. At a very unlikely high price of energy equivalent to, say, $60 per barrel of oil (1992 dollars) there should be enough energy from coal, shale oil, solar power, natural gas, and fossil oil plus oil from biomass - buttressed by the virtually inexhaustible supply of nuclear power - to last so many hundreds or thousands of years into the future, or millions if we include nuclear energy, that it simply does not matter enough to estimate how many hundreds or millions of years. And even if energy would sell at such a most-unlikely high price, rather than the actual

1993 oil price of (say) $15 per barrel, the difference in our standard of living would hardly be noticeable.

From this we may conclude that whatever impact population growth might have upon the energy situation - negative effects through increased demand, positive effects through new discoveries, with a net effect that may be positive or negative - the long-run effect of population growth on the standard of living through its effect on energy costs is quite unimportant. And refined calculations of its magnitude are of no interest in this context.

Are We Running a Ponzi Scheme on Future Generations?

Several writers — among whom the first may have been the Nobel-prize-winning economist Paul Samuelson (in 1975) — have said that population growth constitutes a pyramid game or "Ponzi scheme." Here is how one letter-writer put it:

> "Julian Simon's solution for our economic woes is a pyramid scheme for which our children will pay with a degraded environment and a worsened quality of life. Shame on American Demographics for touting such stuff."
>
> —*Barbara Willin, Summerville, South Carolina*[3]

A Ponzi scheme is a fraud in which each early buyer is sold a franchise for recruiting several additional buyers, each of whom receives a franchise for recruiting additional buyers, and so on. Each person who succeeds in recruiting his/her full quota makes money. But eventually there are no more buyers to be found because the market is saturated. The scheme collapses, and the later franchise buyers lose their money. The scheme is named after Charles Ponzi, who perfected a similar scheme in the securities market in the 1920s.

But population growth does not constitute a Ponzi scheme: there is no reason to expect sources to run out: resources may be expected to become more available rather than more scarce. Hence there is no reason to think that consumption in the present is at the expense of future consumers, or that more consumers now imply less for consumers in the future. Rather, it is reasonable to expect that more consumption now implies more resources in the future because of induced discoveries of new ways to supply resources, which eventually leave resources cheaper and more available than if there were less pressure on resources in the present.

There is a second important difference between a Ponzi scheme and this view of population and resources. As the Ponzi scheme begins to peter out, the price of franchises falls as sellers find it more difficult to induce more buyers to purchase, and the system begins to fall apart. But if a resource becomes in shorter supply in any period, price rises in a fashion that reduces usage (and presumably reduces population growth), and hence it constitutes a self-adjusting rather than a self-destructing system.

Of course this view of population and resources runs against all "common sense" — that is, against conventional belief. But science is only interesting when it gives us knowledge that is not arrived at by common sense alone.

Natural Resources and the Risk of Running Out

You might wonder: Even if the prospect of running out of energy and minerals is small, is it safe to depend on the continuation of technical progress? Can we be sure that technological progress will continue to forestall growing scarcity and even increase the availability of natural resources? Would it not be prudent to avoid even a small possibility of a major scarcity disaster? Would it not be less risky to curb population growth to avoid the mere possibility of natural-resource scarcities even if the chances really are good that higher population will lead to lower costs? A reasonable person may be "risk averse."

Risk aversion is not very relevant for natural resources, for several reasons. First, the consequences of a growing shortage of any mineral - that is, of a rise in relative price - are not dangerous to life or even to the standard of living, as noted above with respect to energy. Second, a relative scarcity of one material engenders the substitution of other materials - say, aluminium for steel - and hence mitigates the scarcity. Third, a scarcity of any mineral would manifest itself only very slowly, giving plenty of opportunity to alter social and economic policies appropriately. Fourth, just as greater affluence and larger population contribute to the demand for more natural resources, they also contribute to our capacity to alleviate shortages and broaden our technological and economic capacity, which makes any particular material ever less crucial. Fifth and perhaps most important, we already have technology in hand - nuclear fission - to supply our energy needs at constant or declining cost forever.

Even so, let's next say a word about the appropriate level of confidence that progress will continue in the future.

Can We Be Sure Technology Will Advance?

Some ask: can we know that there will be discoveries of new materials and of productivity-enhancing techniques in the future? Behind the question lies the implicit belief that the production of new technology does not follow predictable patterns of the same sort as the patterns of production of other products such as cheese and opera. But there seems to me no warrant for belief in such a difference, either in logic or in empirical experience. When we add more capital and labour, we get more cheese; we have no logical assurance of this, but such has been our experience, and therefore we are prepared to rely upon it. The same is true concerning knowledge about how to increase the yield of grain, cows, milk and cheese from given amounts of capital and labour. If you pay engineers to find ways to solve a general enough problem - for example, how to milk cows faster, or with less labour - the engineers predictably will do so. There may well be diminishing returns to additional inventive effort spent on the same problem, just as there are diminishing returns to the use of fertiliser and labour on a given farm in a given year. But as entirely new forms of technology arise and are brought to bear on the old problems, the old diminishing-returns functions then no longer apply.

The willingness of businesses to pay engineers and other inventors to look for new discoveries attests to the predictability of returns to inventive effort. To obtain a more intimate feeling for the process, one may ask a scientist or engineer whether he/she expects his/her current research project to produce results with greater probability than if she/he simply sat in the middle of the forest reading a detective novel; the trained effort the engineer applies has a much greater likelihood of producing useful information - and indeed, the very information that is expected in advance - than does untrained on-effort. This is as predictable in the aggregate as the fact that cows will produce milk, and that machines and workers will turn the milk into cheese. Therefore, to depend upon the fact that technical developments will continue to occur in the future - if we continue to devote human and other resources to research - is as reasonable as it is to depend upon any other production process in our economy or civilisation. One cannot prove logically that technical development will continue in the future. But neither can one so prove that capital and labour and milk will continue to produce cheese, or that the sun will come up tomorrow.

As I see it, the only likely limit upon the production of new knowledge about resources is the occurrence of new problems; without unsolved problems there will be no solutions. But here we have a built-in insurance policy: if our ultimate interest is resource availability, and if availability should diminish, that automatically supplies an unsolved problem, which then leads to the production of new knowledge, not necessarily immediately or without short-run disruption, but in the long run.

I'm not saying that all problems are soluble in the forms in which they are presented.

I do not claim that biologists will make us immortal in our lifetime, or even that the length of human life will be doubled or tripled in the future. On the other hand, one need not rule out that biogenetics can create an animal with most of our traits and a much longer life. But such is not the sort of knowledge we are interested in here. Rather, we are interested in knowledge of the material inputs to our economic civilisation.

A sophisticated version of this argument is that the cost of additional knowledge may rise in the future. Some writers point to the large teams and large sums now involved in natural-science endeavours. Let us notice, however, how much cheaper it is to make any discoveries now than it was in the past because of the existing base of knowledge and the whole information infrastructure. Simon Kuznets could advance further with his research on GNP estimates than could William Petty. And a run-of-the-mill graduate student can now do some things that Petty could not do. Additionally, a given discovery is more valuable now than it was then; GNP measurement has more economic impact now than in Petty's day. I have calculated that the net present value of the invention of agriculture in social terms at the time of discovery was less than the net present value now of something even as trivial as computer games, because of the small population and income then (discounted even at 2 percent per year)[4] in gross social product from the transition to nuclear fission. And agriculture was the only big discovery for thousands of years, whereas the transistor and nuclear power and lots more inventions occurred within just a few recent decades.

Summary: The Ultimate Resource is the Human Imagination in a Free Society

There is no persuasive reason to believe that the relatively larger use of natural resources that would occur with a larger population would have any special deleterious effects upon the economy in the future. For the foreseeable future, even if the extrapolation of past trends is badly in error, the cost of energy is not an important consideration in evaluating the impact of population growth. Other natural resources may be treated in a manner just like any other physical capital when considering the economic effect of different rates of population growth. Depletion of mineral resources is not a special danger for the long run or the short run. Rather, the availability of mineral resources, as measured by their prices, may be expected to increase - that is, costs may be expected to decrease - despite all notions about "finiteness."

Sound appraisal of the impact of additional people upon the "scarcity" (cost) of a natural resource must take into account the feedback from increased demand to the discovery of new deposits, new ways of extracting the resource, and new substitutes for the resource. And we must take into account the relationship between demand now and supply in various future years, rather than considering only the effect on supply now of greater or lesser demand now. And the more people there are, the more minds that are working to discover new sources and increase productivity, with raw materials as with all other goods.

This point of view is not limited to economists. A technologist writing on minerals put it this way: "In effect, technology keeps creating new resources."[5] The major constraint upon the human capacity to enjoy unlimited minerals, energy and other raw materials at acceptable prices is knowledge. And the source of knowledge is the human mind. Ultimately, then, the key constraint is human imagination acting together with educated skills. This is why an increase of human beings, along with causing an additional consumption of resources, constitutes a crucial addition to the stock of natural resources.

We must remember, however, that human imagination can flourish only if the economic system gives individuals the freedom to exercise their talents and to take advantage of opportunities. So another crucial element in the economics of resources and population is the extent to which the political-legal-economic system provides personal freedom from government coercion. Skilled persons require an appropriate framework that provides incentives for working hard and taking risks, enabling their talents to flower and come to fruition. The key elements of such a framework are economic liberty, respect for property, and fair and sensible rules of the market that are enforced equally for all.

We - humanity - should be throwing ourselves the party to outdo all parties, a combination graduation-wedding-birthday-all-rites-of passage party, to mark our emergence from a death-dominated world of raw-material scarcity. Sing, dance, be merry - and work.

But instead we see gloomy faces. They are spoilsports, and they have bad effects.

The spoilsports accuse our generations of having a party - at the expense of generations to come. But it is those who use the government to their own advantage who are having a party at the expense of others - the bureaucrats, the grants-grabbers, the subsidy-looters. Don't let them spoil our merry day.

Note

1. Fischman and Landberg, "Resource and Environmental Consequences of Population Growth and The American Future". In Ronald Ridker, ed., 1972, *Population, Resources, and The Environment. Vol. 3*. The Commission on Population Growth and The American Future. Washington, D.C.

2. U.S. ERDA. 1976. Energy and the Environment. Washington, D.C.

3. *American Demographics*, September 1990, pp 52-53.

4. Simon, Julian L., "Paradoxes and Difficulties in the Evaluation of Progress and Technological Advance, Past and Future," *Technology in Society*, Vol. 10, 1988, pp. 425-432, reprinted in Simon, 1992.

5. Feiss, Julian W. 1963-65. "Minerals". In Scientific American eds., *Technology and Economic Development*. Harmondsworth, Eng.: Penguin, in association with Chatto & Windus. p. 117.

Property Rights, Markets and Sustained Development

3

The Market and Nature

FRED L. SMITH, JR.

Many environmentalists are dissatisfied with the environmental record of free economies. Capitalism, it is claimed, is a wasteful system, guilty of exploiting the finite resources of the Earth in a vain attempt to maintain a non-sustainable standard of living. Such charges, now raised under the banner of "sustainable development," are not new. Since Malthus made his dire predictions about the prospects for world hunger, the West has been continually warned that it is using resources too rapidly and will soon run out of something, if not everything. Nineteenth century experts such as W.S. Jevons believed that world coal supplies would soon be exhausted and would have been amazed that over 200 years of reserves now exist. US timber "experts" were convinced that North American forests would soon be a memory. They would similarly be shocked by the reforestation of eastern North America—reforestation that has resulted from market forces and not mandated government austerity.

In recent decades, the computer-generated predictions of the Club of Rome enjoyed a brief popularity, arguing that everything would soon disappear. Today, sustainable-development theorists from the World Bank's Herman Daly and the United Nations' Maurice Strong to US Vice President Albert Gore and Canadian author David Suzuki, seem certain that, at last, Malthus will be proven right. It was this environmental view on display at the United Nations' "Earth Summit" in Rio de Janeiro in 1992. This conference, vast in scope and mandate, was but the first step in the campaign to make the environment the central organising principle of global institutions.

If such views are taken seriously, then the future will indeed be a very gloomy place, for if such disasters are in the immediate future, than drastic government action is necessary. Consider the not atypical view of David Suzuki:

"There has to be a radical restructuring of the priorities of society. That means we must no longer be dominated by global economics, that the notion that we must continue to grow indefinitely is simply off, that we must work towards not zero growth, but negative growth."

For the first time in world history, the leaders of the developed nations are being asked to turn their backs on the future. The resulting policies could be disastrous for all humankind.

The Environmental Challenge

The world does indeed face a challenge in protecting ecological values. Despite tremendous success in many areas, many environmental concerns remain. The plight of the African elephant, the air over Los Angeles, the hillsides of Nepal, the three million infant deaths from water-borne diseases throughout the world, and the ravaging of Brazilian rain forests all dramatise areas where problems persist and innovative solutions are necessary.

Sustainable development theorists claim these problems result from "market failure": the inability of capitalism to address environmental concerns adequately. Free market proponents suggest such problems are not the result of market forces, but rather of their absence. The market already plays a critical role in protecting those resources which are privately owned and for which political interference is minimal. In these instances there are truly sustainable practices. Therefore, those concerned with protecting the environment and ensuring human prosperity should seek to expand capitalism, through the extension of property rights, to the broadest possible range of environmental resources. Our objective should be to reduce, not expand, political interference in both the human and natural environments.

Private stewardship of environmental resources is a powerful means of ensuring sustainability. Only people can protect the environment. Politics *per se* does nothing. If political arrangements fail to encourage individuals to play a positive role, the arrangements can actually do more harm than good. There are tens of millions of species of plants and animals that merit survival. Can we imagine that the

150 or so governments on this planet—many of which do poorly with their human charges—will succeed in so massive a stewardship task? Yet, there are in the world today over five billion people. Freed to engage in private stewardship, the challenge before them becomes surmountable.

Sustainable Development and its Implications

The phrase sustainable development suggests a system of natural resource management that is capable of providing an equivalent, or expanding, output over time. As a concept, it is extremely vague, often little more than a platitude. Who, after all, favours non-sustainable development? The basic definition promoted by Gro Harlem Brundtland, former Prime Minister of Norway and a prominent player at the 1992 Earth Summit, is fairly vague as well: "Sustainable development is a notion of discipline. It means humanity must ensure that meeting present needs does not compromise the ability of future generations to meet their own needs."

In this sense, sustainability requires that as resources are consumed one of three things must occur: New resources must be discovered or developed; demands must be shifted to more plentiful resources; or, new knowledge must permit us to meet such needs from the smaller resource base. That is, as resources are depleted, they must be renewed. Many assume that the market is incapable of achieving this result. A tremendous historical record suggests exactly the opposite.

Indeed, to many environmental "experts," today's environmental problems reflect the failure of the market to consider ecological values. This market-failure explanation is accepted by a panoply of political pundits of all ideological stripes, from Margaret Thatcher to Earth First! The case seems clear. Markets, after all, are short-sighted and concerned only with quick profits. Markets undervalue biodiversity and other ecological concerns not readily captured in the marketplace. Markets ignore effects generated outside of the market, so-called externalities, such as pollution. Since markets fail in these critical environmental areas, it is argued, political intervention is necessary. That intervention should be careful, thoughtful, even scientific, but the logic is clear: Those areas of the economy having environmental impacts must be politically controlled. Since, however, every economic decision has some environmental effect, the result is an effort to regulate the whole of human activity.

Thus, without any conscious decision being made, the world is moving decisively toward central planning for ecological rather than economic purposes. The Montreal Protocol on chlorofluorocarbons, the international convention on climate change, the convention on biodiversity, and the full range of concerns addressed at the UN Earth Summit—all are indicative of this rush to politicise the world's economies. That is unfortunate, for ecological central planning is unlikely to provide for a greener world.

Rethinking the Market Failure Paradigm

The primary problem with the market-failure explanation is it demands too much. In a world of pervasive externalities—that is, a world where all economic decisions have environmental effects—this analysis demands all economic decisions be politically managed. The world is only now beginning to recognise the massive mistake entailed in economic central planning; yet, the "market failure" paradigm argues that we embark on an even more ambitious effort of ecological central planning. The disastrous road to serfdom can just as easily be paved with green bricks as with red ones.

That markets "fail" does not mean that governments will "succeed." Governments, after all, are susceptible to special interest pleadings. A complex political process often provides fertile ground for economic and ideological groups to advance their agendas at the public expense. The US tolerance of high-sulphur coal and the massive subsidies for heavily polluting "alternative fuels" are evidence of this problem. Moreover, governments lack any means of acquiring the detailed information dispersed throughout the economy essential to efficiency and technological change.

More significantly, if market forces were the dominant cause of environmental problems then the highly industrialised, capitalist countries should suffer from greater environmental problems than their centrally-managed counterparts. This was once the conventional wisdom. The Soviet Union, it was argued, would have no pollution because the absence of private property, the profit motive, and individual self-interest would eliminate the motives for harming the environment. The opening of the Iron Curtain exploded this myth, as the most terrifying ecological horrors ever conceived were shown to be the Communist reality. The lack of property rights and profit motivations discouraged efficiency, placing a greater stress on natural resources. The result was an environmental disaster.

Do Markets Fail—or Do We Fail to Allow Markets?

John Kenneth Galbraith, an avowed proponent of statist economic policies, inadvertently suggested a new approach to environmental protection. In an oft-quoted speech he noted that the United States was a nation in which the yards and homes were beautiful and in which the streets and parks were filthy. Galbraith then went on to suggest that we effectively nationalise the yards and homes. For those of us who believe in property rights and economic liberty, the obvious lesson is quite the opposite.

Free market environmentalists seek ways of placing these properties in the care of individuals or groups concerned about their well-being. This approach does not, of course, mean that trees must have legal standing, but rather is a call for ensuring that behind every tree, stream, lake, air shed, and whale stands at least one owner who is able and willing to protect and nurture that resource.

Consider the plight of the African elephant. On most of the continent, the elephant is managed like the American buffalo once was. It remains a political resource. Elephants are widely viewed as the common heritage of all the peoples of these nations and are thus protected politically. The "common property" management strategy being used in Kenya and elsewhere in East and Central Africa has been compared and contrasted with the experiences of those nations such as Zimbabwe which have moved decisively in recent years to transfer elephant-ownership rights to regional tribal councils. The differences are dramatic. In Kenya, and indeed all of eastern Africa, elephant populations have fallen by over 50 per cent in the last decade. In contrast, Zimbabwe's elephant population has been increasing rapidly. A programme of conservation through use that relies upon uniting the interest of man and the environment succeeds where political management has failed.

The Market and Sustainability

The prophets of sustainability have consistently predicted an end to the world's abundant resources, while the defenders of the free market point to the power of innovation—innovation which is encouraged in the marketplace. Consider the agricultural experience. Since 1950, improved plant and animal breeds, expanded availability and types of agri-chemicals, innovative agricultural techniques, expanded irrigation, and better pharmaceutical products have all combined to spur a massive expansion of world food supplies. That was not expected by those now championing "sustainable development." Lester Brown, in his 1974 Malthusian publication By Bread Alone, suggested that crop-yield increases would soon cease. Since that date, Asian rice yields have risen nearly 40 per cent, an approximate increase of 2.4 per cent per year. This rate is similar to that of wheat and other grains. In the developed world it is food surpluses, not food shortages, that present the greater problem, while political institutions continue to obstruct the distribution of food in much of the Third World.

Man's greater understanding and ability to work with nature have made it possible to achieve a vast improvement in world food supplies, to improve greatly the nutritional levels of a majority of people throughout the world in spite of rapid population growth. Moreover, this has been achieved while reducing the stress to the environment. To feed the current world population at current nutritional levels using 1950 yields would require ploughing under an additional 10 to 11 million square miles, almost tripling the world's agricultural land demands (now at 5.8 million square miles). This would surely come at the expense of land being used for wildlife habitat and other applications.

Moreover, this improvement in agriculture has been matched by improvements in food distribution and storage, again encouraged by natural market processes and the "profit incentive" that so many environmentalists deplore. Packaging has made it possible to reduce food spoilage, reduce transit damage, extend shelf life, and expand distribution regions. Plastic and other post-use wraps along with the ubiquitous Tupperware have further reduced food waste. As would be expected, the United States uses more packaging than Mexico, but the additional packaging results in tremendous reductions in waste. On average, a Mexican family discards 40 per cent more waste each day. Packaging often eliminates more waste than it creates.

Despite the fact that capitalism has produced more environment-friendly innovations than any other economic system, the advocates of sustainable development insist that this process must be guided by benevolent government officials. That such efforts, such as the United States' synthetic fuels project of the late 1970s, have resulted in miserable failures is rarely considered.

In the free market, entrepreneurs compete in developing low-cost, efficient means to solve contemporary problems. The promise of a potential profit, and the freedom to seek after it, always provides the incentive to build a better mousetrap, if you will. Under planned economies, this incentive for innovation can never be as strong, and the capacity to reallocate resources toward more efficient means of production is always constrained.

This confusion is also reflected in the latest environmental fad: waste reduction. With typical ideological fervour, a call for increased efficiency in resource use becomes a call to use less of everything, regardless of the cost. Less, we are told, is more in terms of environmental benefit. But neither recycling nor material or energy use reductions *per se* are a good thing, even when judged solely on environmental grounds. Recycling paper often results in increased water pollution, increased energy use, and, in the United States, actually discourages the planting of new trees. Mandating increased fuel efficiency for automobiles reduces their size and weight, which in turn reduces their crashworthiness and increases highway fatalities. Environmental policies must be judged on their results, not just their motivations.

Overcoming Scarcity

Environmentalists tend to focus on ends rather than processes. This is surprising given their adherence to ecological teaching. Their obsession with the technologies and material-usage patterns of today reflects a failure to understand how the world works. The resources people need are not chemicals, wood fibre, copper, or the other raw materials of concern to the sustainable-development school. We demand housing, transportation, and communication services. How those demands are met is a derivative result based on competitive forces—forces which respond by suggesting new ways to meet old needs as well as improving the ability to meet needs in older ways.

Consider, for example, the fears expressed in the early post-war era that copper would soon be in short supply. Copper was the life-blood of the world's communication system, essential to linking together humanity throughout the world. Extrapolations suggested problems and copper prices escalated accordingly. The result? New sources of copper in Africa, South America, and even the United States and Canada, were found. That concern, however, also prompted others to review new technologies, an effort that produced today's rapidly expanding fibre optics links.

Such changes would be viewed as miraculous if not now commonplace in the industrialised, and predominantly capitalistic, nations of the world. Data assembled by Lynn Scarlett of the Reason Foundation noted that a system requiring, say, 1,000 tonnes of copper can be replaced by as little as 25 kilograms of silicon, the basic component of sand. Moreover, the fiber-optics system has the ability to carry over 1,000 times the information of the older copper wire. Such rapid increases in communication technology are also providing for the displacement of oil as electronic communication reduces the need to travel and commute. The rising fad of telecommuting was not dreamed up by some utopian environmental planner, but was rather a natural outgrowth of market processes.

It is essential to understand that physical resources are, in and of themselves, largely irrelevant. It is the interaction of man and science that creates resources: Sand and knowledge become fibre optics. Humanity and its institutions determine whether we eat or die. The increase of political control over physical resources and new technologies only increases the likelihood of famine.

Intergenerational Equity

Capitalism is ultimately attacked on grounds of unsustainability for its purported failure to safeguard the needs of future generations. Without political intervention, it is argued, capitalists would leave a barren globe for their children. Thus, it is concluded, intergenerational equity demands that politics intervene. But are these criticisms valid?

Capitalists care about the future because they care about today's bottom line. Market economies have created major institutions—bond and stock markets, for example—which respond to changes in operating policies that impact future values. A firm that misuses capital or lowers quality standards, a pet store that mistreats its stock, a mine that reduces maintenance, a farmer that permits erosion—all will find the value of their capital assets falling. Highly specialised researchers expend vast efforts ferreting out changes in management practices that might affect future values; investment houses pay future-analysts very well indeed to examine such questions.

Markets, of course, are not able to foresee all eventualities, nor do they consider consequences hundreds of years into the future. Yet, consider the time horizon of politicians. In most cases, they are concerned with only one thing: getting re-elected, a process that provides them at best a limited time horizon. Politically-managed infrastructure is routinely undermaintained; funds for new roads are more attractive than the smaller sums used to repair potholes; national forests are more poorly maintained than private forests; erosion is more serious on politically-controlled lands than on those maintained by private corporations. If the free market is shortsighted in its view of the future, then the political process is even more so. It is therefore the free market which best ensures that there will be enough for the future.

Warring Paradigms

The two alternative perspectives on environmental policy—free markets and central planning—differ dramatically. One relies upon individual ingenuity and economic liberty to harness the progressive nature of market forces. The other rests upon political manipulation and government coercion. In point of fact, these approaches are antithetical. There is little hope of developing a "third way." Yet, there has been little debate on which approach offers the greatest promise in enhancing and protecting environmental concerns. The political approach has been adopted on a wide scale throughout the world, with more failure than success, while efforts to utilise the free-market approach have been few and far between.

Nevertheless, there are numerous cases where private property rights have been used to complement and supplement political environmental strategies. One excellent example is a case in England in the 1950s where a fishing club, the Pride of Derby, was able to sue upstream polluters for trespassing against private property. Even the pollution issuing from an upstream municipality was addressed. This ability to go against politically-preferred polluters rarely exists where environmental resources are politically managed.

At the heart of the division between statist and free market environmentalists is a difference in moral vision. Free market environmentalists envision a world in which man and the environment live in harmony, each benefiting from interaction with the other. The other view, which dominates the environmental establishment, believes in a form of ecological apartheid whereby man and nature must be separated, thus protecting the environment from human influence. From this view rises the impetus to establish wilderness lands where no humans may tread, and a quasi-religious zeal to end all human impact on nature.

Thus, the establishment environmentalists view pollution—human waste—as an evil that must be eliminated. That waste is an inevitable by-product of human existence is of secondary concern. To the environmentalist that endorses this ideology, nothing short of civilization's demise will suffice to protect the earth.

The view that free market environmentalists endorse is somewhat different. Not all waste is pollution, but only that waste which is transferred involuntarily. Thus, it is pollution to dispose of garbage on a neighbour's lawn, but not to store it on one's own property. The voluntary transfer of waste, perhaps from an industrialist to the operator of a landfill or recycling facility, is merely another market transaction.

Conclusion

The United Nations Earth Summit in Rio de Janeiro considered an extremely important issue: What steps should be taken to ensure that economic and ecological values are harmonised?

The world faces a fateful choice as to how to proceed: by expanding the scope of individual action via a system of expanded private property rights and the legal defences associated with such rights, or by expanding the power of the state to protect such values directly. In making that choice, we should learn from history. Much of the world is only now emerging from decades of efforts to advance economic welfare via centralised political means, to improve the welfare of mankind by restricting economic freedom, by expanding the power of the state, to test out the theory that market forces are inadequate to protect the welfare of society. That experiment has been a clear failure on economic, civil liberties, and even ecological grounds. Economic central planning was a utopian dream; it became a real world nightmare.

Today, the international environmental establishment seems eager to repeat this experiment in the ecological sphere, increasing the power of the state, restricting individual freedom, certain that market forces cannot adequately protect the ecology. Yet, as I've quickly sketched out here, this argument is faulty. Wherever resources have been privately protected, they have done better than their politically managed counterparts—whether we are speaking of elephants in Zimbabwe, salmon streams in England, or beaver in Canada. Where such rights have been absent or suppressed, the results have been less fortunate. Extending property rights to the full array of resources now left undefended, now left as orphans in a world of protected properties, is a daunting challenge. Creative legal arrangements and new technologies will be necessary to protect the oceans and airsheds of the world, but those tasks can be resolved if we apply ourselves. The obstacles to ecological central planning are insurmountable. The need for centralised information and a comprehensive system of controls in order to coerce the population of the world to act in highly restricted ways, as well as that for omniscient decision-makers to choose among technologies, can never be met.

Ecological central planning cannot protect the environment, but it can destroy our civil and economic liberties. There is too much at stake to allow the world to embark upon this course. The environment can be protected, and the world's peoples can continue to reach new heights of prosperity, but it is essential to realise that political management is not the proper approach. Rather, the leaders of the world should follow the path of the emerging nations of Eastern Europe and embrace political and economic freedom. In the final analysis, the free market is the only system of truly sustainable development.

4

The Common Law: How It Protects the Environment

ROGER E. MEINERS
BRUCE YANDLE

"Empowering those most directly affected by pollution, common law property rights protect powerfully, preventing polluters from arbitrarily fouling streams or spewing poisons onto neighboring property."

— Elizabeth Brubaker

The purpose of this paper is to show, by examining specific cases in American and English history, that strong legal traditions enabled ordinary citizens to protect their air, land, and water, often against politically potent parties. Even public law officials, such as attorneys general, used the common law to protect citizens against environmental dangers. Unfortunately, as we will see, statutory regulation has largely supplanted the common-law legal regime that once provided solutions for many environmental problems.

Protecting the Environment

Long ago, before the terminology of environmental degradation evolved and the regulatory machinery began to determine what constitutes illegal pollution, people knew that they and their property could suffer from noxious pollutants. Such pollution was offensive; sometimes it injured people's health; and sometimes it damaged property values. The protection against this invasion came primarily through legal actions for trespass and nuisance. Those who allowed something noxious to escape their control and invade the property of others could be held accountable for their actions through private litigation (In many cases, either trespass or nuisance could apply, since both actions were often involved when pollution reached others' property). While evidence of harm had to be shown for damages to be assessed, the basic notions are commonsensical.

Nuisance actions may be private or public. A private nuisance is a substantial and unreasonable interference with the use and enjoyment of an interest in property. Such interference may be intentional or may be due to carelessness. As US Supreme Court Justice George Sutherland said in a case in 1926, "Nuisance may be merely a right thing in a wrong place like a pig in the parlor instead of the barnyard."[1] Legal actions can lead to recovery for damages to land as well as to recovery for damages to health or any other benefit attached to our interests in property.

A public nuisance is an act that causes inconvenience or damage to public health or order or that obstructs public rights. If a business creates noxious emissions that affect many citizens, a public attorney may bring an action on behalf of all affected citizens to have the activity terminated.

Trespass created rights similar to those against nuisance. If a harmful substance is allowed, intentionally or carelessly, to invade the property of another, whether by land, air, or water, there may be a trespass. If so, the defendant is held responsible for damages.

Since water is often not owned by property owners whose land abuts a lake or a stream, the common law extends protection to water quality through *riparian rights*. Riparian rights to water are user rights that allow water users to sue those who damage water quality to the point where its use and enjoyment are reduced.

On the following pages, we will show how the principles of nuisance and trespass protected people against pollution in the past. We will look at protection of surface water through riparian rights, protection of underground water and land, and protection against air pollution. We will see the strengths of the common-law protections as well as their limitations.

Surface Water

In the late nineteenth century, the Carmichael family owned a 45-acre farm in Texas, with a stream running through it, that bordered on the state of Arkansas. In the 1890s the city of Texarkana, Arkansas, built a sewage system and connected numerous residences and businesses to it. The sewage collected by the city system was deposited "immediately opposite plaintiffs' homestead, about eight feet from the state line, on the Arkansas side."[2] The Carmichaels sued the city in federal court in Arkansas.

The Carmichaels were forced to connect their property to a public water system to obtain water for their family and livestock. The cost of the water hook-up and its use was $700. In addition, they claimed that the value of their property was reduced by $5,000; the enjoyment of their homestead over the previous two years was reduced by $2,000; and the dread of disease was valued at $2,000.

The court found that the

> cesspool is a great nuisance because it fouls, pollutes, corrupts, contaminates, and poisons the water of [the creek], depositing the foul and offensive matter . . . in the bed of said creek on plaintiffs' land and homestead continuously. . . . [thereby] depriving them of the use and benefit of said creek running through their land and premises in a pure and natural state as it was before the creation of said cesspool. . . .

The claims for damages were awarded. The Carmichaels also sought a permanent injunction against the cesspool. Judge Rogers found that the city of Texarkana was operating properly under state law to build a sewer system, but that there was no excuse for fouling the water used by the Carmichaels, regardless of how many city residences benefited from the sewer system.

Citing other cases, the court found that the action at law for damages was proper, as was the request for an injunction. The court cited a leading text on the law of torts:

> If a riparian proprietor has a right to enjoy a river so far unpolluted that fish can live in it and cattle drink of it and the town council of a neighboring borough, professing to act under statutory powers, pour their house drainage and the filth from water-closets into the river in such quantities that the water becomes corrupt and stinks, and fish will no longer live in it, nor cattle drink it, the court will grant an injunction to prevent the continued defilement of the stream, and to relieve the riparian proprietor from the necessity of bringing a series of actions for the daily annoyance. In deciding the right of a single proprietor to an injunction, the court cannot take

into consideration the circumstance that a vast population will suffer by reason of its interference.

Judge Rogers noted: "I have failed to find a single well-considered case where the American courts have not granted relief under circumstances such as are alleged in this bill against the city..."

A 1913 New York high court case illustrates how riparian rights could protect the water-quality rights of one citizen even against large business interests.[3] A new pulp mill polluted a creek. A downstream farmer, Whalen, sued the mill for making the water that passed by his land unfit for agricultural use. He had to obtain an alternative water source for his crops and animals. The trial court awarded damages of $312 and granted an injunction, ordering the mill to end harmful pollution within one year or close operations.

But the appellate division overturned the injunction and reduced the damages to $100. The court noted that the mill was an important economic asset to the area. It cost over $1 million to build and employed about five hundred people. Thus, it was worth far more than the water was to the plaintiff. However, the Court of Appeals (New York's highest court) unanimously reinstated the decision of the trial court:

> Although the damage to the plaintiff may be slight as compared with the defendant's expense of abating the condition, that is not a good reason for refusing an injunction. *Neither courts of equity nor law can be guided by such a rule, for if followed to its logical conclusion it would deprive the poor litigant of his little property by giving it to those already rich.* (italics added)

Even when a government gave an industry preferential water rights, the common law would not allow the preference to expand to include a right to pollute, as a turn-of-the-century case makes clear. The Idaho constitution stated that mining operations "shall have preference over those using the same [water] for manufacturing or agricultural purposes."[4] A farmer downstream from a mining operation sued for damage to his farm caused by acids dumped in the Coeur d'Alene River, which was used by the farm. The federal appeals court held that even if the mining operation was in "the ordinary and usual mode of mining" and even if it had a constitutional preference for access to water, it had no "right to dump injurious and deleterious materials into a stream." The court noted that, with the exception of one case from Pennsylvania, this view represents "an unbroken line of decisions in the United States and England."

So, long before the Environmental Protection Agency (EPA) came into existence, firms knew that if they substantially polluted their neighbors' water, they could expect to be found liable. To minimise liability, water polluters installed pollution control devices. Paper mills in Wisconsin routinely owned miles of downstream river property, knowing that otherwise they would be liable for violation of riparian rights[5].

While such water pollution cases showed up consistently over the years, they did not represent massive numbers of cases. For example, a 1971 survey of all reported American common-law cases involving violation of riparian rights (water pollution) found a total of 445 cases, including cases dating back to the nineteenth century. From 1945 to 1970, there were never more than four common-law water and air pollution case decisions reported in any given year, most of them involving water pollution.[6] The reason, we believe, is that the law was generally understood and therefore only infrequently contested.

Common-law protection of water does not apply just to those who own property that abuts a waterway but to all who have the right to use the water, for purposes including recreation. Those who enjoy sport fishing in England have long protected water quality through private litigation brought by angling associations. For example, a fishing club, Pride of Derby, and its association, the Derbyshire Angling Association, brought legal action against the borough of Derby, British Celanese, Ltd., and the British Electricity Authority.[7] The defendants discharged sewage, industrial waste, and heated effluent which polluted and raised the temperature of the River Derby, damaging the anglers' fishing. The suit by the fishing clubs was joined by the Earl of Harrington, who owned land along the river.

The lower court issued an injunction restraining the three defendants from reducing the quality of the river's water, but suspended the injunction for two years to give the defendants time to alter their operations. The borough also asked the Court to substitute damages for an injunction, since the community could not easily rebuild its sewage treatment plant. But the court rejected this position, noting that damages would be an inadequate remedy for the angling club, whose members want to exercise their right to fish. The injunction required the borough to redesign its sewage system, Celanese to change its discharge practices, and the British Electricity Authority to reduce its discharge of superheated water. This British case illustrates the effectiveness of common law in protecting property rights, in this case the right to fish, and shows how this protection translates to environmental protection. Because the right to fish is a privately owned right in Great Britain, individuals can protect streams effectively through court suits.

Land and Underground Water

The primary problem caused by pollution of land is groundwater pollution due to seepage from improperly disposed wastes. Wastes that are properly contained rarely cause harm. After the famous Love Canal incident in the late 1970s, Congress passed the Superfund law in 1980 (the Comprehensive Environmental Response Compensation and Liability Act) to regulate the cleanup of toxic waste sites. Despite this law, some common-law cases have occurred, and they show how the common law might have dealt with the problem of groundwater pollution over time.

For example, in 1981, the Illinois EPA, backed by the federal EPA, supported the right of a chemical waste landfill to remain in operation. The landfill had been built with state and federal approval, but residents of a nearby village alleged that the landfill was damaging their water supply. The Illinois supreme court found that the landfill was causing groundwater contamination and that there could be a chemical explosion given the disposal technique used.[8] It held that the landfill was a public and a private nuisance. The village residents were there first; their right not to have their property damaged could not be stripped in favour of a "general societal" desire for a landfill. The court noted that toxic landfills are legitimate, but they must be constructed so as not to impose costs on surrounding landowners who have not agreed to the intrusion. The court issued a permanent injunction against the landfill and ordered that the toxic wastes be dug up, moved, and the land restored.

The courts' view of the standards for groundwater contamination has evolved over the decades, as the 1982 case of *Wood vs. Picillo*[9] illustrates. Neighboring property owners sued a farmer who maintained a hazardous waste dump on his property. They claimed that the dump emitted noxious fumes and polluted groundwater. The Rhode Island Supreme Court agreed. In doing so it overturned a 1934 decision that would have supported the defendant's position. The 1934 decision was based on the state of science at that time, when knowledge about the course of groundwater was, as the court stated in 1982, "indefinite and obscure." Since 1934, the court said:

> the science of groundwater hydrology as well as societal concern for environmental protection has developed dramatically. As a matter of scientific fact the courses of subterranean waters are no longer obscure and mysterious. . . . We now hold that negligence is not a necessary element of a nuisance case involving contamination of public or private waters by pollutants percolating through the soil and traveling underground routes.

In other words, the common law now imposes strict liability (that is, liability even when there is no negligence) on polluters who cause damage to waters. This standard of care is consistent with old common-law tort rules imposing strict liability in case of hazardous materials, rules recorded in a famous 1868 British case, *Rylands v. Fletcher*. This case is often cited for restating the ancient proposition, "So use your property as not to injure your neighbor's property." Strict liability is imposed if there is evidence of injury or of potential to cause injury in cases of hazardous substances.

Thus, advances in knowledge of the effects of toxic substances, and the ability to track them, means tougher standards today than in years past. As a New York court noted in 1983, "One who creates a nuisance through an inherently dangerous activity or use of an unreasonably dangerous product is absolutely liable for resulting damages, regardless of fault, and despite adhering to the highest standard of care."[10]

In 1994, the supreme court of Florida, in a common-law case, held that a firm that damaged an underground water supply by negligent disposal of toxic wastes was responsible for the cost of restoring the groundwater to make it fit for human consumption.[11] The company argued that it should only be responsible for the change in the value of the land that covered the groundwater. The court rejected this damage measure, holding the firm responsible for the $3 million spent for alternative water supplies and $5.6 million for the estimated cost of restoration of groundwater quality.

Air

In comparison with surface and groundwater pollution cases, few common-law air pollution cases are found. One reason is that most air pollution comes from multiple sources, making it difficult to identify defendants. But it is clear that the courts have long recognised common-law liability for air pollution when liability can be assigned to a polluter causing harm.

An early case illustrating common-law protection against air pollution was *Georgia vs. Tennessee Copper Co.*[12] The state of Georgia, on behalf of its citizens, sued two companies that operated copper smelters in Tennessee near the Georgia border. The court noted that a public nuisance had been created because the "sulphurous fumes cause and threaten damage on so considerable a scale to the forests and vegetable life, if not to health, within [several counties in Georgia]. . . ." Defendants argued that they had recently constructed new facilities that reduced the scope of the problem, but the Supreme Court held for Georgia. The Court gave the companies a reasonable time to build more emission-control equipment, but held that if such equipment did not reduce emissions enough to protect plant life in Georgia, the state could ask the court for an injunction to shut down the smelters.

A similar strong support of the right to clean air was found in a case before the Arizona Supreme Court in 1931.[13] A heavy-metal smelter nine miles from the plaintiff's farm caused $1,300 worth of damage to his crops in 1926. In 1927, because the smoke continued to deposit harmful levels of chemicals on the farm, the plaintiff did not plant a crop but sued for the value that would have been created had the smoke not continued. The high court upheld the verdict for the farmer, holding that he could recover the profits that he could have earned from planting. The fact that the pollution resulted

> from the carrying on of a perfectly lawful business in the most approved way made no difference. The landowners were deprived of their legal right to farm unmolested by the poisonous fumes none the less by the fact that the discharge thereof was a necessary incident to a business of this character.

A more modern air pollution case is *R. L. Renken vs. Harvey Aluminum*.[14] An aluminum plant in The Dalles, Oregon, employed 550 people, making it the largest employer in town. It was sued by several orchard owners who claimed that their crops had been damaged by fluoride emissions. Finding the pollution to be a trespass and a nuisance, the court awarded the orchard owners approximately $10,000 each in damages for the crop losses and ordered the plant to install state-of-the-art emission-control equipment, which was estimated to cost over $2 million. If the equipment was not in place within a year, the plant could be ordered closed.

The parties ended up in court again several years later over a dispute about a settlement they had agreed to after the previous case. The court enforced the agreement, requiring the company to compensate the orchard owners more than $940,000 "for past or future economic losses in their respective orchards." Clearly, the orchard owners had a right to air that was not polluted in a way that harmed their cultivation.

In spite of this history, it is true that air pollution is difficult to solve through common law. Air pollution has been a problem for centuries. Smoke from coal was long a bane of population centres. Cities and states began to pass ordinances and statutes to deal with their smoke problems over a century ago[15]. In other words, local communities responded to the limitations of the common law in dealing with pollution, first by regulating the quality of coal-fired boilers and setting restrictions on locomotives. As coal use declined, automobiles gradually became the larger problem. California took the lead in investigating how to deal with auto emissions. Other states, including New York and Texas, also imposed ambient air-quality standards before the federal standards emerged.

As a result of these forces—common law augmented by local regulations—air quality improved steadily during the 1950s and 1960s before the passage of the Clean Air Act of 1970. Indeed, the rate of improvement in air quality *slowed* after the EPA became the air quality czar[16]. It may be that the common law alone could not have solved all air pollution problems, but there seems little doubt that common law, combined with state and local controls, would have taken major strides in controlling air and water pollution.[17] However, it was largely supplanted by politically engineered federal controls.

Since passage of the Clean Air Act of 1970, the Environmental Protection Agency has been the primary controller of air quality in the United States. The agency, and various state pollution agencies operating under EPA supervision, issue pollution permits to major pollution sources and determine pollution limits on vehicles and other emission sources. There have been few common-law air pollution cases since 1970.

Limitations of the Common Law

The cases reported here represent the majority view, but there was a minority view, also. Some courts would rule for polluters, holding that the economic benefit of a factory that employed many people outweighed the damage to a few property holders. Some courts held that pollution was just a fact of modern life and necessary for progress to occur. The courts were not always consistent in their decisions.

Legal action is always costly. Even the wealthy do not sue everyone who violates their rights. Low-income people may find some litigation beyond their reach, but the existence of contingency fees enables even the poor to bring suits when their cases are good ones. And, as we saw in the case of farmer Whalen and the pulp factory, the common law does not "deprive the poor litigant of his little property by giving it to those already rich." There is no such principle in the legislative process, which is dominated by special interests.

Multiple polluters that each inflict low levels of damage are unlikely to be held liable—especially when the damage is shared by many. For that reason, problems with air pollution caused by automobiles cannot be handled effectively through common-law courts.

Injuries and harms that come after long gestation periods present another challenge. Parties who can show evidence of injury or imminent harm may have a common-law cause of action. However, efforts to obtain injunctions for speculative harms are not generally successful. Regulation may be the only answer to limit actions that may cause future harms such as cancer. But we cannot know how the law might have evolved had it not been pushed to one side by regulation. It is not difficult to imagine environmental courts— ordinary courts assisted by special masters trained in environmental science—and other arrangements evolving to satisfy the needs of people concerned about their environmental rights.

The Statute Law Substitute

The common-law concept of property rights, with its requirements for proof of damages and its associated rights and duties, formed the foundation for environmental protection in the United States for many decades. This concept was changed fundamentally by the flurry of federal environmental statutes passed in the late 1960s and 1970s. The new statutes give standing to all citizens, whether they are harmed or not. In contrast, common-law requires well-identified plaintiffs and evidence of damages to obtain standing to sue. Under statutes, evidence of a technical violation of a statute-based regulation, rather than evidence of harm or the threat of harm, can spur legal action. Some federal statutes also provide funding for groups who can monitor and then sue polluters who fail to meet any one of a host of technical rules.

Unlike common-law remedies, which provide payment for damages to those who are harmed or enjoin polluters to stop their harmful actions, statute law calls for penalties and fines to be paid to the U.S. Treasury or to "public interest" organisations that seek to raise awareness about the environment. Put simply, the statutes provide subsidies to encourage environmental suits that may be totally unrelated to actual environmental harm.

The record of common-law cases is clear: It imposes tough liability on those who damage the environment. Yes, mistakes are made. Sometimes liability is imposed when it should not be; sometimes the damages assessed are too high; sometimes liability is not imposed when it should be. But it seems far better for such individual mistakes to be made (some of which are rectified on appeal) than for politically inspired policies to impose costs on all.

The Lure of Central Planning

Why, if environmental quality is the goal, was the common law abandoned in favour of central planning and political control?

Political control of the environment cannot be based on evidence of superior performance by governmental bodies. The evidence continues to mount that governments are poor environmental stewards and that government regulation is wasteful, cumbersome, and sometimes ineffective. "For all its accomplishments, we conclude that the pollution control regulatory system is deeply and fundamentally flawed," wrote Clarence Davies and Jan Mazurek[18] of Resources for the Future after conducting a comprehensive three-year assessment of U.S. environmental regulation. Expressing concern about failed federal efforts to build an efficient pollution control system, they concluded: "The United States does not need to wait for a consensus to act: to do so would be to wait forever. Failure to make the changes will be costly to the economy, to the environment, and to every citizen".

Yet the idea that the law and the market have failed, rather than the regulatory system, is today's received wisdom, and the claim is routinely made that insightful political leaders, aided by economists and other theoreticians, can resolve these difficult problems. This idea, which the great economist Friedrich Hayek called the fatal conceit, explains the lure of central planning, which is the alternative to common-law protection.

Politicians can advance their careers by convincing voters of their superior ability to manage problems. Only on rare occasions do politicians yield control, as in the case of transportation deregulation. Such instances are usually driven by special-interest politics that happen to produce results consistent with consumer interests, or else by market competition.

While politicians can perhaps be excused for supporting notions that further their careers, less defensible are the members of academia and other "learned professions" who provide intellectual respectability for destructive central planning. Yet they too have fallen under the sway of misguided theory. Submitting to the "nirvana" fallacy, they often believe that markets will operate better if they are "planned" than if they operate freely under a consistent rule of law. Nowhere is this fallacy more prevalent than with environmental protection.

While it is appropriate to recognise the limitations of the common-law in the field of environmental protection, supporters of centralised environmental controls act as if today's statutory approach has no comparable limitations. For example, one critic accuses free market environmentalism advocates of "wishful thinking" about the common-law[19] while failing to address any of the shortcomings of today's political and bureaucratic system.

It should not be surprising that many policy experts are disdainful of free market environmentalism, which is built on voluntary action and common-law rights supplemented by local rules. Economists and other experts preached for decades that electricity and telephones were natural monopolies. Competitive markets simply could not exist in the face of such monopolies, they said; there was a "market failure" that had to be remedied by government control. Now that the deregulation of transportation and communication has led to dramatic reductions in costs, and the deregulation of electricity is promising to do the same, the argument has switched to environmental problems. Surely, they must be caused by "market failure."

Would-be central planners argue that when it comes to the environment, markets and the law on which the nation was founded cannot protect us. Instead, we must have knowing, wise leaders with the "institutional expertise to fully evaluate complex scientific and technical evidence" and the ability to "make important policy decisions"[20]. Needless to say, we disagree.

Conclusion

Several years ago we were asked by reporters from *Forbes* if we believed that the EPA should be abolished[21]. It seemed at first like a radical notion. But it is a logical extension of the history reported here. In the United States, the Interstate Commerce Commission and the Civil Aeronautics Board have disappeared, and the railroads, trucks, and airlines still function—substantially more efficiently than when regulation "saved" us from exploitation. We are seeing the benefits of communication and electrical service deregulation. Perhaps the environment could be next.

If so, we would be able to return to the regime that served us well in the past and that has shown signs of evolving as knowledge and environmental concerns began to change. The common law provides harsh penalties against firms that disregard the rights of citizens by exposing them to harms. Indeed, when real harm is inflicted, citizens get far better relief through common-law suits than they do from appeals to the Environmental Protection Agency.

Ideas have consequences. Eventually, citizens will recognise that the common-law, bolstered by local regulation, can protect the environment more effectively and fairly than can congressional statutes and bureaucratic regulations.

Notes

1. *Village of Euclid vs. Ambler Realty Co.*, 272 U.S. 365 (1926).

2. *Carmichael vs. City of Texarkana*, 94 F. 561 (W.D. Ark., 1899).

3. *Whalen vs. Union Bag & Paper Co.*, 208 N.Y. 1 (Ct.App., N.Y. 1913).

4. *Bunker Hill & Sullivan Mining & Concentrating Co. vs. Polak*, 7 F.2d 583 (9th Cir., 1925)

5. Davis, Peter. 1971. "Theories of Water Pollution Litigation". *Wisconsin Law Review* 1971: 738-816.

6. Authors' count of cases recorded and reported by Westlaw.

7. Brubaker Elizabeth. 1995. *Property Rights in the Defense of Nature*. Toronto: Earthscan Canada.

8. *Village of Wilsonville vs. SCA Services*, 426 N.E.2d 824 (1981).

9. *Wood vs. Picillo* 443 A.2d 1244 (1982).

10. *New York vs. Schenectady Chemicals, Inc.*, 459 N.Y.S.2d 971 at 976 (Sup. Ct., N.Y., 1983).

11. *Davey Compressor Co. vs. City of Delray Beach*, 639 So.2d 595 (Sup. Ct., Fla., 1994).

12. *Georgia vs. Tennessee Copper Co.* 27 U.S 618 (1907).

13. *United Verde Extension Mining Co. vs. Ralston*, 296 P. 262 (Sup. Ct., Az., 1931).

14. *Renken vs. Harvey Aluminum*, 347 F.Supp. 55 (D. Ore., 1971).

15. Stern, Arthur. 1982. "History of Air Pollution Legislation in the United States". *Journal of the Air Pollution Control Association*

16. Goklany, Indur M. 1995. "Richer is Cleaner: Long-Term Trends in Global Air Quality". In *The True State of the Planet*, ed. Ronald Bailey. New York: Free Press

17. States were active in pollution reduction efforts prior to the establishment of the EPA. For example, in the early 1960s there was an average of three *arrests* per month in Wisconsin for violations of water standards.

18. Davies, J. Clarence, and Jan Mazurek. 1997. *Regulating Pollution: Does the U.S. System Work?*

19. Thompson, Andrew. 1996. "Free Market Environmentalism and the Common Law: Confusion, Nostalgia, and Inconsistency". *Emory Law Journal* Fall.

20. *Ibid.*

21. Brimelow, Peter, and Leslie Spencer. 1992. "Should We Abolish the EPA?", *Forbes*, Sept. 14.

5

The Theory of New Resource Economics

RANDY T. SIMMONS
JOHN BADEN

Proposals to privatise portions of the public lands have been misunderstood and misrepresented by academics, politicians, and some environmentalists. The confusion and ignorance about the New Resource Economics (NRE), which incorporates concepts about the "Tragedy of the Commons," the role of government, the importance of voluntary association, and the function of property rights, have distorted explanations of the likely consequences of privatisation. Our purpose is to provide understanding and to correct misrepresentations by explaining the NRE. We argue that there are firm theoretical and empirical justifications for claiming that private management of many of the lands currently held by the government will more efficiently and equitably meet the needs and wants of people.

The NRE is firmly rooted in classical economics and builds on theories of Austrian economics, property rights, and public choice. It assumes that individuals act on information and incentives and that institutions generate information and structure incentives. These assumptions allow the NRE analyst to explain the causes of and to suggest the solutions to pollution, depletion of natural resources, extinction of plants and wildlife, and inefficient use of resources.

Underlying Assumptions

The NRE assumes that consumers maximise utility when they exercise private choice in the market. When those consumers become voters, we cannot assume that they have been lobotomised or that they magically shift psychological gears; they remain self-interested. It is also assumed that producers who seek to maximise profits in the private sector will also attempt to maximise market advantage and subsidies in the public sector. Government is not an impartial entity that resolves value conflicts; it is an entity composed of self-interested individuals who are politicians and bureaucrats. Politicians are assumed to maximise votes, not some nebulous concept of the public welfare. Bureaucrats are not efficient computers seeking the public interest, but rather are self-interested, just like politicians, consumers, voters, and producers. They are not a special subset serving as vestal virgins of the public interest. They want bigger discretionary budgets in order to obtain greater job security, perquisites of office and income, professional satisfaction, and prestige. They may pursue ideological commitments as well. Because individuals are self-interested in both their private and public roles, the incentives created by institutional arrangements are of fundamental importance when analysing a polity.

Social institutions locate a society on a continuum between anarchy and Leviathan. They are primary determinants of the degree that self-centred individuals will be encouraged to consider the preference of others. The NRE relies on the principles of limited government, secure property, personal liberty, individual enterprise, and voluntary association.

Property Rights and Responsibility

Linking authority with responsibility is the key to capturing social benefits from individual self-interest. As Garrett Hardin pointed out, there is a difference between the *word* "responsibility" and the *fact* of responsibility. He offered Charles Frankel's definition in explanation:

"A decision is considered responsible when the man or group that makes it has to answer for it to those who are directly or indirectly affected by it."[1] Responsibility, then, causes people to consider an action's costs and benefits to themselves and to others. When people are responsible they act as if the wants and values of others matter—not just out of benevolence, but also because of the potential gain or loss from ignoring those considerations.[2]

One of the best ways to establish responsibility and to take advantage of self-interest is through the establishment and enforcement of transferable private property rights. When a resource is privately owned, the owner typically protects it from misuse in order to keep its value (as measured by himself and others) from falling and attempts to increase its value through wise management. A farmer, for example, pays careful attention to the combination of tilling, fertilisers, and water inputs necessary to sustain the productivity of his or her farmland. When the productive capacity falls, due, for example, to erosion, the farm decreases in value. Private ownership creates responsibility by providing a solid link between action and result.

A substantially different result emerges when people produce and consume without paying the social costs of their actions—that is, when authority is divorced from responsibility. Garrett Hardin's parable of the "Tragedy of the Commons" is one of the more forceful illustrations.[3] He described a pasture for which there were no secure, transferable, or enforceable property rights. On the pasture, or commons, people were not protected from the effects of others' actions. The dilemma is twofold. First, the cost of one user's actions is dispersed among the community of users, and only that user enjoys the benefit—a clear violation of Frankel's definition of responsibility. This activity is known as "free-riding."[4] Since any single individual action has a small impact and the individual share of any ill produced by the action is minimal compared to the personal benefits, the dominant strategy is for users to overexploit unowned resources at the expense of the community. *Because benefits are privatised and costs are socialised, there is little incentive to conserve the resource,* to use it wisely, or to manage it as if future generations mattered. Consider the near extinction of the American bison, the rapid decline of beaver populations in North America, and the over-harvest of the Great Whales.

The second part of the dilemma is not often recognised, possibly because the term "free rider" implies that a deliberate effort has been made to benefit oneself at the expense of others. In many situations there is no sensation of riding free because there is no personal interaction between those who create costs and those who pay them.[5] Each individual is simply reacting to the choices presented by the situation. He is not consciously acting against fellow citizens. When people create costs not personally felt, the information link between action and result is severed. Information about the wisdom of the action is more difficult to obtain. People who create social costs, therefore, often see only the personal benefits of their actions, not the costs imposed on others. Conversely, people whose actions benefit others sometimes do not recognise that benefit, and such actions are often not rewarded.

Under these conditions, good intentions easily go astray. Poor decisions do not confront reality checks and wise decisions are not rewarded. Simply being concerned and willing to donate time, effort, and money to resolving perceived problems is inadequate. Good intentions will not suffice. Without the positive feedback provided by responsibility, people who desire to do good will not know if their actions are having the intended effects. This is a fundamental problem with bureaucratic management. When the government owns and manages a resource, the linkage between decision and result is made by the bureaucrat's charitable and sometimes professional instincts, and by the pressures of those with political influence, not by the rationally ignorant general citizen.

Creating responsibility where little or none exists does require a form of coercion. Agreements must be enforced, and a way must be found to protect private property from theft—the ultimate free ride. Without enforcement, property rights cannot create responsibility. As Thomas Hobbes explained: "Covenants without the sword are but words, and of no power to secure a man at all."[6]

A society of responsible individuals is one in which people agree to enact laws and to provide means of enforcement in order to achieve the closest possible approximation to the ideal free society. A form of management is called for that links the authority to act with the responsibility for actions taken.

Emphasising the benefits of protecting property rights and enforcing contracts does not negate tendencies to act altruistically with a community in mind. If establishing and maintaining peace depended solely on the force of law and not on such incentives as norms, customs, and a sense of community, it is doubtful that peace would be lasting or that large groups could be moved toward achieving it. Christian, Kantian, and humanist principles are important to the structuring of societies and should not be ignored or obscured by the search for responsibility.

Even (perhaps especially) allowing for humanitarian virtues, it is not wise to trust in them as motivating agents in situations where people can avoid costs of their actions. Given the capacity of a very few to render useless the contributions of many, it is foolhardy to expect other-interest to supersede self-interest. Rousseau warned of the dangers of even a single self-seeker in a society of perfectly other-regarding persons and argued that the self-seeker "would certainly get the better of his pious compatriots."[7]

Responsibilities and Government

The NRE identifies a legitimate role for government, especially in the areas of natural resources and environmental management. Property rights and contracts must be defined and enforced, there are public goods to be provided, and there are common pool resources to be managed where entrepreneurs have not yet discovered a way for private provision and management. But there is little justification and great danger when government goes beyond these activities.

When government rejects the private property approach, it must choose some form of regulation. In Hardin's commons, an entity, possibly a Pasture Protection Agency, must regulate behaviour through a system of permits, fines, and supervision. The agency must decide how many and whose cows can be allowed on the commons, whether sheep and goats should be accommodated, and whether the commons should be opened for some other use, such as recreation. Except in the case of a pure democracy where the median voter rules, regulatory management is "top down."

The major flaw in governmental management is that those people making decisions are separated from the effects of their decisions. When conflicts arise they decide whose values will prevail. They decide which uses are acceptable, and they manage with the financial resources of third parties. Public hearings, elections, public participation processes, and court actions are all ways to curb governmental irresponsibility, but even these tools are often used to provide benefits to one group through the inefficient use of taxpayers' money.

Public lands management has proved to be highly inefficient in terms of providing an economic return to the nation of owners, but it has effectively used the taxpayers' resources to benefit local constituencies. The national forests are a prime example. Logging continues to be carried out in forests where the costs of logging exceed the value of the timber logged, while causing needless environmental degradation.[8]

Thus economic inefficiency is compounded by environmental atrocity. By subsidising the harvest on these environmentally fragile and uneconomic sites, the Forest Service helps local timber firms stay in business, keeps certain local politicians happy, and most importantly, preserves its saw-timber management budget. None of the politicians, bureaucrats, company executives, or timber workers pays the full cost of his actions, so there is little incentive to use the forest resource or the taxpayers' money more efficiently.

Other government agencies produce the same kinds of results. In nine of the eleven Western states of America, the federal government spends more on the public lands than it collects, even though the region is prosperous and holds vast mineral, timber, forage, and recreational value. Dams are built using benefit/cost ratios smaller than one. The process the government uses to study its grazing program for a specific unit commonly costs more than the value of the grazing permits for that unit. These perverse results are the predictable consequences of management by policymakers and bureaucrats who are insulated and buffered from the results of their decisions.

As should be clear, the NRE analyst would predict not necessarily overuse but misuse. When managers are irresponsible, as they must be in government, appropriate use will occur only by serendipity. Certainly those wilderness lands that contain high-quality deposits of minerals are not being overexploited for their mineral potential.

One of the most consistent findings in the area of resources policy is that public management occurs without regard for internalised costs and benefits at the margin. What does matter is the distribution of these costs and benefits. Marion Clawson's characterisation of multiple use management as "a little of everything everywhere, regardless of costs and results" supports this conclusion.[9] Management questions become political questions of who receives how much of the benefits and who incurs how much of the costs. Given that in all societies wealth and political power are positively related, the relatively wealthy are commonly subsidised. This process is not likely to change until the public lands are removed from a system in which everyone chooses for everyone else and everyone spends monies belonging to others on still others.

To the NRE analyst, governmental management of resources is highly suspect; and as the examples cited above indicate, there are good empirical and theoretical reasons for the suspicion. Continued political management of the public lands means continued reliance on individual voters who find it rational to remain ignorant on most issues, since one vote has an insignificant impact on the outcome. Although politicians, bureaucrats, and producers are informed about specific issues, they will find it rational and perhaps even necessary for success in the highly competitive political arena to suppress and distort information.

Even when voters are informed, interested, and involved there is no guarantee that good policies will result. Political incentives run counter to the public interest, whether measured by efficiency, equity, or careful use of natural resources. Because public management implies an inherent separation of authority and responsibility, citizens seek governmentally provided goods and services for which others will pay. Wants are systematically exaggerated and inflated to "needs." Special interests, programme administrators, and politicians cooperate to concentrate benefits on themselves and diffuse the costs. Rarely will an individual voter or politician find it rational to organise effective opposition to these activities, because the high cost of organising is concentrated on the individual and his or her share of the benefits is small.

There is little wonder that many local people have opposed privatising the public lands. With public ownership, they have been free-riding on other taxpayers; with private management, they would have to start paying for such rides. There is even the possibility that uses would change once property rights were established and people had to determine the true value of present uses. Once the lands were in private hands, however, the benefits of private ownership would become apparent. If a voluntary group wanted to protect a particular marsh or piece of forest, it could approach the owners with an offer to buy it outright or they could propose a more creative protection scheme, such as restrictive covenants. Entrepreneurs would begin searching for ways to provide the most valued uses of the lands. Under a system of property rights, managers would be held accountable for their actions. Potential future users would be represented by speculators—owners who hold the resource in anticipation of its appreciation in value; they withdraw resources from current consumption and postpone their use to a period when their social value is higher. In such a setting, timber harvest would no longer occur where the costs of management and harvest exceeded the value of the timber.

Not Perfection But Improvement

The arguments for privatisation that are based on the NRE are made with the full recognition that the private market is not perfect and that people have motivations beyond obtaining material goods. The market is assumed to have many imperfections, including transaction costs, imperfect information, externalities, and difficulties in producing adequate supplies of public goods. Goal-oriented, self-interested people are assumed to make mistakes, and individuals outside the government are assumed to be no more infallible than those inside.

The traditional justification for governmental management of resources is market failure. The NRE justification for privatisation is governmental failure—the result of inherent irresponsibility. The two failures come together at privatisation. The cure for market failure need not extend beyond establishing and enforcing property rights when ways are found to do so. The cure for governmental failure, at least in resource management, is privatisation. When these analytical concepts are coupled with the normative concept of individual freedom, they make a compelling case for privatising much of the federal estate.

Notes

1. Garrett Hardin, "An Operational Analysis of Responsibility," in *Managing the Commons*, ed. Garret Hardin and John Baden (San Francisco: W. H. Freeman, 1977). p. 67.

2. The classic statement of this is Adam Smith's: "it is not from the benevolence of the butcher, the brewer or the baker that we expect our dinner, but from their regard to their interest. We address ourselves, not to their humanity but to their self love, and never talk to them of our own necessities but of their advantage" (from his An Inquiry into the Nature and Causes of the Wealth of Nations [New York: Oxford University Press, 1976, pp. 26-27).

3. Garrett Hardin, "The Tragedy of the Commons," *Science*, No. 162 (1968)~. 1243—48.

4. The free-rider problem is addressed in the theory of public finance and the theory of externalities in economic; see, for example, James M. Buchanan and Marilyn R. Flowers,

The Public Finances (Homewood, Ill.: Richard D. Irwin, 1975). A large body of literature on the prisoners' dilemma also addresses the free-rider problem; see Duncan R. Luce and Howard Raiffa, Games and Decisions: Introduction and Critical Survey (New York: Wiley, 1957). The United States conducts a large group free rider experiment every session of Congress. The result is a budget containing projects for which every dollar of expenditures paid by the taxpayers from all 435 congressional districts generates benefits worth less than a dollar for a small number of people in a few districts. If each congressman refrained from promoting uneconomical but vote-generating projects, the federal budget would be more easily balanced. Few refrain. The result is that most items in national budgets are favoured by few voters. Public land use allocations give those in political power the chance to promote special interests without even the need for budgetary expenditures. Mischief is cheaper to them, and thus more frequent. For an NRE proposal to deal with this problem see Rodney D. Fort and John Baden, "The Federal Treasury as a Common Pool Resource and the Development of a Predatory Bureaucracy," in *Bureaucracy vs. Environment: The Environmental Costs of Bureaucratic Governance*, ed. John Baden and Richard L. Stroup (Ann Arbor, Mich.: University of Michigan Press, 1981), pp. 9-21.

5. James M. Buchanan, *Freedom in Constitutional Contract: Perspectives of a Political Economist* (College Station, Tex.: Texas A&M University Press, 1977), p. 164.

6. Thomas Hobbes, *Leviathan* (London: J. NI. Dent, 1947), p. 57.

7. Jean Jacques Rousseau, *The Social Contract and Discourses*, trans. G. D. H. Cole (New York: Dutton, 1974), pp. 135—36.

8. See William F. Hyde, "Compounding Clear Cuts: The Social Failures of Public Timber Management in the Rockies," in Baden and Stroup, eds., pp. 186—202.

9. Marion Clawson, *Economics of National Forest Management*, Working Paper EN—6 (Washington, D.C.: Resources for the Future, 1974), p. 108.

6

Reconceptualising Sustainable Development

JULIAN MORRIS

Two colleagues and I were recently driven from Delhi to Agra to see the Taj Mahal. As we crossed the border of Uttar Pradesh we saw long lines of trucks simply waiting in line, doing nothing. Their drivers were hanging around smoking *bidis*. We were shocked. Then it struck me: sustainable transport. The truck drivers must have been given an edict that they should drive less, perhaps to conserve fuel or to reduce nasty emissions to the environment. I asked the driver. "No sir, not sustainable," he said, laughing at the ridiculous Englishman. "The truck drivers must wait to pay a toll to cross the border." "But we're still in India" I protested, "You means they have *internal* tariffs here?" "Yes, sir."

The term 'sustainable development' has been around for about 30 years but has only recently been popularised. It derives originally from the biological concept of 'sustainable yield' – that is to say, the rate at which species such as cod and elephants may be harvested without depleting the population. Starting in the late 1980s, environmentalists and government officials began applying the terms 'sustainability' and 'sustainable development' when discussing environmental policy. Thus, numerous measures aimed at conservation and pollution prevention have been justified on the grounds that they are necessary to promote sustainable development. More recently, and in light of the AIDS crisis in Africa, the interpretation of sustainable development has been broadened to include issues such as healthcare and education, the lack of which are seen as constraints on economic development.

The most common definition of the term derives from a report prepared for the World Commission on Environment and Development, which stated in 1987:

"Sustainable development is development that meets the needs of the present without compromising the ability of future generations to meet their own needs."[1]

So defined, sustainable development is, like motherhood and apple pie, not a concept to which many would object. It would take a perverse outlook indeed to support the idea that people's needs should *not* be met both now and in the future. But, unobjectionable as it is in principle, the concept is sufficiently broad to allow various different interpretations. Indeed, a voluminous literature has sprung up debating its interpretation in periodicals with fancy names such as the *Journal of Sustainable Development* and the *International Journal of Sustainable Development*, which is not to be confused with the *International Journal of Sustainable Development and World Ecology*, or *Sustainable Development International*.

Most of this literature has focused on identifying specific outcomes, such as controlling the climate or saving periwinkles, and then setting about developing policies intended to achieve those outcomes. Usually these policies require stronger systems of global governance (periwinkles are important).

An alternative interpretation focuses less on specific outcomes and more on increasing the chances of superior outcomes. It is the purpose of this chapter, first, to critique the conventional outcome-oriented vision of sustainable development and then to offer an alternative vision.

Unintended Consequence of Outcome-Oriented Sustainable Development Policies

Some environmental groups have claimed that rich countries are burning too many hydrocarbons (coal, oil, gas) and that this harms poor countries. In order to redress the balance, these groups demand a reduction in the consumption of hydrocarbons by rich countries. While such a policy would almost certainly reduce the differential in income and wealth between people in rich and people in poor countries, it would do so in the main by destroying wealth and reducing income of those in rich countries. The reason for this is twofold. First, energy is a basic factor of production, so increasing the cost of energy by mandating a shift to lower-carbon forms will reduce output. Second, hydrocarbons are used by consumers in all manner of applications, both directly, for example in cars and gas stoves, and indirectly, when they turn on their lights. So, reducing the availability of hydrocarbons will create energy poverty.

Although some middle-income countries might benefit from a shift in the location of industrial production, for the most part people in poor countries would suffer – those in the poorest countries, especially, because they have little industrial capacity. The reason is that reducing income in rich countries will reduce demand for all products, including agriculture, textiles, and apparel, which are the main products currently exported from poor to rich countries.

The likely adverse effect on the world's poorest people of constraining consumption of hydrocarbons by the rich is one instance of a more general phenomenon: the unintended effects of outcome-oriented policies that are justified on the grounds that they promote 'sustainable development'.

Consider the effect of the Basel Convention, an international agreement intended to prevent the illegal dumping of hazardous waste in poor countries. As Alan Oxley points out:

"... restrictions on the export of used lead to India has undermined the formal lead recycling industry in that country with perverse environmental consequences. The formal recycling industry, which operates under strict environmental and health and safety regulations, requires high throughput of lead. Because of the relatively low levels of lead use locally, the industry requires imported lead in order to operate at a profit. The restriction on exports of used lead to India has led to the closure of a number of formal-sector lead recycling facilities, which became unprofitable. As a result, more of the locally-produced waste lead is now being recycled in the informal sector. These are unregulated backyard operations, which typically cause contamination of water and air, with adverse health consequences."[2]

Likewise, the ban on trade in elephant ivory, enacted under the Convention on International Trade in Endangered Species, probably does more harm than good by undermining incentives to conserve elephants locally. In Southern Africa, governments have, to a greater or lesser extent, decentralised management of wildlife. In many places in Botswana, Namibia, South Africa, Tanzania and Zimbabwe, local people receive a proportion of the income from hunting, eco-tourism and other economic activities associated with the wildlife with which they share their land. As a result, local people have become stakeholders in the wildlife management system and instead of seeing elephants and other wildlife as a threat they realise that by conserving them they are able to benefit. The ban on international trade in elephant ivory reduced the value of elephants to these people and thereby reduced their incentives to conserve them.

The UN Commission on Sustainable Development is outcome-oriented. One need only search for a few minutes on its website to find a hundred outcomes that it seems to favour – most of which seem, mysteriously, to involve global governance by ... the UN.[3] In 2000, the UN issued *The Millennium Declaration*, which included a commitment to halve by the year 2015 the proportion of the world's population whose income is less than $1 per day.[4] Now, of course, such an outcome may seem laudable; reducing poverty and its accompanying misery *is* a good thing.

But several objections to this proposition could be raised. First, one could object that the outcome is far too modest. How can we accept that in 2015 there will still be 400 million people living on only a dollar a day? Second, one could object that the simply reducing the number of poor people is not enough. What happens to the people who were poor is important. Merely ensuring that they have $1.10 per day for a few years by sending them food packages or other forms of emergency aid is hardly much use in the long-run. Increases in wealth must be sustained if they are to be considered sustainable. This points to a general problem of starting with a desired outcome and then attempting to formulate an appropriate policy to achieve that outcome: all sorts of mischief becomes acceptable because the ends justify the means.

Good Intentions are not Enough

Good intentions are laudable but if good intentions were enough to alleviate poverty, malnutrition and disease, these dreadful problems would no longer plague us. Indeed the facts that more than 10 million people each year die of preventable or curable diseases and that 800 million people survive on less than $1 per day, are testament to the failure of good intentions – and the many billions of dollars spent in their pursuit.[5]

In July 1944, officials representing 44 of the world's nation states gathered in Bretton Woods, New Hampshire, for the United Nations

Monetary and Financial Conference. At this conference, under the chairmanship of John Maynard Keynes, officials agreed to use money taken from taxpayers in the wealthier countries there represented, to subsidise loans to the governments of the war-ravaged countries of Europe. To facilitate this operation, the International Bank for Reconstruction and Development (IBRD) and the International Monetary Fund (IMF) were set up with a mandate to make loans to governments. In 1949, following Harry S.Truman's Point Four Programme, the IBRD and IMF began lending to the governments of 'developing' countries. Since then, hundreds of billions of dollars have been spent on 'aid', yet a balanced assessment indicates that, although there may have been a few benefits, on average it has caused harm. Graham Hancock, former East Africa correspondent for *The Economist*, noted over a decade ago:

'Garnered and justified in the name of the destitute and the vulnerable, aid's main function in the past half century has been to create and then entrench a powerful new class of rich and privileged people. In that notorious club of parasites and hangers-on made up of the United Nations, the World Bank, and the bilateral agencies, it is aid – and nothing else – that has permitted hundreds of thousands of "jobs for the boys" and that has permitted record breaking standards to be set in self-serving behaviour, arrogance, paternalism, moral cowardice, and mendacity. At the same time, in the developing countries, aid has perpetuated the rule of the incompetent and venal men whose leadership would otherwise be utterly non-viable; it has allowed governments characterised by historic ignorance, avarice, and irresponsibility to thrive; last but not least, it has condoned – and in some cases facilitated – the most consistent and grievous abuses of human rights that have occurred anywhere in the world since the dark ages. 'In these closing years of the twentieth century the time has come for the lords of poverty to depart.'[6] Since then, there have been few signs that 'aid' is being put to better use.

Institutions for Sustainable Development

The reason transfers of financial resources from the governments of rich countries to the governments of poor countries have been largely unsuccessful in stimulating economic development is that lack of resources is not the primary problem in poor countries. Take the case of Nigeria, which happens to contain one of the largest oil deposits on the planet. The oil wealth in Nigeria has been controlled by government officials – until recently it was in the hands of the murderous kleptocrat General Sani Abacha – who used it to line their pockets and keep the politically important elite happy, rather than to promote development. There is little point to pouring money into a country whose government has no intention to encourage economic development. Indeed, as Mengistu showed in Ethopia, Mobutu in Zaire, Pol Pot in Cambodia, and Idi Amin in Uganda, dictators will happily accept 'aid' if it helps to prop up their regime. In such cases, government-to-government transfers are not merely counterproductive, they are murderous.

The more fundamental problem is that 'aid' is based on a largely false premise, namely that poverty itself is a barrier to development. In general this is simply not true. Economic development in Western Europe did not require massive redistribution from the rich to the poor. Rather, it required a change in the structure of Europe's institutions; a move away from the Feudal system of the early middle ages to a trading economy.[7] Whilst Sub-Saharan Africa now appears to be facing a genuine crisis, in the form of disease that is destroying the economically productive sector of society, it is probably unique in the world (if not world history) in requiring external assistance to escape from such a quagmire. And even then, such assistance is unlikely to lead to significant growth; rather, it might prevent total economic collapse. If countries are to develop sustainably, institutional reform, not aid, is the solution.

But what do we mean by 'institutional' reform? Institutions are the framework within which people act and interact – they are the rules, customs, norms, and laws that bind us to one another and act as boundaries to our behaviour. Institutions reduce the number of decisions that we need to take; they remove the responsibility to calculate the effect of each of our actions on the rest of humanity and replace it with a responsibility to abide by simple rules. In a system in which rules emerge spontaneously and rules are selected by evolutionary processes, good rules will tend to crowd out bad rules. That is to say, over time, rules that result in better outcomes will be preferred to rules that result in worse outcomes.

Some rules are, of course, essentially arbitrary — which side of the road to drive on, for example. Clearly a rule is required here, or the consequences would be fatal. But whether one follows the English rule, which is based on the fact that (right-handed) jousters would pass on the left, or the French rule, which is based on the fact that it is not English, is of no great consequence.

Other rules are not so arbitrary. For example, the rule that contracts should, generally, be upheld in a court of law has a very distinct consequence. It grants people greater certainty in their transactions with one another and thereby encourages such transactions to take place. If the rules were that contracts were not legally binding, the effect on commerce would be devastating. In *The Other Path*, Hernando de Soto documented the plight of his fellow countrymen in Peru, most of whom were and are denied the formality of such law.[8] They must, instead, rely upon informal mechanisms to enforce contracts, property rights and other relationships.

Whilst such informal mechanisms – customs and norms, for example – work well for groups that are relatively homogenous and where there is little trade with outsiders, they impose significant constraints on the ability for groups to improve their lot. Societies that have adopted formal institutions – such as property rights, markets, contract law, tort law, trademarks, patents, copyright, and so on – have tended to do much better economically and socially than societies that have relied primarily on informal institutions.

Property Rights[9]

It is the institution of private property that, more than any other, has enabled people to escape from the mire of poverty. Property rights are capital; they give people incentives to invest in their land and they give people an asset against which to borrow, so that they might become entrepreneurs. As in Peru, the 700 million rural poor in India are not oppressed by multinational companies. Most of them have never even heard of multinational companies and those that have probably dream of working for them. No, the 700 million rural poor in India are oppressed by tenure rules which make it difficult for them to rent, buy or sell property formally. Land transactions typically involve paying large bribes to local officials, who have a vested interest in maintaining the status quo.

Property rights are created in order to resolve competing claims over resources. Thus, if 10 men all graze their cattle on the same piece of land and there are no rules governing how much each man can graze, then each man has a strong incentive to graze as much of the available land as possible. Under such a system – known as 'open access' – the cattle will quickly denude the land and, in the absence of free land on which to move the cattle, will die.

Historically, open access has been a rarity, occurring only when land is so plentiful that ownership is not necessary, or when people are prevented from owning property. In most cases, before the tragedy occurs, the users of the land would see the advantage of either dividing it up into individual plots or creating rules for using it that reduce the likelihood of denudation. In either case, the land has been privatised – made the exclusive property of one or more people. If the land is split into plots it becomes 'several' or 'individual' property; if the owners agree to common rules it is called 'common' property.[10]

Privatisation is expected to occur when the costs of exclusion (that is, the costs of limiting access to a piece of previously open land, for example by fencing and policing) are equal to or less than the external costs (which, in this case, means the costs associated with the denudation of the land).[11]

But the costs of exclusion will depend upon the exclusion technologies available. Wherever there are externalities present (such as the threat of encroachment by cattle belonging to other ranchers), users would have an incentive to produce new and cheaper exclusion technologies. So, over time, we would expect more and more land to become privately owned and the sum of external costs to decline precipitously. Land that is owned privately (whether individually or in common) will in general be better managed than land that is unowned or owned by the state. This is because the owner(s) know that they will reap the benefits from any investments made in the land, so they have stronger incentives to make those investments. Going back to our cattle-rancher example: those who graze more cattle than their land can support will soon cease to be cattle ranchers. That creates a strong incentive to discover the 'carrying capacity' of the land – and to increase it through new technologies. Entrepreneurial peasants constantly introduce new crops and production methods, creating an environment of diverse agriculture.[12]

Technological innovation not only enables peasants to improve their lot, it also benefits those with whom they trade by lowering the cost of purchasing food and other goods and reducing the risk of famine. But such innovations will be stifled if those who might innovate new technologies are not allowed to benefit from the investments they make through the ownership of property. The individual's *incentive* to *invest* in his land and *innovate* new methods of production will be greater when he can *own* and *exchange property*. Thus, Michael Stahl concludes: "At the farm level, the presence or absence of clearly defined property rights makes the difference between active interest in investing in soil conservation measures or apparent indifference to environmental degradation"[13]

Individual property rights also encourage pollution prevention. In English common law for example, if the owner of property A emits a substance that causes damage to property B, then the owner of A must compensate the owner of B for the harm caused.[14] Thus, even at the height of the industrial revolution a smelting works in an industrial area was enjoined for causing damage to shrubs and trees on a nearby property.[15] There remains an ancient maxim, *sic utere tuo ut in alienum non laedas*, roughly translated as "so use your own as not to harm another." Widely applied in the courts of law this rule would protect not only the property of the owner but also neighbouring properties and even the environment – and society – as a whole. However, the maxim has not received sufficiently general application, mostly because states have stepped in and asserted that the polluting activities should continue because they are in the general interests of mankind.

Property also begets wealth. Once a person owns property, he or she can use it as collateral against a loan. Because such collateral gives the lender security, they will be willing to offer loans at lower rates. So property reduces the cost of becoming an entrepreneur. Of course, some of these entrepreneurs will fail and they may have their property repossessed by the lender. But at least they will have had the opportunity to try to escape from poverty and, of course, most will be no worse than if they never had the property in the first place. In any case, many will succeed in their endeavours. Some may become wealthy; most will simply be less poor.

Intellectual Property Rights[16]

Another institution that encourages innovation is intellectual property. In 1474, the Venetian Republic enacted a law that entitled inventors to a temporary exclusive right to profit from their inventions. Later such laws were adopted widely throughout Europe and North America. In some cases the laws have been abused, but for the most part they have been very beneficial to humanity.

Strong, readily enforceable intellectual property is particularly important for those products and processes that require large investments in research, development and marketing but for which the costs of copying are relatively low. Chemicals, pharmaceuticals, and biotechnology each rely heavily on patents. The music, film, book, art and software industries each rely heavily on copyright. Meanwhile, all manufacturers and sellers of brand goods (which is most manufacturers and most sellers) rely on trademarks and servicemarks to guarantee the identity (and hence brand-associated characteristics) of products.

There are of course drawbacks to IP, including temporarily higher prices of the protected goods, a reduction in the number of goods directly derived from those that are patented,[17] the legal and administrative costs involved in enforcement, and so on.[18] These drawbacks have led several commentators to conclude that patents and other forms of intellectual property are not desirable. However, the problem with focusing on these drawbacks is that in doing so one often forgets that the inventions and creative works might never have come about but for the existence of IP.

It is all very well to criticise the excessively broad application of patent to the internal combustion engine or aeroplane wing control (which was patented by Orville Wright) but if there was no patent protection how much longer would we have had to wait for the car and the aeroplane? Perhaps more importantly, without the stimulus of patent protection, would we have had all the wonderful synthetic chemicals and pharmaceuticals that make our lives and the lives of our loved ones so much better? Without copyright protection, would we have enjoyed the explosion of music, art, literature and film that we have experienced over the course of the past century? Without trademarks and servicemarks, how much more complicated would our lives be, constantly battered with confusingly similar marks?

In sum, were we to abandon or significantly diminish our system of intellectual property rights, we might gain in the very short term through lower cost products, but the cost in the medium to long term would be felt in terms of fewer products, as well as higher expenditures on trade secrecy and other means of protecting knowledge, which might well increase the cost of products.

Freedom of Contract

Another fundamental institution for sustainable development is freedom of contract. This includes both the freedom *to* contract – the freedom to make whatever agreements one desires, subject to fair and simple procedural rules – and the freedom *from* contract – the freedom not to be bound by the decisions of others. Freedom of contract is a fundamental part of the freedom to associate with others. It includes the freedom to transact – to buy and sell property – and as such it is an essential adjunct to the right to clearly defined and readily enforceable property rights.

The freedom to contract enables people to bind themselves to agreements and thereby creates greater legal certainty. This in turn encourages people to engage in trade and investment.

Armed with enforceable property rights and contracts, the peasant becomes a merchant. The freedom from contract prevents others from attempting to interfere with one's right to engage in exchange. Sadly, governments rarely respect the rights of parties to freedom from contract. Restrictions on trade abound in every country in the world. As I pointed out, in India they even have tariffs on the borders, where trucks can wait for days while their drivers pay a small fee (Uttar Pradesh charges 160 rupees – about three dollars). The delay comes from the drivers having to go backwards and forwards filling out ridiculous forms. And as they wait, their loads rot or are stolen. Such interventions are truly unsustainable: they waste capital (the trucks and goods lie idle), and they waste good food, which in turn reduces the value of perishable agricultural goods, and keeps farmers poor.

The Rule of Law

Property rights and contracts are nothing if they are not enforceable. And enforcement is only possible if there are courts wherein disputes over the rights and duties of parties may be resolved and a legal system that will enforce those judgements. There are two key issues of concern here: the first is the willingness of the state to permit competition in the provision of legal services; the second is the enforcement of whatever decision is made.

When the state exerts a monopoly on the provision of law, the costs of resolving disputes through the formal judicial system are typically large. The costs of such dispute resolution can, however, are dramatically reduced if people are able to resolve their disputes privately, for example through arbitration. However, for such arbitration to be enforceable, the courts must accept the rights of parties to settle disputes by arbitration. This is essentially a matter of freedom of contract: if parties contract to have their dispute settled by arbitration then the courts should only get involved if that contract is breached. Sadly, in most countries the courts have resisted this – presumably because they see it as a threat to their power.

The second issue, enforcement of decisions, requires the existence of a credible system of sanctions. One of the reasons rich countries have been able to become rich is that the police and the administrators of criminal justice are generally trusted and are trustworthy. That is not to say that they are free from corruption; rather, that the level of corruption is small in comparison to the levels of corruption in many poor countries. In Indonesia, for example, it is common for criminals to use a combination of threats and bribes to ensure that they stay out of jail. Similar problems exist in many other countries. In Nigeria and South Africa, people have so little trust in the police that they employ private security agencies. In South Africa these seem on the whole to be rather reliable. In Nigeria they are no doubt better than the police – if you can afford them – but they also run scams, stealing from the very houses they are supposed to protect.

Decentralised Decision-Making and Sound Science

If government is to intervene in the actions of the citizens, there must be both a strong justification – the benefits must outweigh the costs – and there must be in place mechanisms to ensure that poor decisions can be changed. Space does not permit a discussion of how in practice costs and benefits would be estimated. Needless to say, however, decisions must at the very least be informed by sound science – that is science which has undergone a rigorous process of peer review.[19]

To prevent mistakes being perpetuated, decisions must be reviewable. Given the difficulties associated with reviewing legislation, that probably means building in redundancy: legislation should perhaps have a sell by date of two or three years after which it must go through the same process of assessment to decide if it is still justified.

In addition, decisions to limit human activities should be taken at the most local level possible. But must be bound by the other principles that prevent abuses of local power.

Those principles are mostly contained within the other principles already adumbrated, such as respect for property and freedom from contract.

Generally speaking, state bureaucracy should be avoided because bureaucracies have a tendency to expand and to create justifications for expansion. One way to deal with this would be to contract out all government services.[20]

Towards Good Governance for Sustainable Development

This combination of property rights, freedom of contract, the rule of law, decentralized decision-making and sound science provides the basis upon which sustainable development can take place. In short, they represent good governance. Sadly few countries have come close to instituting such systems of good governance. The challenge for those eager to see the world become a more sustainable place is clear: stop squealing about the importance of global governance and instead promote good governance.

Notes

1. WCED (1987), p. 43.

2. *Sustainable Development: Promoting Progress or Perpetuating Poverty?* edited by Julian Morris, 2002 Profile Books Ltd. London. pp. 127.

3. http://www.un.org/esa/sustdev/index.html — Accessed 12 June 2002.

4. UN General Assembly Resolution 55/2, para. 19. New York: United Nations.

5. The following description is taken from Morris (1995).

6. Hancock (1989), pp.192-93.

7. North and Thomas (1972).

8. De Soto (1989).

9. Portions of this section are taken from Morris (1995).

10. The meaning of 'private' property should be clarified: in the sense in which it is used here, private property means property for which an identifiable person or group of people is (are) the principal 'residual claimant(s)', see Barzel (1989). Residual

claimants have a right to any good produced by that property, and are liable for any externalities generated. Many fascinating alternative mechanisms for managing property, especially where individual rights are difficult to delineate, are discussed in Elinor Ostrom (1988, 1990) and Schlager and Ostrom (1992).

11. Alchian (1965); Demsetz (1967); Anderson and Hill (1975); Ault and Rutman (1979).

12. Brookfield and Padoch (1994).

13. Stahl (1993).

14. It is necessary only for the owner of B to show that this physical harm was most likely the result of the pollution emanating from A. However, it must be shown that the damage actually exists or is imminent; otherwise there is only the potentiality of an action, not an action as such: *Pemberton v Bright* [1960] 1 WLR 436.

15. *St. Helen's Smelting Co vs. Tipping* (1865) 11 All ER 1483.

16. The following is drawn from Morris *et al.* (2002).

17. "If later innovators cannot freely build on the work of others, or must pay to do so, they may be less likely to engage in inventive activity themselves" (Besen and Raskind, 2001)

18. For example, the legal fees arising out of a battle between Kodak and Polaroid cost Kodak $100 million (Cole, 2000).

19. It is important to distinguish here sound science from consensus science. Often interest groups on both sides of a particular argument will say that so many thousands of scientists have said X. Whilst such statements probably persuade a certain section of the public, they rarely reflect the very real disputes which exist in the scientific literature. Ensuring that these disputes are taken into consideration is not an easy task – but forcing 'consensus' is no solution. In addition, any cost-benefit analysis must consider the opportunity costs of diverting resources towards one action rather than another. So, for example, if one believes that the global climate will warm by two degrees Celsius by the year 2100 and that the cost of this will be 5 per cent of total world output, then before taking action to prevent this harm one should calculate what the cost of taking action will be. If it turns out that no action can be taken that reduces world output by less than 5 per cent, then it would be folly to take any action.

20. Of course, strict rules would be necessary to prevent abuse of power, such as a requirement that all tenders be clearly specified and bids be considered strictly on the basis of conformity with hurdle-type criteria and cost.

References

Alchian, A. (1965), "Some Economics of Property Rights", *Il Politico*, Vol.30, pp.816-29, reprinted in Alchian, A. (1977), *Economic Forces at Work*, Indianapolis: Liberty Press.

Anderson, T.L. and P.J. Hill (1975), "The Evolution of Property Rights: A Study of the American West", *Journal of Law and Economics*, Vol. 18, pp. 163-75.

Ault, D.E. and G.L. Rutman (1979), "The Development of Individual Rights to Property in Tribal Africa", *Journal of Law and Economics*, Vol.22, pp.163-82.

Barzel, Y. (1989), *The Economic Analysis of Property Rights*, Cambridge: Cambridge University Press.

Brookfield, H. and C. Padoch (1994), "Appreciating Agro-diversity", *Environment*, Vol.36, No.5, June, pp.6-11, 37-45.

Besen, S.M. and Raskind, L.J. (1991), "An Introduction to the Law and Economics of Intellectual Property," *Journal of Economic Perspectives*, 3(1), Winter 1991, pp 3-27.

Cole, Julio H. (2000), "Patents and Copyrights: Do the Benefits Exceed the Costs?", paper presented at the *Mont Pélerin Society*, Santiago, Chile, November 12-17.

De Soto, H. (1989), *The Other Path: The Invisible Revolution in the Third World*, New York: Harper and Row.

Demsetz, H. (1967), "Toward a Theory of Property Rights", *American Economic Review*, Vol.57, pp.347-59.

Hardin, G. (1968), "The Tragedy of the Commons", *Science*, Vol.62, 13 December, pp. 1,243-48.

Hancock, G. (1989), *The Lords of Poverty*, London: Macmillan; New York: Atlantic Monthly Press.

Morris (1995), *The Political Economy of Land Degradation*, London: Institute of Economic Affairs.

Morris, J. Mowatt, R., Reekie, W.D., Tren, R. (2002), *Ideal Matter: Globalisation and the Intellectual Property Debate*, Brussels: Centre for the New Europe.

North and Thomas (1972), *The Rise of the Western World*, Cambridge: Cambridge University Press.

Ostrom, E. (1988), "Institutional Arrangements and the Commons Dilemma", in V. Ostrom *et al.* (1988), pp.101-39.

————. (1990), *Managing the Commons*, Cambridge: Cambridge University Press.

Ostrom, V., D. Feeny and H. Picht (eds.) (1988), *Rethinking Institutional Analysis and Development*, San Francisco: ICS Press.

Schlager, E. and E. Ostrom (1992), "Property-Rights Régimes and Natural Resources: A Conceptual Analysis", *Land Economics*, Vol.68, pp.249-62.

Stahl, M. (1993), "Land Degradation in East Africa", *Ambio*, Vol.22, No.8, December, pp.505-8.

Wade, N. (1974), "Sahelian Drought: No Victory for Western Aid", *Science*, Vol.185, 19 July, pp.234-37.

WCED (1987), *World Commission on Environment and Development: Our Common Future*, Oxford: Oxford University Press.

Forests and Wildlife

7

Self-Governance and Forest Resources

ELINOR OSTROM

Introduction

Forest resources share attributes with many other resource systems that make difficult their governance and management in a sustainable, efficient and equitable manner. While some 'forests' are small enough that fencing them or protecting their borders from intrusion is relatively easy, excluding beneficiaries from access and use of most forests is costly. The difficulty of exclusion creates the possibility that individuals who benefit from the use of a forest will not contribute to its long-term sustainability. For many uses of a forest, one person's harvesting subtracts products that are not available to others. Thus, many aspects of forests can be considered as common-pool resources. Common-pool resources are characterised by difficulty of exclusion and generate finite quantities of resource units so that one person's use subtracts from the quantity of the resource available to others (E. Ostrom *et al.* 1994). The ecosystem services generated by forest resources - watershed protection, carbon sequestration, biodiversity enhancement, etc. may be considered as externalities or as public goods. Ecosystem services are, however, closely tied to the sustainability of the forest stock, and are thus threatened by the same set of incentives that tempt users of an unregulated forest resource into a race to use up the timber and destroy the forest itself. Destruction or degradation of forest resources is most likely to occur in open-access forests where those involved and/or external authorities have not established an effective governance regime to regulate the following:

- Who is allowed to appropriate forest products;

- The timing, quantity, location and technology of appropriation;

- Who is obligated to contribute labour or funds to provide or maintain the forest;

- How appropriation and obligation activities are to be monitored and enforced;

- How conflicts over appropriation and obligation activities are to be resolved; and

- How the rules affecting the above will be changed over time with changes in the extent and composition of the forest and the strategies of participants.

A self-governed forest resource is one where actors, who are major users of the forest, are involved over time in making and adapting rules within collective-choice arenas regarding the inclusion or exclusion of participants, appropriation strategies, obligations of participants, monitoring and sanctioning, and conflict resolution. Some extremely remote forests are governed entirely by users and not at all by external authorities.

In most modern political economies, however, it is rare to find any resource system—including the treasuries of private for-profit corporations—that are governed *entirely* by participants without rules made by local, regional, national, and international authorities also affecting key decisions (V. Ostrom 1991, 1997). Thus, in a self-governed system, participants make many, but not all, rules that affect the sustainability of the resource system and its use.

The Conventional Theory
of Common-Pool Resources

Much of the literature on common-pool resources is derived from work on irrigation systems, fisheries and rangelands, but forests are also frequently studied as common-pool resources. Most theoretical studies by political economists have analysed simple common-pool resource systems using relatively similar assumptions (Feeny *et al.* 1996). In such systems, it is assumed that the resource generates a highly predictable, finite supply of one type of resource unit (hardwood timber, for example) in each relevant time period. Users are assumed to be homogeneous in terms of their assets, skills, discount rates and cultural views. They are also assumed to be short-term, profit-maximising actors who possess complete information. In this theory, *anyone* can enter the resource and appropriate resource units.

Users gain property rights only to what they harvest, which they then sell in an open competitive market. The open-access condition is a given. The users make no effort to change it. Users act independently and do not communicate or coordinate their activities in any way. The prediction from this theory is that over-harvesting will result.

Until recently, textbooks in resource economics presented this conventional theory as the *only* theory needed for understanding common-pool resources (for a different approach, see Baland and Platteau 1996).

With the growing use of game theory, use of a common- pool resources is frequently represented as a one-shot or finitely repeated, Prisoner's Dilemma game (Dawes 1973; Dasgupta and Heal 1979). These models formalise the problem differently, but do not change any of the basic theoretical assumptions about the finite and predictable supply of resource units, complete information, homogeneity of users, their maximisation of expected profits, and their lack of interaction with one another or capacity to change their institutions. Nor is the prediction changed. The empirical validity of the theory was not effectively challenged until the mid-1980s since there were many dramatic examples of resources destroyed by users acting independently. The massive deforestation in tropical countries and the desertification of the Sahel confirmed the worst predictions to be derived from this theory for many scholars. Garrett Hardin's (1968) dramatic article in *Science* convinced many non-economists that this theory captured the essence of the problem facing users of most forests in the world. Since users are viewed as being trapped in these dilemmas, recommendations were repeatedly made that external authorities must impose a different set of institutions on such settings. Some recommend private property as the most efficient form of ownership (Demsetz 1967; Posner 1977; Simmons *et al.* 1996). Others recommend government ownership and control (Ophuls 1973).

Implicitly, theorists assume that regulators will act in the public interest and understand how ecological systems work and how to change institutions so as to induce socially optimal behaviour (Feeny *et al.* 1996: 195).

The possibility that the users themselves would find ways to organise had not seriously been considered in much of the policy literature until the last decade. Organising so as to create rules that specify rights and duties of participants creates a public good for those involved. Anyone who is included in the community of users benefits from this public good, whether they contribute or not. Thus, getting 'out of the trap' is itself a second-level dilemma. Further, investing in monitoring and sanctioning activities in order to increase the likelihood that participants follow the agreements they have made, also generates a public good. These investments, therefore, represent a third-level dilemma. Since much of the initial problem exists because the individuals are stuck in a setting where they generate negative externalities for each other, it is not consistent with the conventional theory that they solve a second- and third-level dilemma in order to address the first-level dilemma under analysis.

Until the work of the National Academy of Science's Panel on Common Property (National Research Council 1986), the conclusions derived from this basic theory were applied to all common-pool resources regardless of what kind of institutions local users developed. The growing evidence from many field studies of common-pool resources, however, called for a serious rethinking of the generalisability of the conventional theory (see Berkes 1986, 1989; Berkes *et al.* 1989; E. Ostrom 1990; Bromley *et al.* 1992; McCay and Acheson 1987; Agrawal forthcoming). Empirical studies have challenged the applicability of the conventional theory to forests by showing how forest users in many locations have organised themselves to vigorously protect and, in some cases, enhance local forests (Fortmann and Bruce 1988; Fox 1993; Fairhead and Leach 1996). Often, problems in forest management have emerged when local self-organisation has not been recognised by policymakers, and the autonomy of forest users to continue their own forest use practices has been threatened (Arnold and Campbell 1986; Gillis 1988; Arnold and Stewart 1991).

Loss of ownership combined with substantial increases in population, greater commercialisation of forest products and technological changes that made rapid deforestation possible, have increasingly threatened forests in all parts of the world (Jodha 1986; Poffenberger 1990; Messerschmidt 1993). Evidence has been mounting from a wide variety of sources, however, that local forest users are still capable of managing forest resources in many diverse locations (Ascher 1995; Becker *et al.* 1995; Agrawal and Yadama 1997; Shivakoti *et al.* 1997). Considerable interest has been rekindled in promoting various forms of community forestry institutions (Peluso and Poffenberger 1989; Alcorn 1990; Herring 1990; Gilmour and Fisher 1992; Shepherd 1992; Lynch and Talbott 1995; Ford Foundation 1998).

A Growing Consensus on Variables Enhancing Self-Organisation

Evidence from field research on common-pool resources challenges the generalisability of the conventional theory. While the conventional theory is successful in predicting outcomes in settings where users are alienated from each other or cannot communicate effectively, it does not provide an explanation for settings where users are able to create and sustain agreements to avoid serious problems of over-appropriation. Nor does it predict well when government ownership will perform appropriately or how privatisation will improve outcomes. A fully articulated, reformulated theory encompassing the conventional theory as a special case does not yet exist. On the other hand, scholars familiar with the results of field research substantially agree on a set of variables that enhance the likelihood of users organising themselves to avoid the social losses associated with open-access, common-pool resources (E. Ostrom 1990, 1992a, b; Schlager 1990; McKean 1992, 1998; Tang 1992; E. Ostrom *et al.* 1994; Wade 1994; Baland and Platteau 1996; Arnold 1998). Drawing heavily on Ostrom (1992b: 298-9) and Baland and Platteau (1996: 286-9), considerable consensus exists that the following attributes of common-pool resources and of users are conducive to an increased likelihood that self-governing associations will form. Though further empirical work is needed to test their relevance and explanatory power for self-organised forest management, these attributes do appear to have broad applicability.

Attributes of the Resource:

R1. *Feasible improvement:* The resource is not at a point of deterioration such that it is useless to organise or so under-utilised that little advantage results from organising.

R2. *Indicators:* Reliable and valid information about the general condition of the resource is available at reasonable costs.

R3. *Predictability:* The availability of resource units is relatively predictable.

R4. *Spatial extent:* The resource is sufficiently small, given the transportation and communication technology in use, that users can develop accurate knowledge of external boundaries and internal microenvironments.

Attributes of the Users

A1. *Salience:* Users are dependent on the resource for a major portion of their livelihood or other variables of importance to them.

A2. *Common understanding:* Users have a shared image of the resource (attributes R1, 2, 3 and 4 above) and how their actions affect each other and the resource.

A3. *Discount rate:* Users have a sufficiently low discount rate in relation to future benefits to be achieved from the resource.

A4. *Distribution of interests:* Users with higher economic and political assets are similarly affected by a current pattern of use.

A5. *Trust:* Users trust each other to keep promises and relate to one another with reciprocity.

A6. *Autonomy:* Users are able to determine access and harvesting rules without external authorities countermanding them.

A7. *Prior organisational experience:* Users have learned at least minimal skills of organisation through participation in other local associations or learning about ways that neighbouring groups have organised.

Many of these variables are in turn affected by the type of larger regime in which users are embedded. Larger regimes can facilitate local self-organisation by providing accurate information about natural resource systems, providing arenas in which participants can engage in discovery and conflict-resolution processes, and providing mechanisms to back up local monitoring and sanctioning efforts. The probability of participants adapting more effective rules in macroregimes that facilitate their efforts over time is higher than in regimes that ignore resource problems entirely or, at the other extreme, presume that all decisions about the governance and management of resources need to be made by central authorities.

The key to further theoretical integration is to understand how these attributes interact in complex ways to affect the basic benefit-cost calculations of a set of users utilising a resource (E. Ostrom 1990: Ch. 6). Each user i has to compare the expected net benefits of harvesting from a resource continuing to use the old rules (BO) to the benefits he or she expects to achieve with a new set of rules (BN). Each user i must ask whether his or her incentive to change (D_i) is positive or negative.

$$D_i = BN_i - BO_i$$

If D_i is negative for all users, no one has an incentive to change. If D_i is positive for some users, they then need to estimate three types of costs:

C1: the up-front costs of time and effort spent devising and agreeing upon new rules; C2: the short-term costs of adopting new harvesting strategies; and

C3: the long-term costs of monitoring and maintaining a self-governed system over time (given the norms of the community in which they live).

If the sum of these expected costs for each user exceeds the incentive to change, no user will invest the time and resources needed to create new institutions.

Thus, if $D_i < (C1_i + C2_i + C3_i)$ for all i in the entire set of users, no change occurs.

In field settings, not everyone is likely to expect the same costs and benefits from a proposed change. Some may perceive positive benefits after all costs have been taken into account, while others perceive net losses. Consequently, the collective-choice rules used to change the day-to-day operational rules related to harvesting, planting, monitoring and other management activities affect whether an institutional change favoured by some and opposed by others will occur. For any collective-choice rule, such as unanimity, majority, ruling elite or one-person rule, there is at least one minimal coalition of users that must agree prior to the adoption of new rules.

If there is an individual k in all minimal coalitions for whom $D_k \leq (C1_k + C2_k + C3_k)$, no new rules will be adopted. And if for at least one minimal coalition $D_k > (C1_k + C2_k + C3_k)$, for all members, it is feasible for a new set of rules to be adopted. If there are several such coalitions, the question of which coalition will form, and so which rules will result, is a theoretical issue beyond the scope of this discussion. This analysis is applicable to a situation where a group starts with an open-access set of rules and contemplates adopting its first set of rules limiting access. It is also relevant to the continuing consideration of changing rules over time.

Thus, the existing collective-choice arrangements in a location for making operational rules related to the entry and use of forests affects whether self-organisation will occur. Existing institutional arrangements also have a major effect on the distributional impacts of any rule changes that may occur. The rule used to modify institutional arrangements in field settings varies from reliance on the decisions made by one or a few leaders, to a formal reliance on majority or super-majority vote, to reliance on consensus or close to unanimity. If there are substantial differences in the perceived benefits and costs of users, it is possible that a minimal coalition will impose a new set of rules that strongly favours those in the winning coalition and imposes losses or lower benefits on those in the losing coalition (Thompson *et al.* 1988). If expected benefits from a change in institutional arrangements are not greater than expected costs for many users, however, the costs of enforcing a change in institutions will be much higher than when most participants expect to benefit from a change in rules over time. Where the enforcement costs are fully borne by the members of a minimal coalition, operational rules that benefit the other users lower the long-term costs of monitoring and sanctioning for a governing coalition.

Where external authorities enforce the rules agreed upon by a coalition of users, the distribution of costs and benefits is more likely to benefit that coalition and may impose costs on the other users (see Walker *et al.* 1997). The attributes of a resource (listed above) affect both the benefits and costs of institutional change. If resource units are relatively abundant (R1), there are few reasons for users to invest costly time and effort in organising. If the resource is already substantially destroyed, the high costs of organising may not generate substantial benefits.

Thus, self-organisation is likely to occur only after users observe substantial scarcity. The danger here, however, is that exogenous shocks leading to a change in relative abundance of the resource units occur rapidly and users may not adapt quickly enough to the new circumstances (Libecap and Wiggins 1985).

The presence of frequently available, reliable indicators about the conditions of a resource (R2) improves the capacity of users to adapt relatively quickly to changes that could adversely affect their long-term benefit stream (Moxnes 1996). A resource flow that is highly predictable (R3) is much easier to understand and manage than one that is erratic. In the latter case, it is always difficult for users (or, for that matter, for scientists and government officials) to judge whether changes in the resource stock or flow are due to overharvesting or to random exogenous variables (see Feeny *et al.* 1996 for a discussion of these issues related to the collapse of the California sardine industry). Unpredictability of resource units in microsettings, such as private pastures, may lead users to create a larger common-property unit to increase the predictability of resource availability somewhere in the larger unit (Netting 1972; Wilson and Thompson 1993). The spatial extent of a resource (R4) affects the costs of defining reasonable boundaries and then of monitoring them over time.

The attributes of the users themselves (listed above) also affect their expected benefits and costs. If users do not obtain a major part of their income from a resource (A1), the high costs of organising and maintaining a self-governing system may not be worth their effort. If users do not share a common understanding of how complex resource systems operate (A2), they will find it extremely difficult to agree on future joint strategies. Given the complexity of many forest resources—especially multispecies or multiproduct resources—understanding how these systems work may be counterintuitive even for those who make daily contact with the forest. In forests that are highly variable (R3), it may be particularly difficult to understand and to separate those outcomes stemming from exogenous factors and those resulting from the actions of users. Of course, this is also a problem facing officials as well as users. Users with many other options, who thus discount the importance of future income from a particular resource (A3), may prefer to 'mine' one forest without spending resources to regulate it. They simply move on to other forests once this one is destroyed, assuming there will always be still more forests available to them. Users who possess more substantial economic and political assets may have similar interests to those with fewer assets (A4) or they may differ substantially on multiple attributes. When the more powerful have similar interests, they may greatly enhance the probability of successful organisation if they invest their resources in organising a group and devising rules to govern that group. Those with substantial economic and political assets are more likely to be a member of a winning coalition and thus have a greater impact on decisions about institutional changes. Mancur Olson (1965) long ago recognised the possibility of a privileged group whereby some were sufficiently affected to bear a disproportionate share of the costs of organising to provide public goods (such as the organisation of a collectivity). On the other hand, if those with more assets also have low discount rates (A3) related to a particular resource and lower salience (A1), they may simply be unwilling to expend inputs or may actually impede organisational efforts that might lead to them having to cut back on their productive activities.

Users who trust one another (A6) to keep agreements and use reciprocity in their relationships face lower expected costs for monitoring and sanctioning over time.

Users who lack trust at the beginning of a process of organising may be able to build this form of social capital (Coleman 1988; E. Ostrom 1992a) if they initially adopt small changes that most users follow before trying to make major institutional changes. Autonomy (A7) tends to lower the costs of organising. A group that has little autonomy may find that those who disagree with locally developed rules seek contacts with higher-level officials to undo the efforts of users to achieve regulation. With the legal autonomy to make their own rules, users face substantially lower costs in defending these rules against other authorities. Prior experience with other forms of local organisation (A7) greatly enhances the repertoire of rules and strategies known by local participants as potentially useful to achieve various forms of regulation. Further, users are more likely to agree upon rules whose operation they understand from prior experience, than upon rules that are introduced by external actors and are new to their experience. Given the complexity of many field settings, users face a difficult task in evaluating how diverse variables affect expected benefits and costs over a long time horizon. In many cases, it is just as difficult, if not more so, for scientists to make a valid and reliable estimate of total benefits and costs and their distribution.

The growing theoretical consensus does *not* lead to a conclusion that most users utilising common-pool resources will undertake self-governed regulation. Many settings exist where the theoretical expectation should be the opposite. Users will overuse the forest unless efforts are made to change one or more of the variables affecting perceived costs or benefits. Given the number of variables that affect these costs and benefits, many points of external intervention can enhance or reduce the probability of users' agreeing upon and following rules that generate higher social returns. But both social scientists and policymakers have a lot to learn about how these variables interact in field settings and even how to measure them so as to increase the empirical legitimacy of the growing theoretical consensus. Many aspects of the macroinstitutional structure surrounding a particular setting affect the perceived costs and benefits.

Thus, external authorities can be effective in enhancing the likelihood and performance of self-governing institutions. Their actions can also seriously impede these developments as well. Further, when the activities of one set of users, A, have 'spillover effects' on others beyond A, external authorities can either facilitate processes that allow multiple groups to solve conflicts arising from negative spillovers or take a more active role in governing particular forests themselves. Researchers and public officials need to recognise the multiple manifestations of these attributes in the field. Users may be highly dependent on a forest (A1), for example, because they are in a remote location and few roads exist to enable them to leave. Alternatively, they may be located in a central location, but other opportunities are not open to them due to lack of training or a discriminatory labour market. Users' discount rates (A3) in relation to a particular forest may be low because they have lived for a long time in a particular location and expect that they and their grandchildren will remain in that location, or because they possess a secure and well-defined bundle of property rights to this forest (see Schlager and Ostrom 1992). Reliable indicators of the condition of a forest (R2) may result from activities of users themselves—such as regularly harvesting of medicinal plants or because of efforts to gather reliable information by users or by external authorities (Blomquist 1992). Predictability of forest resources (R3) may result from relatively constant weather conditions or because of actions taken by users to even out the availability of forest products over time. They may have autonomy to make their own rules (A6) because a national government is weak and unable to exert authority over forests that it formally owns, or because national law formally legitimates self-governance of forests as in Uttar Pradesh in India.

When the benefits of organising are commonly understood by participants to be very high, users lacking many of the attributes conducive to the development of self-governing institutions may be able to overcome their liabilities and still develop effective agreements. The crucial factor is not whether all attributes are favourable but the relative size of the expected benefits and costs they generate as perceived by participants. While all of these variables affect the expected benefits and costs of users, it is difficult—particularly for outsiders—to estimate their impacts given the difficulty of making precise measures of these variables and weighing them on a cumulative scale. Further empirical analysis of these theoretical propositions is, thus, dependent on the conduct over time of careful comparative studies of a sufficiently large number of field settings using a common set of measurement protocols (see E. Ostrom and Wertime 1994).

On The Design Principles of Robust, Self-Governed Common-Pool Resource Institutions

Of course, the performance of self-governed common-pool resource institutions varies across systems and time.

Some self-governed resources have survived and flourished for centuries, while others have faltered and failed. As discussed above, some never get organised in the first place. In addition to the consensus concerning the theoretical variables conducive to self-organisation, considerable agreement also exists about the characteristics of those self-governing systems that are robust in the sense that they survive for very long periods of time utilising the same basic rules for adapting to new situations over time (Shepsle 1989).

The particular rules used in the long-surviving, self-governing systems vary substantially. Consequently, it is not possible to arrive at empirical generalisations about the particular types of rules used to define who is a member of a self-governing community, what rights they have to access a resource, and what particular obligations they face. It is possible, however, to derive a series of design principles that characterise the configuration of rules that are used. By design principles, I mean a condition that helps to account for the success of these institutions in sustaining a forest or other common-pool resource and gaining the compliance of generation after generation of users to the rules applied in a location.

Based on considerable research on common-pool resources, robust, long-term institutions are characterised by most of the design principles listed below. Fragile institutions tend to be characterised by only some of these design principles. Failed institutions are characterised by very few of these principles (see, for example, Blomqvist 1996; Morrow and Hull 1996; Schweik *et al.* 1997).

These principles work to enhance the shared understanding of participants of the structure of a resource and its users and of the benefits and costs involved in following a set of agreed-upon rules.

Design principles illustrated by long-enduring common-pool resource institutions:

1. *Clearly defined boundaries*: Individuals or households with rights to withdraw resource units from the common-pool resource and the boundaries of the common-pool resource itself are clearly defined.

2. *Congruence*:

    a. The distribution of benefits from appropriation rules is roughly proportionate to the costs imposed by provision rules.

    b. Appropriation rules restricting time, place, technology and/or quantity of resource units are related to local conditions.

3. *Collective-choice arrangements*: Most individuals affected by operational rules can participate in modifying operational rules.

4. *Monitoring*: Monitors, who actively audit common-pool resource conditions and user behaviour, are accountable to the users and/or are the users themselves.

5. *Graduated sanctions:* Users who violate operational rules are likely to receive graduated sanctions (depending on the seriousness and context of the offence) from other users, from officials accountable to these users, or from both.

6. *Conflict-resolution mechanisms*: Users and their officials have rapid access to low-cost, local arenas to resolve conflict among users or between users and officials.

7. *Minimal recognition of rights to organise:* The rights of users to devise their own institutions are not challenged by external governmental authorities.

For common-pool resources that are part of larger systems:

8. *Nested enterprises:* Appropriation, provision, monitoring, enforcement, conflict resolution and governance activities are organised in multiple layers of nested enterprises.

Design Principle 1 (having rules that clearly define who has rights to use a resource and the boundaries of that resource) ensures that users can clearly identify anyone who does not have rights and take action against them.

Design Principle 2 involves two parts. The first is congruence between the rules that assign benefits and the rules that assign costs. The crucial issue here is that these rules be considered fair and legitimate by the participants themselves (see McKean 1992). In many settings, fair rules are those that keep a relatively proportionate relationship between the assignment of benefits and of costs. Otherwise, those who contribute time, funds and effort to sustaining a forest or other common-pool resource system resent the 'unfair' allocation of benefits to those carrying a lesser load. The whole distribution system can disintegrate if it is not perceived as fair. The second part of this design principle is that both types of rules need to be well matched to local conditions, such as soils, slope, number of diversions, crops being grown.

Design Principle 3 is concerned with the collective-choice arrangements used to modify operational rules—those that directly affect where, when and who can harvest from a resource. If most users are not involved in modifying these rules over time, the information about the benefits and costs as perceived by different participants is not fully taken into account in any efforts to adapt to new conditions and information over time. Users who begin to perceive the costs of their system being higher than their benefits, and who are prevented from making serious proposals for change, may simply begin to cheat whenever they have the opportunity. Once cheating on rules becomes more frequent for some users, others will follow suit. In this case, enforcement costs become very high or the system fails. No matter how high the level of acceptance of an initial agreement, there are always conditions that tempt some individuals to cheat (even when they perceive the overall benefits of the system to be higher than the costs). If one person is able to cheat while others conform to the rules, the cheater is usually able to gain substantially to the disadvantage of others.

Thus, without monitoring of rule conformance—Design Principle 4—few systems are able to survive very long. The sanctions that are used, however, do not need to be extremely high in the first instance. The important thing about a sanction for a user who has succumbed to temptation is that their action is noticed and that a punishment is meted out. This tells all users that cheating on rules is noticed and punished without making all rule infractions into major criminal events.

If the sanctions are graduated (Design Principle 5), a user who breaks rules repeatedly and who is noticed doing so, eventually faces a penalty that makes rule breaking an unattractive option. While rules are always assumed to be clear and unambiguous in theoretical work, this is rarely the case in field settings. It is easy to have a disagreement about how to interpret a rule that limits harvesting or requires inputs. If these disagreements are not resolved in a low-cost and orderly manner, then users may lose their willingness to conform to rules because of the ways that 'others' interpret them in their own favour (Design Principle 6).

Design Principles 7 and 8 are related to autonomy. When the rights of a group to devise their own institutions are recognised by national, regional and local governments, the legitimacy of the rules crafted by users will be less frequently challenged in courts, administrative and legislative settings. Further, in larger resources with many participants, such as many forests, nested enterprises that range in size from small to large enable participants to solve diverse problems involving different scale economies. By utilising base institutions that are quite small, face-to-face communication can be used for solving many of the day-to-day problems in smaller groups. By nesting each level of organisation in a larger level, externalities between groups can be addressed in larger organisational settings that have a legitimate role to play in relationship to the smaller entities.

Theoretical Puzzles

In addition to the consensus concerning the variables most likely to enhance self-organisation and the design principles characterising successful, long-term governance arrangements, many unresolved theoretical issues still exist about the self-governance of common-pool resources. Two major theoretical questions relate to the effect of size and heterogeneity.

Size

The effect of the number of participants facing problems of creating and sustaining a self-governing enterprise is unclear. Drawing on the early work of Mancur Olson (1965), many theorists argue that size of a group is negatively related to solving collective-action problems in general (see also Buchanan and Tullock 1962). Many results from game theoretical analysis of repeated games conclude that cooperative strategies are more likely to emerge and be sustained in smaller rather than larger groups (see synthesis of this literature in Baland and Platteau 1996). Scholars who have studied many user-governed forestry institutions in the field have concluded that success will more likely happen in smaller groups (see, for example, Cernea 1989).

In a systematic study of forest institutions, Agrawal (1998) has found that smaller forest user groups are less able to undertake the level of monitoring needed to protect forest resources than moderately sized groups.

In a study of over 100 irrigation systems, Lam (1998) did not find any significant relationship between either the number of users or the amount of land included in the service area and any of the three performance variables he studied.

One of the problems of focussing on size of group as a key determining factor is that many other variables change as group size increases (Chamberlin 1974; R. Hardin 1982). If the costs of providing a public good related to the use of a forest, say a sanctioning system, remain relatively constant as group size increases, then increasing the number of participants brings additional resources that could be drawn upon to provide the benefit enjoyed by all (see Isaac *et al.* 1993). Marwell and Oliver (1993: 45) conclude that when a 'good has pure jointness of supply, group size has a *positive* effect on the probability that it will be provided.' On the other hand, if one is analysing the conflict levels over a subtractable good and the transaction costs of arriving at acceptable allocation formulae, group size may well exacerbate the problems of self-governing systems. Since there are tradeoffs among various impacts of size on other variables, a better working hypothesis is that medium-sized groups may succeed more often than very small or very large groups.

Heterogeneity

Many scholars conclude that only very small groups can organise themselves effectively because they presume that size is related to the homogeneity of a group and that homogeneity is needed to initiate and sustain self-governance. Heterogeneity is also a highly contested variable. Groups can vary along a diversity of dimensions including their cultural backgrounds, interests and endowments (see Baland and Platteau 1996). Each may operate differently. Further, groups who are relatively homogeneous may also find themselves confronted with high levels of conflict when rules related to access to forest resources are not well defined (see Gibson and Koontz forthcoming).

If groups from diverse cultural backgrounds share access to a forest, the key question affecting the likelihood of self-organised solutions is whether the views of the multiple groups concerning the structure of the forest, authority, interpretation of rules, trust and reciprocity differ or are similar. In other words, do they share a common understanding (A2) of their situation? New settlers to a region may simply learn and accept the rules of the established group, and their cultural differences on other fronts do not affect their participation in governing a forest. On the other hand, new settlers are frequently highly disruptive to the sustenance of a self-governing enterprise when they generate higher levels of conflict over the interpretation and application of rules and increase enforcement costs substantially. When the interests of users differ, achieving a self-governing solution is particularly challenging. This problem characterises some fisheries where local subsistence fishermen have strong interests in the sustenance of an inshore fishery, while industrial fishing firms have many other options and may be more interested in the profitability of fishing in a particular location than its sustained yield. The conflict between absentee livestock owners and local pastoralists has also proved difficult to solve in many parts of the world. Differential endowments of users can be associated with both extreme levels of conflict as well as very smooth and low-cost transitions into a sustainable, self-governed system. Johnson and Libecap (1982) reason that the difference in the skills and knowledge of different kinds of users frequently prevents them from arriving at agreements about how to allocate harvesting quotas. Heterogeneity of wealth or power may or may not be associated with a difference in interests. As discussed above, when those who have more assets share similar interests with those who have less assets (A4), groups may be privileged by having the more powerful take on the higher initial costs of organising, while crafting rules that benefit a large proportion of the users. Users may design institutions that cope effectively with heterogeneities. Thus, when they adopt rules that allocate benefits using the same formulae used to allocate duties and responsibilities (Design Principle 2a), users who differ significantly in terms of assets will tend to agree to and follow such rules.

Even in a group that differs on many variables, if at least a minimally winning subset of users from an endangered but valuable forest are dependent on it (A1), and they share a common understanding of their situations (A2), have a low discount rate (A3), include some with more assets among their members (A4), trust one another (A5), and have autonomy to make their own rules (A6), it is more likely that they will estimate the expected benefits of governing their forest to be greater than the expected costs. Whether the rules agreed upon distribute benefits and costs fairly depends both on the collective-choice rule used and the type of heterogeneity existing in the community. Neither size nor heterogeneity is a variable with a uniform effect on the likelihood of organising and sustaining self-governing enterprises. The debate about their effect is focussing on the wrong variables. Instead of concentrating on size or the various kinds of heterogeneity independently, it is important to ask how these variables influence other variables as they affect the benefit-cost calculations of those involved in negotiating and sustaining agreements. Their impact on costs of producing and distributing information (Scott 1993) is particularly important.

Conclusion

The conventional theory of common-pool resources, which presumed that external authorities were needed to impose new rules on those users trapped into producing excessive externalities for themselves and others, has now been shown to be a special theory of a more general theoretical structure. For users to contemplate changing the institutions they face, they have to conclude that the expected benefits from an institutional change will exceed the immediate and long-term expected costs.

When users cannot communicate and have no way of gaining trust through their own efforts or with the help of the macroinstitutional system within which they are embedded, the prediction of the earlier theory is likely to be empirically supported.

Ocean fisheries, the stratosphere and other global commons come closest to the appropriate empirical referents. If users can engage in face-to-face bargaining and have autonomy to change their rules, they may well attempt to organise themselves.

Whether they organise depends on attributes of the resource and the users themselves that affect the benefits to be achieved and the costs of achieving them. Whether their self-governed enterprise succeeds over the long term depends on whether the institutions they develop are consistent with design principles underlying robust, long-living, self-governed systems. The theory has progressed substantially during the past half century.

There are, however, many challenging puzzles to be solved. Further empirical research is needed to test the relevance and explanatory power for forests of the emerging consensus regarding the conditions conducive to effective self-governance of resources and for the design of more effective public policies. Given the number of variables that can independently and interactively affect the incentives and actions of local users—especially when one focuses on forest resources—systematic research conducted in a large number of locations obtaining data about the same set of variables is also extremely important. Hopefully, future policies will be constructed on the basis of an understanding of both the strengths and limitations of self-governance of forestry resources.

References

Agrawal, A. (1998), *"Group size and successful collective action: A case study of forest management institutions in the Indian Himalayas"*. In: Gibson, C., McKean, M. and Ostrom, E. (eds.) *Forest resources and institutions*. "Forests, Trees and People Programme", *Working Paper 3*. FAO, Rome, Italy.

Agrawal, A. Forthcoming. *Greener pastures: exchange, politics and community among a mobile people*. Duke University Press, Durham, NC.

Agrawal, A. and Yadama, G. (1997), "How do local institutions mediate market and population pressures on resources: forest Panchayats in Kumaon, India". In: *Development and Change* 28: 435-65.

Alcorn, J. (1990), "Indigenous agroforestry strategies meeting farmers' needs". In: Anderson, A. (ed.) *Alternatives to deforestation: steps toward sustainable use of Amazon Rainforest*, 141-151. Columbia University Press, New York.

Arnold, J.E.M. (1997), "Social dimensions of forestry's contribution to sustainable development". Position Paper for programme area F of the *XI World Forestry Congress*, Antalya, Turkey, 13-22 October 1997.

————. (1998), "Devolution of control of common pool resources to local communities: experiences in forestry". Paper presented at the meeting of the UNU/WIDER Project on *Land Reform Revisited: Access to Land, Rural Poverty, and Public Action*, Santiago, Chile, April 26-28.

Arnold, J.E.M. and Campbell, J.G. (1986), "Collective management of hill forests in Nepal: the community forestry development project". In: National Research Council, Proceedings of the conference on *Common Property Resource Management*, 425-54. National Academy Press, Washington, DC.

Arnold, J.E.M. and Stewart, W.C. (1991), "Common property resources management in India". Oxford Forestry Institute, *Tropical Forestry Papers 24*. Oxford University, Oxford, UK.

Ascher, W. (1995), *Communities and sustainable forestry in developing countries*. ICS Press, Oakland, CA.

Baland, J.M. and Platteau, J.P. (1996), *Halting degradation of natural resources. Is there a role for rural communities?* Clarendon Press, Oxford, UK.

Banana, A.Y. and Gombya-Ssembajjwe, W. (1998), "Successful forest management: the importance of security of tenure and rule enforcement in Ugandan forests". In: Gibson, C., McKean, M. and Ostrom, E. (eds.) *Forest resources and institutions*. "Forests, Trees and People Programme", *Working Paper 3*. FAO, Rome, Italy.

Bebbington, A, Kopp, A. and Rubinoff, D. (1997), "From chaos to strength? Social capital, rural people's organisations and sustainable development". Paper prepared for the FAO workshop on *Pluralism, Forestry and Rural Development*, December 1997, Rome, Italy.

Becker, C.D. and Gibson, C. (1998), "The lack of institutional supply: why a strong local community in Western Ecuador fails to protect its forest". In: Gibson, C., McKean, M. and Ostrom, E. (eds.) *Forest resources and institutions*. "Forests, Trees and People Programme", *Working Paper 3*. FAO, Rome, Italy.

Becker, C.D., Banana, A.Y. and Gombya-Ssembajjwe, W. (1995), "Early detection of tropical forest degradation: An IFRI pilot study in Uganda". *Environmental Conservation* 22: 31-38.

Berkes, F. (1986), "Local-level management and the commons problem: a comparative study of Turkish coastal fisheries". *Marine Policy 10*: 215-29.

Berkes, F. (ed.) (1989), *Common property resources: ecology and community-based sustainable development*. Belhaven Press, London, UK.

Berkes, F., Feeny, D., McCay, B.J. and Acheson, J.M. (1989), "The benefits of the commons". *Nature 340*: 91-3.

Blomquist, W. (1992), *Dividing the waters: governing groundwater in southern California*. ICS Press, Oakland, CA

Blomqvist, A. (1996), *Food and fashion. Water management and collective action among irrigation farmers and textile industrialists in South India*. The Institute of Tema Research, Department of Water and Environmental Studies, Linköping, Sweden.

Bromley, D.W., Feeny, D., McKean, M., Peters, P., Gilles, J., Oakerson, R., Runge, C.F. and Thomson, J. (eds.) (1992), *Making the commons work: theory, practice, and policy*. ICS Press, Oakland, CA.

Buchanan, J.M. and Tullock, G. (1962), *The calculus of consent*. University of Michigan Press, Ann Arbor, MI.

Cernea, M. (1989), *User groups as producers in participatory afforestation strategies*. World Bank Discussion Papers 70. World Bank, Washington, DC.

Chamberlin, J. (1974), "Provision of collective goods as a function of group size". *American Political Science Review 68*: 707-16.

Coleman, J. (1988), "Social capital in the creation of human capital". *American Journal of Sociology 91*: 309-35.

Dasgupta, P.S. and Heal, G.M. (1979), *Economic theory and exhaustible resources*. Cambridge University Press, Cambridge, UK.

Dawes, R.M. (1973), *The commons dilemma game: an N-person mixed-motive game with a dominating strategy for defection*. Oregon Research Institute Research Bulletin 13: 1-12.

Demsetz, H. (1967) "Toward a theory of property rights". *American Economic Review 57*: 347-59.

Fairhead, J. and Leach, M. (1996), *Misreading the African landscape. Society and ecology in a forest-savanna mosaic*. Cambridge University Press, Cambridge, UK.

Feeny, D., Hanna, S. and McEvoy, A.F. (1996), "Questioning the assumptions of the 'tragedy of the commons' model of fisheries". *Land Economics 72*: 187-205.

Fisher, R.J. (1988), "Indigenous systems of common property forest management in Nepal". *Environment and Policy Institute, Working Paper 18*. East West Center, University of Hawaii, Honolulu.

FAO (Food and Agriculture Organisation) (1997), "Crafting institutional arrangements for community forestry". *Community Forestry Field Manual 7*. FAO, Rome, Italy.

Ford Foundation. 1998 *Forestry for sustainable rural development*. Ford Foundation, New York.

Fortmann, L. and Bruce, J.W. (eds.) (1988), *Whose trees? Proprietary dimensions of forestry*. Westview Press, Boulder, CO.

Fox, J. (ed.) (1993), "Legal frameworks for forest management in Asia: Case studies of community/State relations". *Programme on Environment, Occasional Paper 16*. East West Center, University of Hawaii, Honolulu.

Gibson, C. and Koontz, T. Forthcoming *When community is not enough: institutions and values in community-based forest management in Southern Indiana*. Human Ecology.

Gibson, C., McKean, M. and Ostrom, E. (eds.) (1998), "Forest resources and institutions". *Forests, Trees and People Programme, Working Paper 3*. FAO, Rome, Italy.

Gillis, M. (1988), "Indonesia: public policies, resource management, and the tropical forest". In: Repetto, R. and Gillis, M. (eds.) *Public policies and the misuse of forest resources*, 43-113. Cambridge University Press, New York.

Gilmour, D.A. and Fisher, R.J. (1992), *Villagers, forests, and foresters: The philosophy, process, and practice of community forestry in Nepal*. Sahayogi Press, Kathmandu, Nepal.

Hardin, G. (1968), "The tragedy of the commons". *Science* 162: 1243-8.

Hardin, R. (1982), *Collective action.* Johns Hopkins University Press, Baltimore, MD.

Herring, R.J. (1990), "Rethinking the commons". *Agriculture and Human Values* 7(2): 88-104.

IASCP (International Association for the Study of Common Property) (1998), *Abstracts of the workshop Crossing Boundaries,* Seventh Common Property Conference, held in Vancouver, Canada, 10-14 June 1998.

Isaac, R.M., Walker, J. and Williams, A. (1993), "Group size and the voluntary provision of public goods: experimental evidence utilising large groups". *Journal of Public Economics* 54: 1-36.

Jodha, N.S. (1986), "Common property resources and rural poor in dry regions of India". *Economic and Political Weekly,* 21(27): 1169-81.

Johnson, R.N. and Libecap, G.D. (1982), "Contracting problems and regulation: The case of the fishery". *American Economic Review* 72: 1005-23.

Lam, W.F. (1998), *Governing irrigation systems in Nepal: institutions, infrastructure, and collective action.* ICS Press, Oakland, CA.

Leach, M., Mearns, R. and Scoones, I. (1997), "Challenges to community-based sustainable development: dynamics, entitlements, institutions". *IDS Bulletin.* 28(4): 4-14.

Libecap, G.D. and Wiggins, S.N. (1985), "The influence of private contractual failure on regulation: The case of oil field unitisation". *Journal of Political Economy* 93: 690-714.

Lynch, O. and K. Talbott. (1995), *Balancing acts: community-based forest management and national law in Asia and the Pacific.* World Resources Institute, Washington, DC.

Malla, Y.B. (1997), "Sustainable use of communal forests in Nepal". *Journal of World Resource Management.* 8:51-74.

Marwell, G. and Oliver, P. (1993), *The critical mass in collective action: a micro-social theory.* Cambridge University Press, New York.

McCay, B.J. and Acheson, J.M. (1987), *The question of the commons: the culture and ecology of communal resources.* University of Arizona Press, Tucson, AZ.

McKean, M. and E. Ostrom. (1995), *Common property regimes in the forest: just a relic from the past?* Unasylva, 46(180): 3-15

McKean, M.A. (1992), "Management of traditional common lands (Iriaichi) in Japan". In: Bromley, D.W., Feeny, D., McKean, M., Peters, P., Gilles, J., Oakerson, R., Runge, C.F. and Thomson, J. (eds.) *Making the commons work: theory, practice, and policy,* 63-98. ICS Press, Oakland, CA.

McKean, M.A. (1998), "Common property: what is it, what is it good for, and what makes it work?" In: Gibson, C., McKean, M. and Ostrom, E. (eds.) *Forest resources and institutions.* "Forests, Trees and People Programme", *Working Paper 3.* FAO, Rome, Italy.

Messerschmidt, D.A. (ed.) (1993), "Common forest resource management: annotated bibliography of Asia, Africa, and Latin America". *Community Forestry Note 11.* FAO, Rome, Italy.

Morrow, C.E. and Hull, R.W. (1996), "Donor-initiated common pool resource institutions: the case of the Yanesha forestry cooperative". *World Development* 24: 1641-57.

Moxnes, E. (1996), "Not only the tragedy of the commons: misperceptions of bioeconomics". *Working Paper,* Foundation for Research in Economics and Business Administration. SNF, Bergen, Norway.

National Research Council (1986), *Proceedings of the conference on Common Property Resource Management.* National Academy Press, Washington, DC.

Netting, R. McC. (1972), *"Of men and meadows: Strategies of alpine land use".* Anthropological Quarterly 45: 132-44.

North, D. (1990), *Institutions, institutional change and economic performance.* Cambridge University Press, New York.

Okoth-Ogendo, H.W.O. (1987) "Tenure of trees or tenure of land". In: Raintree, J. (ed.) *Land, trees and tenure: proceedings of an international workshop on tenure issues in agroforestry,* Nairobi, May 27- 31, 1985, 225-229. Land Tenure Center, Madison, WI.

Olson, M. (1965), *The logic of collective action: Public goods and the theory of groups.* Harvard University Press, Cambridge, MA.

Ophuls, W. (1973), *"Leviathan or oblivion".* In: Daly, H.E. (ed.) *Toward a steady state economy.* Freeman, Oakland, CA.

Ostrom, E. (1990), *Governing the commons: the evolution of institutions for collective action.* Cambridge University Press, New York.

—————. (1992a), *Crafting institutions for self-governing irrigation systems.* ICS Press, Oakland, CA.

—————. (1992b), "The rudiments of a theory of the origins, survival, and performance of common-property institutions". In: Bromley, D.W., Feeny, D., McKean, M., Peters, P., Gilles, J., Oakerson, R., Runge, C.F. and Thomson, J. (eds.) *Making the commons work: theory, practice, and policy,* 293-318. ICS Press, Oakland, CA.

—————. (1996), "Crossing the great divide: coproduction, synergy, and development". *World Development* 24: 1073- 1087.

—————. (1998), "The international forestry resources and institutions research programme: a methodology for relating human incentives and actions on forest cover and biodiversity". In: Dallmeier, F. and Comiskey, J.A. (eds.) *Forest biodiversity in North, Central and South America,* and the "Caribbean: Research and Monitoring". *Man and the Biosphere Series,* 22. Unesco, Paris, France and The Parthenon Publishing Group, Carnforth, UK.

Ostrom, E. and Wertime, M.B. (1994), "International forestry resources and institutions (IFRI) research strategy". Working Paper, *Workshop in Political Theory and Policy Analysis.* Indiana University, Bloomington, IN.

Ostrom, E., Gardner, R. and Walker, J.M. (1994), *Rules, games, and common-pool resources.* University of Michigan Press, Ann Arbor, MI.

Ostrom, V. (1991), *The meaning of American Federalism: constituting a self-governing society.* ICS Press, Oakland, CA.

—————. (1997), *The meaning of democracy and the vulnerability of democracies: a response to Tocqueville's challenge.* University of Michigan Press, Ann Arbor, MI.

Peluso, N.L. (1996), "Fruit trees and family trees in an anthropogenic forest: ethics, of access, property zones, and environmental change in Indonesia". *Comparative Studies in Society and History* 38: 510-48.

Peluso, N.L. and Poffenberger, M. (1989), "Social forestry in Java: Reorienting management systems". *Human Organisation* 48:333-44.

Peters, P. (1994), *Dividing the commons: politics, policy, and culture in Botswana.* University of Virginia, Charlottesville, VA.

Poffenberger, M. (ed.) (1990), *Keepers of the forest: land management alternatives in Southeast Asia.* Kumerian Press, West Hartford, CT.

Posner, R. (1977), *Economic analysis of law.* Little, Brown & Co, Boston, MA.

Richards, M. (1997), *Common property resource institutions and forest management in Latin America.* Development and Change 28: 95-117

Rocheleau, D. and Ross, L. (1995), "Trees as tools, trees as text: struggles over resources in Zambra-Chacuey, Dominican Republic". *Antipode* 27: 407-428.

Sarin, M. (1993), "From conflict to collaboration: local institutions in joint forest management". *Joint Forest Management Working Paper 14.* Society for the Promotion of Wastelands Development and Ford Foundation, New Delhi, India.

Schlager, E. (1990), *Model specification and policy analysis: the governance of coastal fisheries.* PhD dissertation, Indiana University, Bloomington, IN.

Schlager, E. and Ostrom, E. (1992), *Property-rights regimes and natural resources: a conceptual analysis.* Land Economics 68: 249-62.

Schweik, C.M. (1998), "Social norms and human foraging: an investigation into the spatial distribution of Shorea robusta in Nepal". In: Gibson, C., McKean, M. and Ostrom, E. (eds.) *Forest resources and institutions.* "Forests, Trees and People Programme", *Working Paper 3.* FAO, Rome, Italy.

Schweik, C.M., Adhikari, K. and Pandit, K.N. (1997), "Land-cover change and forest institutions: a comparison of two sub-basins in the southern Siwalik hills of Nepal". *Mountain Research and Development 17*: 99-116.

Scott, A.D. (1993), "Obstacles to fishery self-government". *Marine Resource Economics 8*: 187-99.

Shepherd, G. (1992), "Managing Africa's tropical dry forests: a review of indigenous methods". *Agricultural Occasional Paper 14*. Overseas Development Institute, London, UK.

Shepsle, K.A. (1989), "Studying institutions: some lessons from the rational choice approach". *Journal of Theoretical Politics 1*: 131-49.

Shivakoti, G., Varughese, G., Ostrom, E., Shukla, A. and Thapa, G. (eds.) *(1997), People and participation in sustainable development: understanding the dynamics of natural resource systems.* Proceedings of an international conference on Political Theory and Policy Analysis held at the Institute of Agriculture and Animal Science, Rampur, Chitwan, Nepal, March 17-21 1996. Indiana University, Bloomington, IN.

Simmons, R.T., Smith, F.L., Jr. and Georgia, P. (1996), *The tragedy of the commons revisited: politics versus private property.* The Centre for Private Conservation, Washington, DC.

Tang, S.Y. (1992), *Institutions and collective action: self-governance in irrigation.* ICS Press, Oakland, CA.

Thompson, L.L., Mannix, E.A. and Bazerman, M.H. (1988), "Negotiation in small groups: effects of decision rule, agendas and aspirations". *Journal of Personality and Social Psychology 54*: 86-95.

Varughese, G. (1998), "Coping with changes in population and forest resources: institutional mediation in the middle hills of Nepal". In: Gibson, C., McKean, M. and Ostrom, E. (eds.) *Forest resources and institutions.* "Forests, Trees and People Programme", *Working Paper 3*. FAO, Rome, Italy.

Wade, R. (1994), *Village republics: economic conditions for collective action in South India.* ICS Press, Oakland, CA.

Walker, J.M., Gardner, R., Ostrom, E. and Herr, A. (1997), *"Voting on allocation rules in a commons: theoretical issues and experimental results".* Working Paper, *Workshop in Political Theory and Policy Analysis.* Indiana University, Bloomington, IN.

Western, D. and Wright, R.M. (eds.) *(1995), Natural connections: perspectives in community-based conservation.* Island Press, Washington, DC.

Wilson, P.N. and Thompson, G.D. (1993), "Common property and uncertainty: compensating coalitions by Mexico's pastoral Ejidatarios". *Economic Development and Cultural Change 41*: 299-318.

World Bank (1992), *World Development Report 1992: Development and the environment.* World Bank, Washington, DC.

8

Common Pool Resources and the Indian Legal System[1]

NIRMAL SENGUPTA

Economic Prospect

Over the years the state in India has assumed an overwhelming role both as a regulator of some renewable resources like the forests and fisheries and as an actual provider of some others like irrigation services. Privatisation is now being attempted in the case of latter. However, privatisation here means establishment of group management involving users and replacing bureaucratic management, since high transaction costs prohibit individualisation of social irrigation systems. Transaction costs in monitoring and enforcement is formidable also for regulating uses of resources like forests or fishstock, dispersed over a wide area. Participatory regulations are again more desirable. Efforts towards establishment of participatory Common Pool Resource (CPR) management have been made in several parts of the country since the late fifties. We will discuss here the compatibility of the legal set up with this process. Whether this is indeed the best management option is not within the scope of the present study. Our analytical context is that the Indian national policy has come to recognise extensive peoples participation as the desirable mode of CPR management. The Eleventh Schedule added in the 73rd Constitution Amendment (Panchayati Raj Act) lists explicitly the development of CPRs as one of the tasks of the *Panchayats*.

An analysis of legislations in India pertaining to CPR is highly complicated. At present there are three sets of effective legislations: the pre-73rd Amendment Constitutional provisions which are still operative, the user-group models like Joint Forest Management (JFM) tried in some areas and the emerging *Panchayati Raj* (PR) scenario. We discuss them in three sections.

Private Property and CPR in India

In pursuance of the objective of mercantile capitalism the East India Company designed its legal system primarily to pave the way for private development of property. The system (a) defined and protected the private rights, including the property rights. To secure the rights from encroachment by executives a Judiciary independent of the executive institutions of the state and acting as a check on them was created. (b) To facilitate economic relationship between propertied subjects, the public law favoured markets and contracts and developed a number of conventions like the validity of the sale of property, binding nature of contracts in debt etc. To implement the law, records of rights and right-holders were necessary.

In the Permanent Settlement system, this was done through the recording of the *zamindars* and the revenue payable by each. For almost a hundred years there were no records of tenants and little concern about the economic system within the jurisdiction of the *zamindars*. The existing CPR relations did not come under the purview of public law in the Permanent Settlement areas.

In the *Ryotwari* Settlement system, the British proceeded a step further in establishing private property. Settlements were made with individual tenants. During the preparation of the records of rights, it surfaced that no individual could lay claims on the local tanks, woodlots, or grazing grounds. The colonial government declared its ownership over such properties and thus committed itself to their upkeep and maintenance, disregarding the CPR relations (Sengupta, 1991, 2001). In a few years this necessitated the induction of engineers in the Revenue Department and ultimately gave rise to the Public Works Department with numerous personnel and considerable expense but with dismal performance. This is the background to the rise of the 'government as provider' model in CPRs in India. Lately, the staggering maintenance cost along with poor performance of government agencies has caused a sense of alarm, and privatisation, in the form of handing over resources to water users associations, and other local user groups is on the agenda.

Between these two extremes of the government's role, there exist many other ways in which the government exercises property rights. Before enlisting them let us discuss the costs associated with these enactments so that the readers can form a clear idea of the historicity and extent of each. The 'provider' role presupposes considerable investment in construction, renovation and/or operations and maintenance. Permanent Settlement in contrast was probably the *least* expensive kind of settlement. It could be made without any demarcation of boundaries, without any survey of land, without any attempt to value the land in detail or to record rights. Any further development requires considerable investment. Even in the very first *Ryotwari* settlement, measurement and assessment of the land of nearly 80,000 cultivators had to be done. And still the government did not know what erstwhile CPRs it owned. Only after eighty years could the complete list of the thirty two thousand tanks in *Ryotwari* areas be released (Sengupta, 1991, 2001). Until then, the government was the title holder (*res publicae*), but could not exercise its right in any form (*res nullius*). Then began a protracted process of "standardisation of tanks", a programme that has lasted more than a hundred years, to list the physical details like irrigation capacity, heights of embankments etc. This investment is needed for being a competent provider, to do any kind of allocation, repair and improvement works. Comparable is the situation of forests, fisheries and other CPRs. Delimiting forests is an expensive exercise. Implementation of a reservation order on a patch of forest takes ten to twenty years of work. And pursuance of any protective provision needs considerable information about the physical details as well as monitoring, entry regulations, policing, regular estimation of forest appropriation, and so on.

Thus, whatever be the legal status, without the necessary investment it is not effected in any form. Although we will be describing some general patterns, their actual extent varies significantly. India's federal system divides responsibilities for different governmental functions between the centre and states. Natural resource management is largely a state subject. Many states and localities had acts dating to the colonial period pertaining to different aspects of natural resource management. We are able to discuss here only the generalities.

After independence, the *Ryotwari* settlement model covered the whole country. The government became the owner of the erstwhile CPR resources. To begin with, the colonial government did not indulge in anything involving these properties, including investments for establishing property rights of any form. The situation may be described as "poorly defined property rights". In reality this is not a poor definition, but a negation; by not extending any statutory sanction to existing CPR rights these resources were made open-access in physical (*de facto*) property law. This is a very common occurrence all over the world. It was sometimes done in colonial interest, to open up areas occupied by indigenous communities. In the American continent, the vast CPRs under native Indian control were opened up to the settlers by assigning 'common property' status to these lands in the civil codes (McCay, 1990). This nomenclature is at the root of much confusion and a considerable research effort has been needed to clarify the difference between *de facto* common property and open access.

In many of these government owned areas CPR management continues by default. Until others actually intervene, taking advantage of the open-access situation, the traditional self-governance by users may survive. In this way many traditional CPR organisations have survived to date. Management of smaller tanks by community effort is a noteworthy example (Sengupta, 1991). However the characteristic of poorly defined property rights results in unresolved conflicts and harassment by other claimants including petty officials, thus increasing the transaction costs. In sea-fishing, fishers' customary rights have not received any statutory sanction. However, the fishers are not prevented from continuing to use the sea as in the past. Thus, the traditional physical rights perpetuate, but since they are not statutory property rights the fishers have always been vulnerable to encroachments. Intercommunity conflicts on questions of trespass in customary territories are frequent, reaching the courts only when there is a criminal offence. The fishers had to adopt the course of agitation when trawler fishing was allowed by the government in 1970s. They succeeded in imposing a seasonal ban on trawler fishing (Kurien and Achari, 1990). Lately, the Indian government has given international joint ventures a free access to fish in the exclusive economic zone of the country. Again the fishers have begun agitation, along with the trawler owners now. In other cases of poorly defined properties, the open access nature actually brought in many claimants resulting in the tragedy of the commons. Particularly in the post-independence period the forests and grazing grounds have faced this situation extensively.

In course of time the open access nature of the resources may change. Government may undertake certain regulatory measures or may divert it to other uses. The regulatory model of government control of CPR is best expressed in the case of forests (Guha, 1989; Pathak, 1994). Beginning with the forest regulations during the late colonial period several 'forest conservation' measures have been introduced. These programmes, though allowing commercial uses, restricted the access of local people more and more. Gradually, large areas of forests were declared reserved and protected. About 90 per cent of all forests in India are now publicly owned and badly managed. Extensive degradation of forests has caused concern and by now the policy seems to be moving towards participatory management.

Regulatory measures introduced during colonial rule are less well-known. Some of these show startling inefficiencies in resource use due to inappropriateness of the property rights introduced. The following example illustrates what might have happened in extensive tracts under *Ryotwari* settlements. (Since detailed information for that period is not available we proceed through a reconstruction.) Ramnathapuram district was an exception within the *Ryotwari* tract since it was earlier a *zamindari* settlement. Unlike in the rest of south India, here some of the customary practices continued till the time of Independence. One was extensive tank bed cultivation during the dry season when the tanks were empty. Every year the revenue department used to distribute temporary ownership (*patta*) for cultivation of tank beds. This emerges from official reports in the 1950s, which record official objection against the continuation of this practice. After the abolition of the *zamindari* the P.W.D. had taken over the tanks from the revenue department. They were determined to stop the tank-bed cultivation on the grounds that the department was not getting enough time to undertake desilting work. To put an end to the practice was not an easy task; the soil rich in sub-soil moisture had high fertility and yielded very good cash crops. For ten years there was a tug of war between the unrelenting farmers and the P.W.D. The Planning Commission recommended a high penalty tax (Sengupta, 1991: 111-113), and finally the government succeeded. A footnote is in order. The tanks are now in dilapidated condition and very little desilting work has been done.

An important feature of CPR in India is that the users do not have any statutory right over them. The 'public property concept' on natural resources may be upheld by the court or an administrative order. Thus, courts have sometimes decreed that natural streams cannot be diverted by people, derecognising traditional checkdam constructions by the users. Here is a story from the modernisation of Tambraparni river system in the 1870s (Sengupta, 1991: 138). When British engineers decided to construct an eighth anicut (weir) on the river, the potential beneficiaries had collected a sum of Rs. 30,000 on their own for the construction. But the government did not use it on the grounds that the beneficiaries may in future, use this to ask for a reduction of rent. The fund was used instead for building a road. Or consider a recent story. In 1986 the Rural Development Department introduced a Tree Patta Scheme for encouraging the poor to grow fuel, fodder and fruit trees on government lands. Under this scheme the beneficiaries were given legal status in respect of usufructuary rights in trees; no right whatsoever was granted on the land. In about three years the scheme was withdrawn on the ground that granting *pattas* (titles) amount to privatisation of government land and is inconsistent with the existing Tenancy Act and Forest Act.

It is worth noting that water rights of users in India are defined only with respect of irrigation water from government irrigation projects (the Irrigation and Drainage Acts). There is no system of rights granted to the users on 'natural' waters. The provision of prior appropriation rights in the U.S.A. encourages public investment. Customary communal and private rights were given statutory recognition in Japan while modernising the legal system in the Meiji Restoration period (McKean, 1992). This retained the private initiatives. Not so in India. Even the 73rd Amendment does not define the ownership and control rights of Panchayati Raj Institutions over the subjects included in the Eleventh Schedule, leaving considerable scope for dispute.

In subsequent years several acts were passed regulating the use of natural resources. In the post-independence period objectives of social justice have been of great concern and several acts have been passed which have a bearing on the management of CPRs. Their applications lead to many ambiguities and inconsistent judicial interpretations. Finally, we note that the legal procedures are highly inefficient. Transaction costs are so high that it is beyond the reach of many. Thus the legal system became more of an instrument of harassment than a proper conflict resolution mechanism. In case of CPRs many conflicts have lingered in civil courts for decades, ultimately being resolved primarily through self-organisation by the users outside the ambit of the legal system (Michael, 1979; Brara, 1989; Sengupta, 2000).

Customary Rights and CPR

We will now turn to another aspect of Indian legal system. The colonialists were not as committed to private property here as in their homeland. The colonial authority was also fully aware that it would be politically and socially cumbersome to administer English or western law to supplant an already complex set of native rules. It was therefore, decreed that a vast area of residuary law was to be administered by the British courts according to 'justice, equity, and good conscience'. This then became a vehicle for recognising some version of native law rather than imposing western law in India (Galanter, 1989; Washbrook, 1981). Thus parallel to the enunciation of these policies of 'public' law the colonial government had also attempted to define the bases of a 'private' or a 'personal' law of their subjects. The private law was to prescribe the moral and community obligations to which the individual was subject, and was to be made by the 'discovery' of existing customary and religious norms. Several British administrators had worked hard to discover "customs" and pre-existing norms including indigenous common property relations (Coward, 1990; Sengupta, 2000).

What we are concerned with here is rehabilitation of the customary relations within the modern legal system, not the survival of customary rights in general. Customary property rights of CPR could survive longer in settlements granting territorial rights to *zamindars*, princes and also to tribal chiefs. Within the Permanent Settlement for example, the estates were constituted as small autonomous units. Within these units, local regulations and laws were determined by customs and practices and were administered in a highly personalised form. At least in the beginning the estate authorities adhered to unwritten norms and customs including the traditional CPR management norms. But this had no statutory sanction. Instead, the *zamindars* were given paramount rights on CPRs, which they came to exercise much later, at the height of the peasant movements. For example, they forcibly asserted their rights over local tanks and stopped water supply (Sengupta, 2001). Such actions ultimately brought revision of property rights and abolition of privileged owners like the overlords. Indeed it was the acts like the Tenancy Acts that gave some administrative recognition to the communal rights over CPR resources.

Is customary/communal management of CPR viable and desirable in the current social order and the rule of private property? Communal ownership of property was widespread in many traditional societies. Numerous anthropological investigations have been carried out about the CPR management systems. Many countries have incorporated the customary communal rights into the modern legal systems. It is one thing to incorporate pre-existing rights while introducing an act. It is another thing to introduce 'traditional' or 'customary' rights as the pre-existing rights. Often the 'customary' is a reconstruct and had ceased to exist as a consequence of social dynamics.

A particularly interesting case is that of Portugal. Unlike what happened in the rest of Europe, the nineteenth century legal system of Portugal had extended recognition to traditional common property (called baldios) system. However, privatisation and nationalisation of these commons continued and by 1966 the process was considered by the autocratic government as completed. In 1966 the concept of common property was omitted from the Civil Code. After ten years, in 1976 the new left-wing government began a programme of returning the commons (baldios) to the communities. The concept was reintroduced in the legal system.

In an incisive analysis Brouwer (1992) noted that both the abolishers and restorers of commons had relied on a single anthropological image of the *baldios*. Their emphases were different. One group emphasised the backwardness and stagnation of the traditional *baldios* the other, the mutual solidarity, equity and democratic management. The same set of documents depicting traditional *baldios* had supplied the arguments on both sides. Meanwhile the actual *baldios* had ceased to be the old traditional types. There was already considerable dynamism within the *baldio* members and considerable differentiation and disunity. The law therefore, never achieved what it was aimed at. Instead, it was wielded by powerful groups in and out of the *baldios* to further their specific interests.

Whether the colonial records of customary rights in India were always truthful or were improper reconstructions of customs is debated (Sengupta, 2000; Coward, 1990). But the use of the provision of rehabilitating customs was not confined to granting some rights to the users. Even customary responsibilities of self-organisation were brought under the purview of law. As described earlier, following the *Ryotwari* settlement the colonial government became morally responsible for the maintenance of the local tanks. There were above 30,000 tanks in Madras Presidency alone and regular maintenance of these tanks would need a colossal amount of funds. On the other hand the Board of Directors of the East India Company were tight-fisted about allotting funds for this unforeseen and unwarranted contingency. The customary practice was the contribution of voluntary labour by the farmers. It was known as *kudimaramath* (own maintenance). In 1858 an act called the Madras Compulsory Labour Act or Kudimaramath Act was passed by which revenue officials were empowered to summon farmers for unpaid labour for irrigation works. The implication has changed further in the hundred years since. (Sengupta, 1991: 69-77). These days the P.W.D. engages agricultural labour to conduct, as *kudimaramath*, those works which originally used to be done by voluntary labour. Thus customary rights and regulations were so improperly coded that they sometimes reached such ridiculous outcomes. Indeed, the recourse to customary practices was to legislate in favour of public control rather than with a view to an honest coding of customs.

Thus far, the legal support CPR systems have received worldwide is from recognition of customary rights in the modern legal system. But this approach has a couple of very serious weaknesses. One is that of crude simplification, even vulgarisation (Scott, 1995; Sengupta, 2001). For example, a proof of a customary right in a court is often documented evidence - an official note or even a historical or an anthropological account. Such documents are necessarily a simplification of the actual custom. Also, it is prepared with a fossilized notion of the custom; laws enacted to perpetuate them invariably fail to correspond to the dynamic ground reality. The other is the ambiguity between primary (national) property law and the recognition of customary law by local law or administrative orders. Parties in conflict muster whichever is convenient and resolution of conflicts depend on personal leanings of the mediators. The existence of law at normative discourse level means nothing unless it also corresponds to the real incentives of the agents. In designing appropriate legal system therefore, it is better not to start with past but with the future, not from traditions and customs but from rational choice situations.

The concept of the 'ground reality' may be seen as a mixture of customary and rational choice practices. In this formation, the customs are weakening and rational choice is the emerging trend in modern society. In designing a supporting legal system one must therefore, primarily adapt to the rational choices of the agents (North, 1990; Ostrom, 1990; Ostrom *et al.* 1994). Any custom and existing right may then be accommodated only to the extent that it does not contradict the basic provisions decided as above. Only in this approach can statutory property rights promote the developmental goal. Besides, "good management results not always from just distribution...... Even if it is endowed with good management a customary system of distribution may still warrant a change in social environment" (Sengupta, 1991: 32). The tradition in India refused drinking water rights to "untouchables". On the other hand, in a couple of recent efforts towards participatory irrigation management, ownership of land has been disassociated from water rights and even the landless have been given tradable water rights on common irrigation systems.

Perhaps fortunately, Indian policymakers are not keen on espousing traditional virtues. In the current efforts of CPR formation in India through JFM (Joint Forest Management), water-users associations or even in *Panchayats*, no effort is made to depict them as an emulation of the traditional system. These are described as challenges in designing new sets of organisations and institutions. The major misinterpretation therefore, has been a complete negation of existing customary rights in many cases. The institutional designs however, are keen to break the stranglehold of traditional vested interests but have often paid little attention to the pre-existing CPR relations and remnants of customary rights. Failing to adapt to the ground reality, in most cases they too are not able to present the right incentive structure for the participants.

Panchayat Model

The Gandhian economic vision of village self-sufficiency is a hypothetical institutional model whose property rights basis was never explored. The major consideration behind the rejection of this model even by the Congress Party immediately after independence was that unlike Gandhian notion, the village is not a harmonious entity. However, a concession was made in the Directive Principles of the Constitution, favouring village *panchayats* as units of local self-government. This Directive Principle (Article 40) remained one of the most vigorously pursued Principles in India.

The 73rd Amendment has cast a constitutional imperative on all the state governments to come up with appropriate panchayati raj (PR) legislation detailing meaningful democratic devolution of functions, functionaries and funds. Specifically, it empowers states to endow *panchayats* with such powers and authority as to enable them to function as institution of self-government. Articles 243-G anticipates PR institutions to "function as institutions of Self-Government" preparing and implementing plans and schemes for economic development and social justice on subjects including forestry, fuel and fodder, fisheries, animal husbandry, water management, watershed development, drinking water, soil conservation, maintenance of community assets etc. Towards this end the State Governments are asked to endow the *Panchayats* with necessary powers and authority. The Eleventh Schedule which lists these CPR activities is an innovation of the 1993 Act. Thus the PR system aims to promote self-governance and development of CPRs. It brings not merely some protected and productive resources under CPR management but also covers a far wider variety of CPR resources. There are some gaps though. The PR Act does not envisage CPR situations of a different territorial order. Sea fishing is one example. In addition, the inclusion and exclusion rules need to be determined for traditional community knowledge and biodiversity in order to avail the benefits of IPR regime.

Even though 29 subjects have been listed in the Eleventh Schedule the exact functions to be transferred to *Panchayats* were left to be determined by the individual State Governments. Analysts noted that this only makes for uniformity of PR institutions in form but not in substance and functions. Earlier experience of autonomous bodies show that the State Governments either do not transfer additional functions or take back those whenever they so feel (Singh, 1993). Most of the CPRs in villages are now controlled by one or the other of the government departments. They are unlikely to relinquish their powers easily. Different acts make room for conflicting interpretations. Brara (1989: 2251) in her detailed study shows that under the Rajasthan Panchayat Act 1953, all common lands belonged to the *panchayats* while the State Tenancy Act declared that all land in Rajasthan was the property of the state. She cites three judicial awards between 1981 and 1983 that had diametrically opposite views on the issue of ownership.

The Act does not mention anything about the status of the functional groups like JFM. In fact it creates an impression that all the *panchayat* members or *panchayat* officials on their behalf will take charge of governance (Sengupta, 1996). The functional group model of participation has a clear incentive structure. All the villagers, or all the people residing in a *taluk* may not be the beneficiary and cost-sharer of each piece of CPR resource in the locality. Involving all of them violates the basic conditions of congruence of rights and duties. Nor would it be able to work efficiently if all the governance structures of all CPR resources are merged into a single body. Also, it is unlikely that the PR institution on its own will extend sanctions to such efforts like womens' fisheries. Nor does the Act prevent in any way the PR institutions from claiming a lion's share of CPR benefits, leaving functional and user groups with very little incentive.

On the positive side however, the existence of PR institutions between the government and functional groups may permit considerable autonomy to the latter. This will help in accommodating wide variations in local conditions. The local authority is more well informed about the community members and may help in better selection of the functional group. Their understanding of residual customs and pre-existing situations is likely to be better than government functionaries now involved in the extension work. Transaction costs in conflict resolution may be curtailed drastically.

A task force on Devolution of Powers and Functions upon the Panchayati Raj Institutions, appointed by the Rural Development Department pointed out (GoI, 2001) that "The functions devolved upon the PRIs are in the nature of 'subjects' rather than in terms of 'activities' or 'sub-activities'". If the letter and spirit of the 73rd Amendment is to be realised through the respective state Acts then the states are required to supplement the stated 29 functions in the 11th

Schedule of the Constitution with detailed functional responsibilities, and identification of functionaries and funds for it. However, the State Panchayati Raj Acts have largely retained the style of listing out broad functions, instead of formulating relevant rules and guidelines detailing functional responsibilities of each tier of panchayats for each of the subjects. The laws have not reached the *panchayats* and the villagers in most states simply because the operational law does not exist (Upadhyay, 2002). To illustrate, in the state of Rajasthan (like in all other states) power and functions that have been legally assigned to the *gram sabha* or '*panchayats* at appropriate level' in tribal dominated areas have been indeed impressive. They include prior consultation before land acquisition, ownership of minor forest produce, managing village markets, managing institutions affecting social sectors and controlling local plans and so on. However, all these powers have either been made "subject to such rules as may be prescribed" or "to the extent and in the manner as may be prescribed". Six years after the central laws and three years after the above-mentioned state law was passed, neither rules have been made nor have any prescribed orders/guidelines been issued.

The *panchayat* law for scheduled areas in Himachal Pradesh says that planning and management of minor water bodies shall be entrusted to "*gram panchayats, panchayat samitis* or *zilla parishads* as the case may be in such manner as may be prescribed". Here again unless there are rules made which prescribe specifically which tier of *panchayat* should be made responsible for planning and management of minor water bodies, operational law on the subject would not be in place. Such rules, however, have not been made, five years after the law was passed.

The Functional Group Models

Current efforts like JFM or water-users associations, start basically with identification of participants. An incentive structure is created through some arrangement of benefit-sharing between the government and the participants. Policies and departmental orders have already been issued towards this end. The Forest Act too is being revised. It is worth investigating how sound this institutional model is. The arguments that favour such participatory management of resources are that it reduces transactions costs. The objections are that (i) resource management objectives and choices of the participants may differ from the 'social' objective and (ii) there is no guarantee that the beneficiaries will make necessary investments and/ or organise themselves to undertake the task.

The departmental objectives in the participatory programmes are very limited. The JFM programme is directed towards protection of degraded forest lands; the participatory irrigation management programmes demand merely that the people pay larger share of recurring costs. But for these the objectives have remained very narrow, being confined to the departmental needs. Since the participants have only been asked to comply with the departmental objectives, there is no scope for a conflict of objectives. Except in some pockets, people do not show much enthusiasm for these programmes: perhaps a reflection of the difference in bureaucratic and popular objectives.

In pursuance of the limited objectives, only limited rights have been given to the users. The forestry programme of shared governance is of 'right-based management' (terminology as per Townsend and Pooley in Hanna and Munasinghe ed. 1995). The users have been granted some share of the forest produce. In water users associations too the approach is the same - by refusing farmers rights to receive water in case of default. However, in the case of irrigation, some contractual and co-management efforts have been made. An interesting variation was practised in two sterling cases of success, in *Sukhomajori* and *Pani Panchayat*. Landless labourers were given rights on communal irrigation systems somewhat in the manner of individual transferable quotas (ITQs), which contributed towards increasing equity.

User participation reduces transaction costs in the form of costs of information, designing and coordination, monitoring and evaluation. The user group models have used these in varying degrees. However an important principle for economising in participation is that "costs and benefits of monitoring a set of rules are not independent of the particular set of rules adopted.... When appropriators design at least some of their own rules they can learn from experience to craft enforceable rules..."(Ostrom, 1990: 96). In these officially designed programmes however elaborate rules to be obeyed by the associations were made to facilitate not their functioning but the external administrative task. Implementation of these very guidelines increased the transaction costs of the user organisations and consequently, reduced their marginal net benefits. Also, when regulations are acceptable and are considered legitimate by those whose interests are being regulated, compliance with regulations increase and consequently, management costs decline (Hanna in Hanna and Munasinghe ed. 1995). The externally imposed, elaborate guidelines often do not succeed in getting voluntary compliance. The theoretical literature on local management of commons also indicate that significance of informal incentives like already established trust, deserve greater attention in explaining rational choice (Seabright, 1993). The rigid boundary and membership rules of these government efforts probably fail on this count too.

The congruence between rights and responsibilities is another important requirement of successful organisations. This is where the state JFM orders faulted seriously (*viz.* Arora, 1994). Rigorous membership rules were drafted with fanciful ideas of right holders and equitable distribution. Little respect was shown to pre-existing organisations. Of course, the belief was that there were no such existing organisations and people did not do anything for protection of government forests. In consequence, membership of organisations was more to comply with norms than functions. Many of these organisations remained on paper only.

The basic economic principle was that an incentive structure towards participation would be created through sharing of benefits between the government and the participants. The question is whether the incentives in the current programmes are enough (e.g. Chaturvedi, 1994). Limited objectives, limited rights given to the users, complex guidelines increasing transaction costs and sharing of benefits with a large number of members as per the laid-down norms make this a critical question.

The problem of ineffective legal mandate for the *panchayats* is compounded by the creation of a large number of village level user-groups formed under various development programmes of the state governments, all of which function independently of the Panchayati Raj Institutions (PRIs), on the subjects assigned to them under the Constitution. These user-groups include the village forest committees/ forest protection committees, the water users associations, the watershed development committees, self-help groups, amongst others. If an 'activity mapping' of the omnibus subjects vested with the *panchayats* is done it will inevitably show an overlap with the activities presently being carried out by these user-groups. The emergence and proliferation of these bodies has taken place over the last 10 years and is a phenomenon running parallel with the evolution and growth of PRIs as constitutional bodies following the 73rd Amendment. These 'parallel bodies' sharing functions of PRIs, while working at the same time at the same village level, can be a major reason behind the surfacing of inter-institutional conflicts. This underlines the need for a streamlined legal framework that operates at the village level in the wake of the 73rd Amendment and in the wake of the fact that some of these user-groups are also finding footings in law (Upadhyay, 2002).

Note

1. Slightly modified and updated version of an article by Nirmal Sengupta, "Common Pool Resources, Indian Legal System and Private Initiative", *Conference on Law and Economics*, Indian Statistical Institute, New Delhi, January 11-13, 1996.

References

Arora, Dolly (1994), "From State Regulation to People's Participation: Case of Forest Management in India", *Economic and Political Weekly*, March 19.

Bates, Robert H. (1989), *Beyond the Miracle of the Market - The Political Economy of Agrarian Development in Kenya*, Cambridge University Press, Cambridge.

Brara, Rita (1989), "Commons Policy as Process – The Case of Rajasthan, 1955-1985", *Economic & Political Weekly*, October 7.

Brouwer, Roland (1992), "The Commons in Portugal: A Story of Static Representations and Dynamic Social Processes", in F. von Benda-Beckmann and M. van der Velde ed. *Law as a Resource in Agrarian Struggles*, Agricultural University, Wageningen.

Chopra, Kanchan; Kadekodi, G. and Murthy, M.N. (1990), *Participatory Development: People and Common Property Resources*, Sage, New Delhi.

Chopra, Kanchan and Kadekodi, G. (1999), *Operationalising Sustainable Development*, Sage, New Delhi.

Chaturvedi, A.N. (1994), "Sustainability of short rotation Coppice System in Arabari", *Wasteland News*, May-July.

Coward, E. Walter Jr. (1990), "Property Rights and Network Order: The Case of Irrigation Works in The Western Himalayas", *Human Organisation*, 49(1).

Galanter, Marc (1989), *Law and Society in Modern India*, Oxford University Press, Delhi.

GoI (Govt. of India), (2001), 4 *Report of the Task Force on Devolution of Powers and Functions upon the Panchayati Raj Institutions*, Department of Rural Development (Panchayati Raj Division), Ministry of Rural Development, Government of India.

Guha, Ramchandra (1989), *The Unquiet Woods: Ecological Change and Peasant Resistance in the Himalayas*, Oxford University Press, Delhi.

Hanna, Susan and Mohan Munasinghe ed. (1995), *Property Rights and The Environment*, The Beijer International Institute of Ecological Economics and The World Bank.

Iyer, K. Gopal (1993), "Government and Community Land in Bihar", in B.N.Yugandhar and K. Gopal Iyer ed. *Land Reforms in India (vol. I): Bihar - Institutional Constraints*, Sage Publications, New Delhi

Jodha, N.S. (1992), "Common Property Resources: A Missing Dimension of Development Strategies", *World Bank Discussion Papers*, No. 169

Kurien, J. and Achari, T.R.T. (1990), "Overfishing along Kerala Coast: Causes and Consequences", *Economic and Political Weekly*, September 1-8

McCay, Bonnie J. (1990), "The Culture of the Commoners - Historical Observations on Old and New World Fisheries" in McCay, Bonnie J. and Acheson, James M. ed., *The Question of the Commons - The Culture and Ecology of Communal Resources*, Univ. of Arizona Press, Tucson

McKean, Margaret (1992), "Management of Traditional Common Lands (Iriaichi) in Japan", in D. W. Bromley ed. *Making the Commons Work - Theory, Practice and Policy*, International Centre for Self- Governance, San Francisco

Michael, L.H. (1979), *Water Resource Management in South India 1878-1938: Irrigation and Hydro-electric Power in the Cauvery Basin*, Ph. D. thesis, University of Wisconsin, Madison, unpublished

North, Douglass C. (1990), *Institutions, Institutional Change and Economic Performance*, Cambridge Univ. Press, Cambridge

Ostrom, Elinor (1990), *Governing the Commons: The Evolution of Institutions for Collective Action*, Cambridge Univ. Press, New York

Ostrom, Elinor; Roy Gardner and James Walker ed. (1994), *Rules, Games & Common-Pool Resources*, Ann Arbor, The University of Michigan Press

Pathak, Akhileshwar (1994), Contested Domains - *The State, Peasants and Forests in Contemporary India*, Sage, New Delhi

Scott, James C. (1995), "State Simplifications: Nature, Space, and People", paper presented at the *Agrarian Questions Conference*, Wageningen Agricultural University, 22-24 May, mimeo

Seabright, Paul (1993), "Managing Local Commons: Theoretical Issues in Incentive Design", *Journal of Economic Perspectives*, 7(4), Fall

Sengupta, Nirmal (1991), *Managing Common Property: Irrigation in India and the Philippines*, Sage, New Delhi.

————. (1995), "Common Property Institutions and Markets", *Indian Economic Review*. 30(2), July-December.

————. (1996), "Common Pool Resources, Indian Legal System and Private Initiative", *Conference on Law and Economics*, Indian Statistical Institute, New Delhi, January 11-13, unpublished.

————. (2000), "Negotiation with an Under-informed Bureaucracy", in Bryan Bruns and Ruth Meinzen-Dick ed. *Negotiating Water Rights*, International Food Policy Research Institute, Washington D.C. and Vistaar Publications, New Delhi.

————. (2001), *A New Institutional Theory Of Production – An Application*, Sage Publications, New Delhi, London and Thousand Oaks.

Singh, S.K. (1993), "73rd Constitution Amendment: An Analytical Framework", *The Administrator*, vol 38, Oct Dec.

Upadhyay, Videh (2002), "Panchayats and Paper Laws - Simmering Discontent on 73rd Amendment" *Economic and Political Weekly*, July 20.

Washbrook, David (1981), "Law, State and Agrarian Society in Colonial India", *Modern Asian Studies*, 15(3).

9

Keepers of Forests:
Foresters or Forest Dwellers?

TRUPTI PAREKH
PARTH J. SHAH

On May 3, 2002, the Ministry of Environment and Forests (MoEF) of India issued a circular to all the states and union territories to summarily evict all forest dwellers and users—the "encroachers." In the Forest Act of 1865, the British government had obliterated centuries old customary rights of local communities and established exclusive state control over forest resources. Legally minded as the British were, their justification was that under Indian custom all property held in common belonged to the king, and since *they* were the conquerors, that property now belonged to them.

Has much changed after independence for the communities that live inside or depend on forests? The Indian state, like its British predecessor, views them as encroachers, without any legitimate claim on forest resources. The British wrested control in the name of scientific management and the Indian state continues that tradition in the name of sustainable management, wilderness conservation, or biodiversity preservation. Who should be the custodians of our forests: foresters—agents of the state in its executive, legislative, or judiciary branch—or forest dwellers, the communities that have kept them all these centuries?

The Conflict of Visions

The genesis of the problems of encroachment, deforestation, and degradation lies in the process of expanding state control over forests and alienation of forest dwelling communities from forests, initiated by the British and continued with added vigour by the state in independent India. There is a mindset—shared not only by the forest administration and the preservationist environmentalists, but also by many who have the interests of native communities foremost on their minds—that sees the well-being of forests and that of forest dwellers as two different and mutually exclusive options. This is based on a premise that the forests can be well protected only if local forest using communities are excluded, and that the needs of the forest dependent communities can be met only if society is ready to suffer the loss of forests. One must choose between these two alternatives: local communities are enemies of the forests and the forests have to be 'protected' from them and the best protection can be ensured by tight control of the state.

There is a conflict of two visions. One is a vision of wilderness and the other of wise use. One views humans as outsiders in the natural ecosystem and the other as integral to the ecosystem.

We argue in this paper that the wilderness vision does not mesh with the ground reality. It creates a false dichotomy, with far-reaching consequences for people as well as for forests. The problem of encroachment is inherently linked with the basic issue of forest (mis)management. Encroachment is the result and not the cause of degradation. And degradation has been caused by state-dominated forest management, which has caused alienation of forest dwellers from their social and economic base.

We contend that mass eviction of millions of tribals from their natural habitat is not a solution to the problem of deforestation and degradation. The focus instead should be on devolving rights to forest dwellers, who are the only people who can become good stewards of forest resources.

The guns-and-guards approach will not work, whether it is practised by the machinery of the ministry or the judiciary. In fact it will not work even if it were enforced by the ever-efficient private sector. Both corporatisation and collectivisation are premised on the wilderness vision; they are two sides of the same coin. Forest dwellers are integral to the forest ecology; they are no encroachers.

Demands for forest and species protection are made by the urban educated class but the costs of protection are imposed on forest dwellers since they are compelled to vacate the area that is declared as a national park or sanctuary. Their displacement however is rarely highlighted. Have your ever seen NGOs and celebrities standing up against their evacuation? Green oustees get little sympathy or support. Brown oustees—those displaced by development projects—get all of it. Could we be any more hypocritical?

The Supreme Court and the "Encroachers"

The Supreme Court (SC), in Writ Petition No. 202/1995, which has come to be known as the "forest conservation case," while dealing with the problem of deforestation and its causes, reviewed the issue of forest encroachment, that is, illegal or unauthorised occupation or cultivation of forestlands. The problem of encroachment was highlighted with reference to some ecologically-fragile regions in the Andaman and Nicobar islands, West Bengal, Karnataka, Madhya Pradesh, Chhattisgarh, Tamil Nadu, and Assam. The SC instructed the Chief Secretaries of these states to indicate the steps to be taken by them in this regard. Taking a cue from this, the MoEF immediately sent a circular on May 3, 2002 to all states and union territories to evict all encroachers by September 30, 2002, even though the SC had not ordered eviction of encroachers.

There were disturbing reports of huts being razed to ground with the help of elephants, in states like Assam and Maharashtra. (Prabhu P, 2002). In states like Andhra Pradesh, "at least some of the socially conscientious officials seemed to be veering round to the civil society and tribal groups' unanimous view that the circular was 'inhuman and impractical' and its implementation, a sure recipe for a tribal upheaval, and that the state government has resolved to bring to the notice of the apex court 'the ground realities' and the practical problems in implementing the order" (Venkateshwarlu K, 2002).

Meanwhile, the report of the SC-appointed Central Empowered Committee (CEC) (CEC, 2002), consisting of three forest officers and two environmentalists, thoroughly condemned the encroachments and recommended their immediate evictions. The Committee treats encroachment as a law and order problem. It recommends a strong contingent of police force and presence of a magistrate (in case of firing). It asks for immunity to the police staff under section 197 of Criminal Procedure Code. With these dictatorial powers bestowed on the state governments, the Committee expects immediate compliance. If the states still fail, it further demands that the state government pay Rs. 1,000 per hectare per month as compensation for 'environmental losses' caused by continuing encroachment and imposes a possible fine of Rs. 100 per month on defaulting officials.

After the Ministry received much protest on the circular, it again issued another circular on October 30 emphasising that the recent directive did not overrule the guidelines issued in September 1990 for regularisation of eligible cases of encroachment. This takes us to the circulars issued by the MoEF in September 1990, which were in line with the trend set in the Forest Policy of 1988. The June 1990 circular on Joint Forest Management was the beginning of a slow policy shift in favour of forest dwellers.

Forest Encroachment: What, Why, and How Much?

The term "encroachment" as used by the SC and MoEF prejudges the issue and criminalises all without any distinction. Ashish Kothari of *Kalpavriksh* aptly states: "We agree with the MoEF and the CEC that encroachment on forestlands by powerful vested interests is a serious issue and must be dealt with strictly. But to label tribal/*adivasi* communities that have traditionally and customarily cultivated lands but do not have the title deeds to prove this as 'encroachers,' and to club them in the same category as powerful vested interests who have indeed eaten up our forests, is an unjust step to take, and in the long run detrimental to ecological conservation itself." (Kothari A, 2002).

Despite its avowed policies, the government has consistently failed to identify genuine forest dwellers, demarcate the land to which they have rights, and grant them proper legal title to those lands. If the government had carried out this process earlier, even by its own inadequate guidelines, forests and forest dwellers both would have been in far better shape today.

The forest administrators have created the impression that encroachments are the major cause of deforestation and degradation, and that large chunks of forestland have been used up for regularisation of encroachment. It should be clearly understood that the extent of these encroachments is hardly the cause of the degradation of our forests. The noise around the encroachment issue has silenced discussion about the performance of our state foresters in sustaining the forests. The recorded forest area of the country is 76.52 million hectares (mha), whereas the forest cover is 63.72 mha, out of which 38.79 mha is degraded and 24.93 mha is dense (FSI,

1999). Thus, the degraded forest area in the country is as high as 60 per cent of the total forest cover. As against this, the total encroachment in forest areas in the country is 1.25 mha (MoEF, 2002), which is merely 1.9 per cent of the total forest area. Out of this total encroachment, the area used by the forest dwellers would be even smaller. It is thus clear that the extent of encroachment is minuscule and marginal in the context of degradation of forests, and is essentially played up to divert attention from real problems.

Evolution of State Forestry: Alienation of Forest Dwellers

British Era of Sovereign Claims

In the late eighteenth century, with the supply of oak falling short in England, the British turned their eyes on the teak of India for shipbuilding for the Royal Navy. The demand for teak from the railways was rising too. At the time of the advent of the British, ownership of forests was with the then princes or local chiefs. The local communities, however, had largely unhindered access and use of forest resources to meet their grazing, firewood, and timber requirements. Many of these communities had evolved informal norms and customs for protection and proper use of forests. In the early years, the British also considered forests and other wastelands to be the property of village communities under whose boundaries they lay, and did not interfere much with the customary usage.

In the beginning, the commercial potential of the forest wealth was largely unrecognised. The British, like the earlier rulers, were interested in the revenue that the land earned. The forests were considered more "as a necessity for the people, [but] as a revenue-earning resource, they were considered insignificant...This being the case...forests were considered as an obstruction to agriculture rather than otherwise, and consequently a bar to the prosperity of the Empire. It was the watchword of the time to bring...forest areas under cultivation, and the policy tended in that direction. The direct and indirect value of forest was underestimated, as is clearly exhibited by the provisions of many of the earlier settlements, especially in Bengal and Punjab, which transferred large forest areas forever to landholders or to the cultivators of the country, who at that period had neither a right to them nor in many instances even appreciated the boon granted to them, as they valued the areas as little as the Government which gave them" (Ribbentrop B, 1900).

With the growing demand for timber, in the first half of the nineteenth century, steady and uninterrupted supply of timber became necessary. A long and heated debate ensued amongst British bureaucrats as to how procure the timber—whether to buy at the market rate, or to enter into lease contracts with local princes for the forest lands with exclusive rights to grow and cut timber, or to outright take over the forests and manage them scientifically so as to stop what they considered wasteful cutting of teak by local contractors and to be able to get uninterrupted timber supply.

For the first time in 1807, a proclamation announced that "the royalty rights in teak claimed by former governments were vested in the Company, and all unauthorised felling of teak by private individuals was prohibited." The proclamation "contained no definition of the term 'sovereignty', nor had those forests been specified over which the sovereignty extended" (Stebbing, 1923, p. 70).

This was the beginning of what would become the Forest Service. In the name of scientific management, powers were given to the Conservator to sanction teak felling and selling. "The private timber trade was annihilated: for even if they bought timber with the Conservator's permission timber merchants could not market it, save by a Government agency." This led E P Stebbing to comment, "...the new regime was far too drastic to be continued as a method of permanent administration. The privilege of cutting fuel for private use, which had been practised at will by all from time immemorial, was also invaded and prohibited, a short-sighted step of amazing folly" (Stebbing, 1923, p. 71). It is surprising that such a step of amazing folly, after temporary suspension for nearly four decades, was resumed by the British in 1860s, claiming "sovereignty" not only over teak, but also over all forested lands.

The British government, keen to promote 'scientific conservancy measures', enquired in 1846 whether the Conservator's power of sanction was to be made applicable only to government forests or to other forests too. The general consensus was to cover "all such forests as could not be clearly established to be private property" (Stebbing, 1923, pp. 118-123). This was the first step in the era of 'efficient management' of forests by the state in India. The logical consequence of the Conservatorship was to gradually take over rights to every piece of land it could put its foot on.

Debate Over Proprietary Rights:
Munro, Brandis, and Baden-Powell

There were debates within the British bureaucracy about the take-over of forests without considering the then existing proprietary rights. The Madras government, in fact, totally rejected state intervention in forests in the belief that tribals and peasants should exercise complete control over forest areas and state should at best play a subsidiary role. When the Forest Act of 1878 was under consideration, the Madras government declared that it could not be extended there, on the grounds that the reserve forests that the act called for could not be established there. "The rights of the villagers over the waste lands and jungles were considered to be of such a nature as to prevent the government from forming independent state property" (Ribbentrop, 1900, p. 100).

Sir Thomas Munro, the governor of Madras, who had abolished the Conservatorship in 1823, had in fact, said in his minutes that the merchants and agriculturists were "too good traders not to cultivate teak or whatever wood is likely to yield a profit. They are so fond of planting...To encourage them no regulation is wanted, but a free market. Restore the liberty of trade in private wood: let the public be guarded by its ancient protector, not a stranger, but the Collector and Magistrate of the country, and we shall get all the wood the country can yield more certainly than by any restrictive measures. Private timber will be increased by good prices, and trade and agriculture will be free from vexation" (Ribbentrop, 1900, PP 84-85).

German forester, Dietrich Brandis, who came to be known as the founder of the forestry service in India, supported the idea of creating government forests, but strongly urged to restrict them to areas of compact valuable blocks in the interiors that could be obtained without impinging on forest rights of communities. Brandis, in fact, advocated leaving aside rest of the areas under the control of village communities as village forests (Guha R, 1998).

All the voices of dissent and reason were, however, defeated by hard-liners like Baden-Powell. Citing the precedent of Indian rulers having claimed rights of absolute ownership, he argued for the absolute control and ownership right of the state on all common land, whether inhabited or not. In order to rationalise his argument, he invoked the rights of the conqueror, who obtained automatically all the rights as sovereign from the oriental sovereigns, i.e., native chiefs. He conveniently set aside the accepted law in England that no property could be taken over from the citizens by the state.

The Imperial Forest Department was created in 1864 to consolidate state control on public forests and to put forestry operations on a scientific footing. Deitrich Brandis, a German 'expert', was appointed the first Inspector General of Forests. The fist attempt to create legal mechanisms to assert and safeguard state control over forests was made through the Indian Forest Act of 1865. This was replaced by a far more comprehensive piece of legislation in 1878.

This act obliterated the centuries old customary use of forest resources by rural communities all over India. It provided for formation of three classes of forests: "Reserved forests," "Protected forests," and "Village forests." Reserved forests consisted of compact valuable areas to be brought under full state control. All private rights were extinguished, transferred elsewhere, or in exceptional cases allowed, for limited exercise. In Protected forests, rights were recorded but not settled and state control was to be firmly maintained by detailed provisions for reservation of valuable trees and by demarcation of areas for grazing and firewood collection. Most of the protected forests were gradually converted to the category of reserve forests to bring them under state control. The third category of Village forests, which were to be earmarked to meet the needs of local communities, remained on paper only, as this option was never exercised in practice. The act also greatly enlarged the punitive sanctions available to forest administration, closely regulating the extraction and transit of forest produce and prescribing a detailed set of penalties for transgressions of the act. The same act, with minor modifications in 1927, is still operational in independent India.

Transporting an Alien Model of Forestry:
Environmental Imperialism

Thus, by bringing the forestry management model alien to communities, in the words of Dietrich Brandis, "an exotic plant, or a foreign artificially fostered institution," (Guha R, 1998, p. 95) the

British government, by a stroke of an executive pen, expropriated the customary rights of local communities and established exclusive state control over forest resources. This inevitably led to total alienation of the local communities from forest management and generated a strong feeling of resentment against the forest department. Thus the very people, who were and could have been the best friends of forests, were turned into their worst enemies. Not only the people were not seen as the original proprietors of forests, they were perceived as being "erratic, unsystematic, and unmindful" of long-term sustainability by forests.

That independent India continued with this alien model speaks volumes of apathy and lack of concern not only for the forest dwelling communities, but also for the forests, and lack of respect for property rights. There are, thus, question marks on the appropriateness of such a model in the land where people have been living deep into the forests and having occupancy rights for generations. Artificially excluding the whole community from their very habitat is not the best way of protecting the forests. Hence, in countries like India, the model of reserving and thus excluding the communities was and still is unworkable.

It is also true for the "Protected Areas," that is, Sanctuaries and National Parks. The idea of unilaterally declaring an area as sanctuary/ national park and, with a fiat, excluding the people from either the enjoyment or evicting them from the area, respectively, with a view to preserving the area in its pristine form, is equally faulty and unworkable in countries like India. This approach ignores the ground realities of our country: that people are an integral part of the forests and without involving them in their management, wildlife or bio-diversity cannot be protected.

The Historical Process of Land-titling
Denied to Forest Dwellers

The history of acquisition of tenure and property rights, not only in India but also throughout the world, has been the history of taking possession (*occupatio* in Roman law), followed by peaceable enjoyment (*usucapio*), and being perfected with the passage of time (Baden-Powell, 1898). A prolonged tenure gives a prescriptive right of ownership. Before and in the early part of the British regime, forests were considered as a non-revenue generating resource. The official policy was to encourage expansion of agriculture in forest areas. Many of princely states made attractive offers of free land to encourage farmers from other areas to come and settle in their territory and start agriculture. Acquisition of a title to the land of long occupation has, thus, remained a norm.

Some people did not clear the area for occupation and use, and remained in the forests. These forest dwellers were later refused the same process of land titling when their forestlands were declared as reserve forests. The people who kept the forests intact are now penalised for not clear-cutting them for agricultural land!

The Post-Independence Era of Clear Cutting

One of the most important causes of the loss of forest cover has been the type of silviculture practices adopted in the post-independence era. After the passage of the Forest Act of 1878, the colonial government started bringing more and more areas under reserve forests and initiated a system for systematic harvesting of forests based on working plans. The plans relied on selective cutting of mature teak tress on rotational basis with natural regeneration from coppices as well as seeds. This ensured that no part of the forest was devoid of vegetative cover.

After independence, however, a new strategy of intensive commercial forestry was adopted from the sixties onwards. Large areas, usually the most productive areas, were brought under plantation working circles where all trees were clear felled to replace them with artificial plantations of fast growing and high yielding teak species. Thus the system of selective felling was replaced by clear cutting of all trees in selected coupes that were to be replaced by teak plantations. As a result thousands of hectares of natural forests were clear felled during the sixties, seventies, and the early eighties. As could have been expected, the teak plantations never came up, except in a few isolated cases. Most of the forest areas have still not recovered from the effects of years of clear cutting.

Recent Policy Changes to
Improve Forest Management

In the late 1980s, for the first time in the history of forest management, there was an acceptance of local communities' claims on the forests. This was a revolutionary break from the past. Even independent India's Forest Policy of 1952 had not recognised local peoples' claims. In fact, it stated categorically that "neighbouring areas are entitled to a prior claim over a forest and its produce". It continued, "the accident of a village being situated close to a forest does not prejudice the right of the country as a whole to receive the benefits of a national asset."

The first policy, advocating local communities' claims on forests, though harsh on the encroachers, is the National Forest Policy of 1998. It states: "having regard to the symbiotic relationship between the tribal people and forests, a primary task of all agencies... should be to associate the tribal people closely in the protection, regeneration, and development of forests as well as to provide gainful employment to people living in and around forests" (MoEF, 1998). It emphasised safeguarding the customary rights and interests of these people.

The MoEF carried forward this concept of involving local communities in the regeneration of forests and initiated a policy of Joint Forest Management (JFM) in June 1990. It states: "the National Forest Policy of 1988 envisages people's involvement in the development and protection of forests.... It [is] one of the essentials of forest management that the forest communities should be motivated to identify themselves with the development and protection of forests from which they derive benefits" (MoEF, 2000). The benefits to the individual members of the forest protection committees under the JFM policy are usufruct rights on grass, lops and tops of branches, minor forest produce and also a stipulated share (which varies in different states from 25 per cent to 100 per cent) in the sale of timber.

After two months, on September 18, 1990, the MoEF brought out six circulars, regarding settlements of disputed claims, *pattas*, leases, grants involving forestlands, guidelines regarding regularisation of encroachments, conversion of forest villages into revenue villages, settlement of other old habitations, payment of compensation for loss of life and property due to predation/depredation by wild animals and payment of fair wages on forestry works. These circulars taken together give a good package for the resolution of old disputes over claims on forestlands and other problems and thus have the potential to reduce the deep distrust of people for the forest department. One of the circulars regarding disputed claims over forestlands states: "It is being felt that even bona fide claims are persistently overlooked causing widespread discontentment among the aggrieved persons. Such instances ultimately erode the credibility of the Forest Administration and sanctity of the forest laws, especially in the tracts inhabited by tribals" (MoEF, 2000b).

No action, however, has been taken on the circulars. In March 1984, the Ministry of Agriculture had suggested that the state and union territory governments may confer heritable and inalienable rights on forest villagers if they were in occupation of the land for more than 20 years. The 1990 MoEF circular concludes: "But this suggestion does not seem to have been fully implemented." These circulars remain on paper only.

Again, these circulars have not altered the age-old mindset of the forest department. For example, the directive on the encroachment does suggest that the respective state governments may provide alternate economic base to such persons by associating them collectively in JFM programmes. Its main thrust, nonetheless, remains immediate evictions of the encroachers. It does not analyse the genesis and causes of this ongoing problem, nor does it take into account the ground reality that forest dwellers are the best steward of those resources.

Very little has been accomplished with regard to the policy on JFM too. One, there are many deficiencies in the policy, like unequal partnership in matters of rights, power and authority between the participating communities and the forest administration, lack of legal and statutory backing to the policy, inadequate benefit-sharing, and so on. Two, the lack of enthusiasm of forest officials towards implementation is glaring and is reflected in the actual forestland covered under JFM. The area under JFM in 22 states is 10.25 million ha, 16 per cent of the total forest area in India (FSI, 1999). More than 60 per cent (7.43 mha) is found in only three states of Madhya Pradesh, Chhatisgarh, and Andhra Pradesh. The performance of rest of the states has been extremely poor. Even the slight shift in the policy away from centralised management to somewhat decentralised management of forests has remained on paper. Necessary statutory and procedural changes have not been followed and the colonial mindset has not changed. These policy changes could have paved the way towards people-friendly solutions.

In the late 1990s, the MoEF and the Planning Commission constituted several committees to examine various forestry issues like afforestation policies and rehabilitation of wastelands, steps to confer ownership rights of minor forest produce to *Panchayats*, increase people's participation in forest management. "While the large number of committees constituted by the government in recent years indicates its keenness for policy change, the actual process of change has been somewhat slow," comment Saigal, Arora, and Rizvi in their book, *The New Foresters*. They conclude, "[t]here has been limited progress on the implementation of the recommendations of different committees" (Saigal *et al.*, 2002, pp. 112-113).

Community Rights over Community Commons

The proponents of exclusive state control on forests base their case on the claim of the "tragedy of the commons." Biologist Garret Hardin first articulated the idea that commonly held open access resources like forests and grazing lands inevitably suffer over-exploitation as no individual has an incentive to stop his use of the resource as long as others are also able to use it. Each individual strives for quick and maximum exploitation of the resource since all the benefits are accrued to him, but the costs are borne by the whole community. It further states that village communities without cohesion do not have the necessary knowledge and expertise to manage the forests in a scientific manner on a long-term basis.

But these proponents fail to recognise that common ownership does not mean that it is a "free for all" resource and would inevitably suffer the tragedy of the commons. "Communally held open access resources" was largely a theoretical construct of Hardin as in practice such resources are never free for all but are controlled by host of intricate rules and regulations for their use. This was indeed the case with communal management of forests before the advent of British control. Local communities were actually managing forest resources in sensible and sustainable ways through informal rules and practices, as evidenced by the existence of widespread network of sacred groves. Bringing these resources under the state control actually created the tragedy of open access rather than solving it, as local communities lost all incentives and interest in the proper management of forests. The forests no longer belonged to them and they started acting irresponsibly.

It is beyond doubt that best way to put forest management on a sound footing is to re-establish the rights of local communities on forest areas of the country. This approach is now followed in several places around the world.

Illustrative Case Studies of Community Rights

CAMPFIRE, Zimbabwe

CAMPFIRE (The Communal Areas Management Programme for Indigenous Resources) involves rural communities in conservation and development by returning to them the stewardship of their natural resources thus harmonising the needs of rural people with those of the ecosystem. It emerged with the recognition that as long as wildlife remained the property of the state, no one would invest in it as a resource. Since its official inception in 1989, CAMPFIRE has engaged more than a quarter of a million people in the practice of managing wildlife and reaping the benefits of using wild lands.

Since 1975, Zimbabwe has allowed private property holders to claim ownership of wildlife on their land and to benefit from its use. Under CAMPFIRE, people living on Zimbabwe's impoverished communal lands, which represent 42 per cent of the country, claim the same right of proprietorship. Many of the communal lands have too little or unreliable rainfall for agriculture, but provide excellent wildlife habitat. Conceptually, CAMPFIRE includes all natural resources, but its focus has been wildlife management in communal areas, particularly those adjacent to national parks, where people and animals compete for scarce resources.

CAMPFIRE begins when a rural community, through its elected representative body, the Rural District Council, asks the government's wildlife department to grant them the legal authority to manage its wildlife resources, and demonstrates its capacity to do so. By granting people control over their resources, CAMPFIRE makes wildlife valuable to local communities because it is an economically and ecologically sound land use. The projects these communities devise to take advantage of this newfound value vary from district to district.

Most communities sell photographic or hunting concessions to tour operators, under rules and hunting quotas established in consultation with the wildlife department. Others choose to hunt or crop animal populations themselves, and many are looking at other resources, such as forest products. The revenues from these efforts generally go directly to households, which decide how to use the money, often opting for communal efforts such as grinding mills or other development projects. The councils, however, have the right to levy these revenues.

Zimbabwe has set aside, in perpetuity, more than 12 per cent of its land as protected wildlife areas. Most of these are surrounded by communal lands. CAMPFIRE helps prevent the protected areas from becoming islands in a sea of development by making wildlife valuable for nearby communities. CAMPFIRE uses economic incentives to encourage the most appropriate management system for these fragile areas.

Who Runs CAMPFIRE

No single organisation runs CAMPFIRE. The members of the Collaborative Group are responsible for co-ordinating various inputs, including policy, training, institution building, scientific and sociological research, monitoring and international advocacy.

The original members of the Group include the *CAMPFIRE Association* representing rural district councils and therefore the interests of the rural communities involved in CAMPFIRE. The Association is the lead agency and co-ordinator of the programme. It chairs the CAMPFIRE Collaborative Group. The other members are the Department of National Parks and Wildlife Management, Ministry of Local Government, Rural and Urban, Zimbabwe Trust, Africa Resources Trust, and World Wide Fund for Nature (WWF).

Critique

In some instances the "decentralisation" of CAMPFIRE has become the "recentralisation" of a district-level elite resulting in ignorance of or hostility to the CAMPFIRE Programme, mistrust of the councils concerned, and increasing intolerance of wildlife.

The integrity of CAMPFIRE's conceptualisation rests on the self-definition and voluntary participation of local people in the resource management. This aspect is compromised by the designation of pre-existing, administratively designed wards as the communal production units. These wards are often internally differentiated, socially and ecologically and lack the cohesion to motivate consensual entry into the Programme.

The high and escalating value of the wildlife resource have had the effect of intensifying political conflict over the appropriation of these values at community, district, and national levels. Within communities and districts, the programme has brought into sharper focus competing interests drawn on class, status and ethnic lines. At the national level the economic performance of the industry has attracted the attention of the political elite and their private sector allies, who seek to appropriate a higher share of its value through patronage, shrewd negotiation or bureaucratic re-centralisation.

Conclusion

All in all, Zimbabwe's CAMPFIRE programme has had the distinction of being the first major project to recognise the importance of providing both benefits and a meaningful role to the people who lived with wildlife and its habitat. The programme decentralises political and administrative powers to people at the grassroots level, distributes millions of dollars to the barefoot masses in communal areas, and has resulted in the adoption of eco-friendly views on wildlife and other natural resources by the people of Zimbabwe. It has also been of significance in reviving the cultural well being of the people in Zimbabwe. The programme has been widely accepted by people because it does not contradict the traditional wisdom about the environment.

Nature Conservancy

The species that have direct human use like elephants, tigers, and crocodiles can be protected by giving local communities an economic stake, as demonstrated by CAMPFIRE, Zimbabwe. How can one protect species that have no human use value? The answer is Nature Conservancy. It is a private environmental organisation set up in 1951 with "a mission to preserve plants, animals, and natural communities that represent the diversity of life on earth by protecting the lands and waters they need to survive."

With the help of members' contributions, Nature Conservancy purchases areas that have a high biodiversity value. It has developed a strategic, science-based planning process, called **"Conservation Design,"** which they use to identify the highest-priority places—

landscapes and seascapes that, if conserved, promise to ensure biodiversity over the long term. Since a single organisation can neither buy all those high priority places, nor protect them single-handedly, it therefore joins together with communities, businesses, governments, partner organisations, and people to arrive at solutions that preserve lands and waters for prosperity. Ecologically sound management techniques developed by Nature Conservancy do not exclude people living in the area nor reject all economic development as antithetical to the goal of biodiversity preservation.

With over 1 million members, it manages the largest system of private nature sanctuaries in the world and has over 20,000 species under its watch. Over 90 million acres of land in the United States, Canada, the Caribbean, South America, and Asia are protected by Nature Conservancy. A US$1 billion campaign has recently been launched to save 200 of the world's 'Last Great Places'.

The approach taken by Nature Conservancy is rather different from that of run-of-the-mill 'green' organisations. The latter usually lobby the government and get the sensitive area declared as a national park or sanctuary or bioreserve. Communities living in that area are then ousted. Species protection is demanded by urban educated class but the burden of protection is born by some of the poorest communities of the country. It is striking that the greens do not see this basic injustice in their approach. Nature Conservancy on the other hand puts its money where its mouth is. The members themselves pay for the species they want to protect. The people whose land is acquired are given full compensation; they sell their land voluntarily to the Conservancy.

Community Forestry in Nepal

During the Rana regime (1850-1950), there were attempts to formalise exploitation of forests through legal process. Ownership rights over big chunks of forests were awarded to private individuals. In the last decades of the Rana rule, the British sought good timber of *sal* from the *Terai* for the railways and military purposes. The Hill forests were inaccessible to outside forces and the local community had rights over the forests, even on the forests officially given over to the private individuals. In some cases, people used to keep *Ban heralu* (caretaker of forest), paying them in the form of paddy.

In 1957, all forests, including private, were nationalised through the Private Forest Nationalisation Act. This led to large-scale felling of timber (by landlords) to prevent the land being classified as forestland and therefore becoming government-owned. Another major fall-out was that people lost their rights and control over the forests. The first Forest Ministry was established in 1959 and the Forest Act, 1961 was formulated on the lines of the Indian Forest Act, 1927. The act initiated handing over management of government forests to the newly formed *panchayats*.

In the late 1970s, forest officers got police powers on the Dehra Doon forestry model. Local people's bona fide use of government forests to meet their basic needs was deemed illegal. Afforestation programmes with the international aid during 1960s and 70s did not succeed, partly because the local communities were not involved in the management. Massive deforestation continued throughout 1950-1980. The crisis led the World Bank to predict in 1978 that by 1993 the hills, and by 2000 the *Terai*, would be totally denuded.

Policymakers began to realise that the objective of arresting the rapid degradation was unachievable without active and substantial involvement of the local people.

The sixth five-year Plan (1981-85) and the Decentralisation Act, 1982 were the first steps towards decentralisation of powers of the Forest Department (FD). In 1989, His Majesty's Government of Nepal (HMGN), after deliberations for four years, jointly with Asian Development Bank (ADB) and Finnish International Development Agency (FINNIDA), brought out the Master Plan for forestry sector. Community forestry thus was a culmination of various experiments initiated in Nepal, as well as of crucial experiences of other countries.

The Master Plan recognised users-groups in place of *panchayat*. It allowed both natural forests as well as degraded forests to be handed over as Community Forests (CF). It emphasised that "Private management and control (if not ownership of forest land) could be the most effective strategy, in the long run, for obtaining maximum production." It was categorical on the required change in the law and mind-set of the forest administration to be "directed away from policing and towards supporting the efforts of the people" and "to allow people to have full control over the forests that they develop and to utilise forest products without too many administrative difficulties" (Master Plan, 1989).

The Forest Act, 1993 gives detailed provisions for community forests to be managed by user-groups. As per the Act, any part of national forests (looking into the distance between the forest and the village, and the wishes as well as the management capacity of local users) are to be handed over with perpetual succession rights to forest user-groups (FUGs). FUGs are an autonomous and corporate body, with legal and statutory status and have perpetual succession rights to develop, conserve, use and manage the forests and sell and distribute the forest products independently by fixing their prices according to a Work Plan. The FUGs were entitled to 100 per cent of the revenue (Forest Act, 1993). Many FUGs have set rules and penalties for those who do not comply with the rules or who are caught felling trees or grazing.

The FD's role is to provide technical guidance and other co-operation, if required by the concerned FUG. Nevertheless, the District Forest Officer (DFO) has power to cancel the registration of the group and take back the community forest, if he finds non-compliance or irregularity on the group's part after giving it reasonable time to submit clarification. The FUG has a right to file a complaint to the Regional Forest Director, whose decision shall be final. If the DFO's decision is disapproved, the CF has to be re-handed over. If his decision is approved, the DFO has to reconstitute the users-group and hand over the community forest to the reconstituted group.

Out of total forest cover of 5.83 million hectares, about 900,000 hectares of forests (21 per cent) has been handed over as CF to about 11,400 user-groups of about 1.3 million households (Department of Forests, 2002).

It has been found that vegetative cover has dramatically improved in the CFs even on degraded forestland. One glance at the forest and one knows whether it is a community forest or national forest. "One can see the difference between the community forests and other forests," admitted K B Shrestha, Director-General of Community and Private Forests, Ministry of Forests and Soil Conservation (Mahapatra, Richard, 2000).

According to a study by Nepal Australia Community Research Management Project, in 1988-99, five FUGs earned $34,445 (Rs 1.6 million) and generated employment worth $6,571. Kakitar Village of Lalitpur, Nepal has already spent $2044 for irrigation purposes and getting potable water (Paudel, Keshab, 2000).

International private and government donor agencies are allowed to establish their offices in the interior areas at district level, even though they cannot directly work with the FUGs. Nevertheless, they remain constantly in direct contact with the community and the ground level reality and in close range of feed back mechanism.

The success of these programmes shatters a widespread myth that poor people have a short time horizon and cannot undertake projects with long gestation periods. People of Kande in Pokhara, Nepal articulated this very well: "Yes the cattle population of our village is now almost half. We have voluntarily disposed off our cattle because now we are confident that whatever more will be produced in the forest would belong to us. So now we are more responsible towards growth of forest. We have seen with our own eyes the wonderful results of natural regeneration" (Mehta, Trupti P, 2002).

Community Rights in India:
The Case of the North-East

Widespread degradation of forests has persisted in the northeastern parts of India. This is *despite* the fact that unlike much of the Indian sub-continent, where forest departments have functioned as state landlords for over a century, in the northeast communities still retain control over much of the regions natural forest ecosystems: either through the District Council (as in the case of Meghalaya, Mizoram, Tripura and the Karbi – Anglong district of Assam) or within the control of the clan, village or tribe (as in the case of Nagaland or Arunachal Pradesh.)

In the northeast the much acknowledged panacea of communal control over the forest resources (as opposed to state control) *appears* to have failed to safeguard the forests.

A possible reason for this may be that while on the face of it, the proportion of forest under the direct control of the state may be miniscule, in reality the people do not have absolute rights over the forests. Even unclassed forests in the hands of private individuals or the community are subjected to the regulatory powers of the state in relation to the use and disposal of forest produce, though the actual pattern of regulation varies from state to state and is mediated by institutions of self governance.

In fact, both in the colonial and post-colonial period the notion of rights over forest resources has been a heavily contested issue between the local communities and the state.

The Constitution and the Northeast

As a result of the historical, social, economic and cultural factors that distinguish the life and outlook of the tribes of the northeast from the rest of the country, this region of India has been provided with a special political and administrative structure.

According to the Sixth Schedule of the Constitution, 'Autonomous District' areas are Constitutionally recognised as areas that need special protection and an administration responsive to the needs and levels of development of tribal people.

The Autonomous District Councils, which are democratically elected institutions, are thus meant to implement these basic policy guidelines.

The administration of the district council is three tiered with:

1. Traditional village administration at the grassroot
2. The 'Elka' administration at the middle level
3. Constitutional District Council at the apex

All three are democratically elected institutions. The Sixth Schedule states that a District Council is to consist of not more than thirty members out of which not more than four shall be nominated by the Governor and the rest to be elected on the basis of adult suffrage. The term of the elected members of the District-Council is five years, while the term of the nominated members is at the pleasure of the Governor.

Legislative Powers of The District Council

Sec 3(1) of the Sixth Schedule deals with the powers of the District-Council to make laws with respect to (among other things):

(a) The *allotment, occupation or use, or setting apart,* of land, other than any kind which is a reserved forest for the purpose of agriculture or grazing as for residential or other non-agricultural purposes or for *any other purposes* likely to promote the interest of the inhabitants of any village or town: provided that *nothing in such laws shall prevent the compulsory acquisition of any land, whether occupied or unoccupied, for public purpose in accordance with the law*

(b) The *management* of any forest not being a reserved forest

(c) The use of canal or watercourse for the purpose of agriculture

(d) The *regulation* of the practice of *jhum* or other forms of shifting cultivation

(e) The establishment of village or town committees or councils and their powers

(f) Any other matter relating to village or town administration including village or town police and public health and sanitation

(Emphasis added.)

However, The laws made by the District Council shall have no effect unless assented to by the Governor [Sec 3(3)]. Also, the President of India may direct that any Act of Parliament shall not apply to an autonomous district. (The above discussion is based on Dutta, Ritwick, 2002. "Community Managed Forests: Law, Problems, and Alternatives.")

So, why, despite being a region where the *community* rather than the State controls forests, are forests so badly depleted in the northeast?

While this arrangement was designed to strike balance between tribal participation and the controlling power of the State, even a cursory look at the wide ranging powers vested in the elected district councils (especially clause (a) above) is enough to undo the myth of tribal autonomy with respect to use and management of forests.

Economic theory asserts that the best way to ensure sustainable use of a resource is to make sure that it is the users of the resource who own, control and manage it. The only way to do this is to assign well-defined, *secure* and enforceable property rights over the resource to specific users.

While the traditional system of community ownership prevalent in the northeast (with various categories of forests, viz. sacred groves; privately owned, clan, community, or village forests) may *seem* to do exactly that, a closer look at the functioning and powers of the district councils reveals that ever since colonial times, the ownership rights of locals over their resources have become more and more uncertain and insecure.

The forest policies that govern the region have been such that increasingly, State control and regulation has been undermining the traditional ownership pattern, via the District Council (which after all, is an elected, representative, *political* body, separate from the actual forest users.)

It is then, not at all surprising that forest cover has dwindled rapidly in the northeast – if the owner(s) of a resource are not secure in their ownership rights, anticipating increasing State control and decreasing autonomy over their own property, then their 'planning horizon' shortens: the approach becomes one of 'making hay while the sun shines'. Perhaps this is what happened in the northeast – the 'hay' in this case was the quick revenue that could be earned by leasing out land to clear-felling timber contractors, as huge demand for timber and wood products arose from the rest of the country.

Lack of a Separate Forest Policy for the Northeast

As noted above, in the northeast, special constitutional provisions have been created to protect the rights and interests of local tribes. Notwithstanding the special administrative arrangements provided for the region and the wide-ranging powers given to the District Council under the Sixth Schedule, the Northeast lacks a *separate* forest policy.

All seven states in the region operate under the guidelines of the national forest policy that applies to the country as a whole. Even the laws passed by various state governments to regulate the forest produce are mere adaptations and extensions of the national laws, specifically the *Indian Forest Act, 1927* and the *Indian Conservation Act, 1980*.

Similarly, forest laws passed by the District Councils remained largely within the framework of these general rules.

When the state of Meghalaya was created in the early 1970s the legislative power enjoyed by the Autonomous District Council was severely curtailed by the insertion of the 'repugnancy clause' under paragraph 12-A in the Sixth Schedule. The paragraph reads:

> 'If *any* provision of *any* regulation made by a District Council or a Regional Council in that state... is repugnant to any provision of a law made by the *legislature* of the state of Meghalaya with respect to that matter, then the law or regulation made by the District council... shall, to the extent of repugnancy, be void and the law made by the Legislature of Meghalaya shall prevail.'

This principle also applies to the laws passed by the District Councils of the states of Tripura and Mizoram. Thus the state governments and even the District Councils operate pretty much as extensions of the Central Government. "The centralising character of the Indian political structure renders the District Council subordinate to the state government while the latter in turn is subservient to the Central Government." (Nongbri 1999)

This arrangement renders completely ineffective the ideas of 'self management' and autonomy that underlie the Sixth Schedule.

Conclusions

(1) The District Councils have been constitutionally given the power to manage all forests other than Government Reserved Forests, and thus the security of ownership enjoyed previously by local user-groups has become tenuous.

(2) Most of the laws enacted by the District Councils for the management of forests are not comprehensive and adequate to deal with the unique circumstances prevailing in a particular Autonomous District, and follow the general rules on the forest policy that applies to the nation as a whole.

(3) In many cases, the District Council has modified some customary laws on forests so that more revenue can be generated, in total disregard to its consequence on the forests.

(4) Finally, the entire administrative structure of the District Council is highly bureaucratic in nature and not much different from the State Forest Department. Thus an elaborate hierarchy of posts exists such as Chief Forest Officer, Assistance Forest Officer, Forest Ranger, Deputy Forest Ranger, Forest Guard etc.

Thus, whereas the Constitution makers had given the District Council the right to make laws and manage forests in the manner best suited for the tribals, the District Council has created an administrative structure which was alien to the tribals and similar to the administrative structure of the Government.

Community Rights in India: The Way Ahead

It is clear that forest management and "encroachments" by local communities are inseparable issues and the attempts to dissociate the two can only add fuel to the fire. The issue of encroachment cannot be merely treated as a law and order problem. Nor can it be treated, as the CEC observes, a "cancer in the forests spreading without pausing and spreading into vitals of the life supporting systems of nature destroying all upon which the life, including the human life itself depends" (CEC, 2002). The cancer, in fact, is not encroachments. The cancer is the exclusion of the local communities from the management of the resources.

Forest resources have suffered greatly because of state control. "Scientific management of forests' in India has survived more than 100 years. Unfortunately, the forests have not. Erosion of people's control over their own resources and decline in the resources' health are not unconnected" (Khare, A, 1992). The myth of "scientific management" of forests, introduced to take forests away from the hands of the people, has been shattered.

To tackle the problem of encroachment, a two-pronged strategy, a long-term and a temporary one should be evolved. The long-term strategy is handing over rights on forest resources to the local communities. This presupposes devolving substantial stakes and rights and economic incentives to them. In the meantime, as a temporary measure, all pre-1980 cultivation by the forest dwellers should be regularised forthwith.

From JFM to Village Forests

While the present policy of JFM encourages participation of local communities in forest management, it falters badly in terms of establishing well-defined community rights over forest areas. Moreover the implementation of existing policy too has been lacklustre. There is thus an urgent necessity to establish the legal framework for moving towards "community" or "village" forests with full rights and autonomy to manage forests on the basis of their knowledge and wisdom. The village communities should have full and perpetual rights on major as well as minor forest products in such forests including that of marketing of products and forest department should play a supportive role in form of providing technical assistance. A clear legislative basis should be provided for this arrangement in the Forest Act.

Section 28 of the Indian Forest Act for the constitution of "village forests" can provide the necessary starting point in this regard. This provision of the Forest Act has never been implemented and has by and large remained dormant. This section should be amended to clearly define the rights and responsibilities of village communities. The government of UP has already done this by notifying the guidelines for JFM as rules under Section 28 of the Indian Forest Act and expressly giving the village communities the rights of forest officials (GOUP, 1997). But, this is not enough. Clear provisions need to be made in the Forest Act itself with procedural details elaborated in the rules as is found in Nepal. This would provide tremendous incentives to the village communities in forest regeneration. Long-term security of tenure and autonomy in decision-making are some of the vital elements in providing incentives to local community organisations to engage themselves in the gigantic task of forest protection and regeneration.

Moving Towards Joint Protected Area Management

The principles underlying community rights on forests are equally applicable in case of protected areas of sanctuaries and national parks. Hence steps should be taken to move towards joint protected area management to ensure that local communities and the wild life can exist together in harmony. This is how the princely states managed their game reserves, which apart from providing habitat to the game animals also met the needs of local communities. Hence innovative programmes need to be initiated which would give local communities vital stakes in the protection of the wildlife.

The main difficulty in initiating such programmes comes from the Wild Life Protection Act, which is exclusionary in nature and does not allow any activities, which are expressly not necessary to protect the interests of the wildlife alone. Section 29 of this Act gives authority to the Chief Wild Life Warden to allow activities that he considers to be in the 'interest' of the wildlife. This Section is too vague and arbitrary, and is liable to be misused. Instead, the Section should be made abundantly unambiguous by specifically providing for activities benefiting local communities, which would in turn ease the conflict between them and the wildlife and thus also be in the interests of the protected wildlife.

Regularising the Existing Encroachments

It is true that cultivation on steep slopes is both harmful for the soil conservation and forest growth as well as inconvenient and economically non-paying proposition for the people, who toil on these slopes. But, to jump from this to outright and immediate evictions is not a solution at all. The improvement in the JFM policy and change in the attitude and mind-set of the forest department, pre-requisites for the people to earnestly participate in the management, would require some time. Also people would require time to trust the forest department and to perceive that forest produce can give them good income. It is also true that most of the encroached lands of the poor communities are of poor quality. But, at present, without any alternatives, the people strive hard on such lands. Once they start getting substantial income from non-timber and timber products, and once they are sure of the permanent or long-term contract, irrevocable at the whim of the forest department, they would be least interested in continuing with the hard toil on the encroached land.

Before the long-term gains start accruing to the people, thus, the encroachments should be regularised or at least not be disturbed. The first requisite step is that the encroachment is regularised, and a clear-cut programme of making unambiguous and legally-binding contracts with the people is initiated.

Norms for Regularisation: Fine Receipts or Field Verification?

The September 1990 circular of the MoEF distinguishes two types of encroachments—one, pre-1980 "eligible" encroachments, liable to be regularised and two, pre-1980 "ineligible" encroachments and all post-1980 encroachments, not liable to be regularised and liable to be evicted. There is, however, no clear definition of which pre-1980 encroachments are eligible and which are not. The Commissioner of Scheduled Castes and Scheduled Tribes has said: "If the claims of the tribal people are to be determined on the basis of the records of the Forest Department or, at best, records of other government departments, his claim may be as good as lost. It is the fact of possession of land, its cultivation and actual reclamation in some cases by his ancestors, which is common knowledge in the village, which is the basis of his claim. These facts may or may not have been brought on record. The reasons for this dissonance can be many. For example, the official may not have visited the area or may have preferred not to take note of the cultivation, or may not have bothered to bring it on record, and such like. They are of no concern to the tribal people. They cannot be expected to know what is there in government records. In these circumstances if the records were to be insisted as evidence, the disputes about land can never be expected to be resolved" (Sharma B D, 1997, p. 36). Thus not only documentary evidence, but physical evidence or testimonies of villagers should also be legitimised as proof of ownership.

Save Natural Forests: Promote Private Forestry

It is interesting to note that 97 per cent of forest lands are owned and managed by the government, but most of the raw material for our wood-based industries is imported from outside.

Wood can become a major industry and combined with its ancillary activities has potential to provide substantial economic benefits to a large number of people, who can grow timber either on government forestlands under JFM or on private lands. Domestic needs of wood for the town and village population have remained a problem, and this often prompts massive government efforts to unsuccessfully promote among the population fanciful alternatives, which are usually non-viable. Tree growing offers a very good economic opportunity to private individuals, co-operatives, and companies.

Until 1980s, most of the large wood-based industries, including paper and pulp industries, were receiving supplies of their raw material, usually at subsidised rates, from government forests. Several of them had long-term raw material supply agreements with the state governments (Saigal and Kashyap, 2002).

However, there was no move either on the part of the government or the industry to grow plantations of timber/bamboo on the leased out forestlands. This has been one of the main reasons behind massive deforestation.

The Forest (Conservation) Act of 1980 was in response to the continuing deforestation. The ban on the lease of forests to industries and moratorium on clear felling by the Forest Department was envisaged to halt the process of deforestation. Meanwhile, the Supreme Court also ordered ban on tree felling in the mid-1990s.

As a consequence of the changes brought about by the ban on felling of trees and their transportation, supply of raw material from government forests to wood based industries has gradually declined, forcing the latter to look for other alternatives. The option of raising own captive plantations on a large scale is not available to these companies due to ceiling limits on agricultural land and restrictions on leasing of government forestland (Saigal and Kashyap, 2002). This has taken away a good opportunity for private parties to grow timber/ bamboo/eucalyptus to supply to industries.

Strict restrictions on the sale of timber grown on private land, without permission of the forest department, is a major deterrent against timber plantations on private lands. Private plantations on non-forestlands suffer from lack of suitable policy support. There is also a separate set of regulations for all individual species found on any private land. Permission from the Divisional Forest Officer or a designated tree authority is required to fell trees and to transport the produce. The private sector's participation in forestry activities is determined by policies at the central and state levels, not only those directly related to forests but also policies and legislation introduced for other sectors e.g. land ceiling on agricultural lands, export-import policies, tax laws etc. (Singh 2002).

Private plantations can supplement the product of the community-managed forests, if the restrictive policies of the government are removed. Moreover, there are many forest areas, which could provide good quality timber, if proper natural regeneration is allowed. This is possible only if the local communities are entrusted with the stewardship of the forests. The same restrictive policies hamper community initiatives of growing these precious species in these areas too.

Forests to Forest Dwellers:
Efficient and Ethical Resolution

Experiences the world over and at home have made it amply clear that resources in the hands of private parties—be that of individuals, communities, or corporations—are better managed than in the hands of government. Who should be entrusted with these resources depends upon the types of resources, circumstances, and local customs and traditions. For resources that have generally been in the commons, like water and forests, the best stewards are the local communities who have been managing those commons historically. Entrusting these resources to any other entity would mean keeping the communities out forcefully—by guns and guards. Whether these guns and guards are employed by the government or a private corporation, they would not able to withstand the battles for bare survival by the communities. Neither corporatisation, nor collectivisation is an option.

The notion that the protection of forests requires them to be separated from humans springs from the Western vision of wilderness. Having lived for centuries in the forests, communities have the requisite traditional know-how to best manage these resources. The traditional knowledge is not infallible, but it is not a giant leap of faith to assume that the communities will learn from outsiders new scientific developments that are appropriate to their concerns. This appraisal by communities and the required meshing of the new with the old knowledge is an effective deterrent to unnecessary scientism—worship of the new just because it is new. Once communities are given clearly demarcated and legally enforceable rights in forests, their management will prove to be the most optimal.

The history of state forestry, from the British to our government, has been of replacing the diverse species of a natural forest with mono species. Both the scientific and sustainable forestry management has led to the same results. Communities are more likely to find economic and social benefits from the existing diversity of resources that the forests offer. There is higher probability of a natural fit between diverse needs of communities and diverse offerings of forests.

In addition to all the utilitarian or efficiency arguments, it must be remembered that local communities have a prior claim—a moral claim—on the forests. They have been living there and using the resource for generations. It is on the premise of prior use that all resources have been settled in any civilised society. It is gross injustice not to recognise the rights of forest dwellers.

Community ownership and management solve two problems simultaneously: the protection of forests and of dignified livelihood to the poorest communities in the country. They build their future from the natural asset of forests. The most efficient as well as moral resolution is to take our forests from the foresters and put them in the hands of forest dwellers.

References

Bethell, Tom (1998), *The Noblest Triumph: Property and Prosperity through the Ages*, New York: St Martin's Press.

Cabinet Circular for Land and Forest Protection Ministry (2000), Nepal, January.

CEC (2002), Recommendations of the Central Empowered Committee (in IA 703/2001 in Writ Petition (Civil) 202/95, July 25.

Deacon, Robert T. (1994), "Deforestation and the Rule of Law in a Cross Section of Countries," *Land Economics*, November, pp. 414-30.

Department of Forests, Ministry of Private and Community Forests (2002), Nepal, May 8.

Dutta, R. (2002), "Community Managed Forests: Law, Problems, and Alternatives." Presented at *The Commons in an Age of Globalisation*, the Ninth Conference of the International Association for the Study of Common Property, Victoria Falls, Zimbabwe, June 17-21, 2002.

Federation of Community Forestry Users, Nepal (FECOFUN) (1999/2000), Annual Report.

Forest Act, 2049 (1993), Published by His Majesty's Government of Nepal (HMGN).

Forest Survey of India (FSI) (1999), *The State of Forest Report, 1999.*

Guha, Ramchandra (1998), "Dietrich Brandis and Indian Forestry," *Village Voices, Forest Choices* Mark Poffenberger and Betsy McGean (ed).

Hobley, M., J.Y. Campbell, and A. Bhatia (1996), "Community Forestry in India & Nepal: Learning from each other," *MNR Discussion Paper No. 96/3*, ICIMOD.

Jan-Willem, Oosthoek (1999), "A Walk through the Forest," Paper presented at a Research Seminar of the Department of History of the University of Stirling on February 24.

Kasere, Stephen, "Campfire–Zimbabwe's Tradition Of Caring" http://www.unsystem.org/ngls/documents/publications.en/voices.africa/number6/vfa6.08.htm.

Khare, Arvind (1992), *Joint Forest Management-Regulations Update*, Society for Promotion of Wastelands Development.

Kothari, Ashish (2002), *The Forest Encroachment Issue: A Briefing Note.*

Lal, J.B. (1989), *India's Forests: Myth or Reality?*

Mahapatra, Richard (2000), *Down To Earth*, February.

Marshall, W. Murphree (1996), "Congruent Objectives, Competing Interests and Strategic Compromise: Concepts and Processes in the Evolution of Zimbabwe's Campfire Programme", *Community Conservation Research in Africa, Principles and Comparative Practice* Working Papers, Paper No 2. Centre for Applied Social Sciences, University of Zimbabwe.

Master Plan (1988), Jointly published by His Majesty's Government of Nepal (HMGN), Asian Development Bank (ADB) and Finnish International Development Agency (FINNIDA).

Mdzungairi, Wisdom (2002), "Rural Folk Need Incentives to Preserve Wildlife", *The Herald* (Harare), http://allafrica.com/stories/200204040313.html

Mehta, Trupti Parekh (2002), "Community Forestry in India and Nepal" *PERC Report*, June. Montana

Mehta, Trupti (2001), "Investigations into Institutional Arrangements and Incentives of Two Respective Forest Management Policies of India and Nepal," Paper presented at Kinship Conservation Institute, PERC, June, unpublished.

Ministry of Environment and Forests (1998), *National Forest Policy*.

——————. (2000), *Guidelines for Joint Forest Management*.

——————. (2002), circular dated May 3.

Nathan, D. (2000), "Timber in Meghalaya". *Economic and Political Weekly (Perspectives)*, 22 January

Nayak, P. (2002), "Community-Based Forest Management In India: The Issue of Tenurial Significance". Paper presented at *9th Biennial Conference of the IASCP* (International Association for the Study of Common Property). June, Victoria Falls, Zimbabwe.

Nongbri, T. (1999), "Forest Policy in North East India", *Indian Anthropologist*, Vol. 29, No. 2

——————. (2001), "Timber Ban in Northeast India: Effects on Livelihood and Gender". *Economic and Political Weekly*, 26 May

North, Douglass (1990), *Institutions, Institutional Change and Economic Performance*.

Paudel, Keshab (2000), "Government Vs Community," *Spotlight*, Vol. 19: No. 47, June 9-June 15.

Poffenberger, M. and B. McGean (ed) (1998), "Upland Philippine Communities: Guardians of the Final Forest Frontiers," *Research Network Report No. 4*, Center for Southeast Asia Studies/ International and Area Studies: University of California.

Powell, Baden (1898), *Forest Law*.

Prabhu, Pradip (2002), "Tribals Face Genocide," *Combat Law*, October-November, A Letter of National Commission for SC and ST to the Prime Minister, dated September 6, 2002.

Ribbentrop, B. (1900), *Forestry in British India*.

Ridley, Matt (1997), *The Origins of Virtue*, USA: Viking Penguin.

Roy, Tirthankar (2000), "Forests and the State," *EPW Reviews*, September 16.

Saigal, S., Hema Arora, Rizvi (2002), *The New Foresters*, IIED.

Saigal, Sushil and Divya Kashyap (2002), *Instruments for Sustainable Private Sector Forestry*, India Case Study No: 5, IIED.

Shah, Anil C. (2003), *Fading Shine of Golden Decade: The Establishment Strikes Back*, March.

Sharma, B.D. (1997), "Resolution of Conflicts Concerning Forest lands-Adoption of a Frame by the Government of India" *Forests and Fallows-The Flickering Hope?*

Singh, Daman (2002), *Policies affecting Private Sector Participation in Sustainable Forest Management*, India Case Study No: 1, IIED.

Stebbing, E.P. (1923), *The Forests of India*, Volume I.

Tripathi, R. and S. Barik. *North-East Ecoregion Biodiversity Strategy and Action Plan* Report Submitted to: Ministry of Environment and Forests Government of India, New Delhi

Venkateshwarlu, K. (2002), "State to go slow on Union Ministry circular," *The Hindu*, September 24.

10

Developing Markets for the Ecosystem Services of Forests

IAN POWELL, ANDY WHITE
NATASHA LANDELL-MILLS

Introduction

The many valuable ecosystem services provided by forests — including watershed protection, biodiversity conservation and carbon storage – are gaining increasing attention from industry and government, as well as private citizens. These individuals are also increasingly aware of the dangers and costs of allowing forest services to be degraded or lost. This degradation can have local impacts, such as floods and landslides, or broader impacts, like global climate change.

This awareness is drawing attention to the economic benefits of healthy forest ecosystems — benefits that until recently have been taken for granted. Indeed, as human demands increase and natural resources become scarcer, those who bear the costs of degradation — such as downstream water utilities, local governments, private insurers and society as a whole — are exploring opportunities to reduce costs by financing forest conservation. At the same time, some forest owners are seeking compensation for the costs of maintaining healthy forests. Interest in reducing costs, increasing incomes and expanding conservation is moving markets for ecosystem services toward cantre stage in the debate about forest conservation.

The growing prominence of markets comes at a time when traditional models of government financed protected areas and conservation are under strain. Growing public deficits and increasing frustration with governmental inefficiencies are spurring action by a broad range of stakeholders. This action is increasingly backed up by willingness to put finances toward environmental services. Private companies, individuals, Non-Governmental Organisations (NGOs) and communities are all getting involved, driven by the need to reduce costs, capture new income, improve public relations, manage risks and ultimately, to protect current well-being.

Many promising investments and programmes already have been established, with more under development (See Box 1). For example, a Costa Rican utility company voluntarily pays into a fund that provides money for private upstream landholders to increase forest cover. This reduces sedimentation, thus providing sufficient water flow for hydro-electricity generation. In Paraguay, AES, an international power company, paid $2 million to form a protective reserve for one of South America's last remaining areas of undisturbed dense tropical forest. This helps to offset carbon emissions. In Karnataka, India, farmers have formed a fund with the assistance of an NGO, the Government of India and the Swiss Development Co-operation to help other local farmers with watershed protection activities, such as regenerating and maintaining fallow land. Often, public authorities – particularly when faced with budgetary crisis — require those who most obviously benefit from ecosystem services to provide financing. This is the case in Colombia, where hydroelectric and water utilities are required by law to allocate a fixed percentage of revenues to an ecosystem fund. The fund pays private landowners for watershed management and purchases hydrologically sensitive lands for management by government agencies.

In addition to direct-investment approaches of this nature, some governments are experimenting with new fiscal approaches. In Brazil, a few states have pioneered a new tax allocation system, under which a percentage of state tax goes directly to municipalities that actively protect watershed areas.

These many innovations are generating important lessons, yet they are limited in scale, scope and impact. Given the tremendous social benefit of forest services, and the many private and public stakeholders who would gain – both socially and economically — from greater protection, it is vital to tap the potential of market approaches.

The purpose of this paper is to help innovators — be they forest owners, investors, policymakers or those who are suffering the off-site effects of degradation — understand the basic opportunities and issues posed by direct investment in forest services. The paper complements the work of others who have reviewed innovative tax policies and other fiscal approaches to forest services. It draws on the combined experience of The Katoomba Group, a collection of international experts engaged in developing markets for forest services.

Box 1

The Growing Number of Markets for Environmental Services

In a recent global review of emerging markets for forest environmental services, over 280 cases of actual and proposed payments for four sets of environmental services were uncovered. These include 75 deals for carbon sequestration deals, 72 for biodiversity conservation, 61 for watershed protection, 51 for landscape beauty and 28 for sales of "bundled services." Far from being concentrated in the developed world, these cases were drawn from a range of countries in the Americas, the Caribbean, Europe, Africa, Asia and the Pacific. While the study suggests impressive expansion in markets, it also highlights the tremendous variety of market structures. Schemes differ according to the number and type of participants involved, the payment mechanisms employed, the degree of competition and their level of maturity. They also often have very different impacts for local and global welfare. These variations reflect local socio-economic and environmental factors, drivers and ultimately local variations in the process of market development.

Source: Landell-Mills, N., J. Bishop, I. Porras. Forthcoming. "Silver bullets or fools' gold? Developing markets for forest environmental services and the poor". Instruments For Private Sector Forestry Series. IIED, London.

The paper first describes the principle ecosystem services provided by forests and how market approaches can assist in conserving these services. The types of financial mechanisms currently in practice, the ways in which these mechanisms develop and the key questions used to evaluate these markets are then presented. The conclusion provides perspective on the steps needed to fully develop and expand markets for forest services.

Background: The Environmental Services of Forests

Forests perform significant services that maintain conditions for all life on earth. The environmental services of forests are those ecological processes from which humans directly benefit. Some of the key environmental services are described below.

Carbon Storage and Sequestration

Simply by existing, forests keep carbon out of the atmosphere. In addition, forests can be managed to actively sequester even more carbon and lock it up in biomass. Forests contain about 40 per cent of total terrestrial carbon. Around half of the total dried biomass of forests is carbon. Plants absorb carbon through photosynthesis from atmospheric carbon dioxide. Topography, soil and species composition and climate are all factors that influence the rate at which carbon sequestration occurs in forests. As forests grow and develop, the amount of carbon tied up in living and dead biomass increases. Once mature, the biomass of a forest stabilises. Therefore, the longevity of trees makes them particularly suited to the sequestration and storage of carbon while conversely, forest clearing and degradation accounts for 15 to 30 per cent of all carbon emissions to the atmosphere. These simple relationships underpin the emerging investments in forest carbon sequestration. As a result, forests are elemental to national and international efforts to address global climate change.

Hydrological Services

Forests have major effects on hydrological processes, although the extent and value of these services varies with each watershed's circumstances. Transactions in hydrological services are, therefore, site-specific. These transactions depend upon local physical, social and environmental characteristics. The three main beneficial hydrological services may be identified:

Flow Regulation: A forest intercepts rainfall and, with a generally large capacity for water absorption and retention, may in some situations help convert irregular precipitation into a more even flow of water from a catchment area. The risk of flooding due to extreme weather can, therefore, be reduced. A forest may also act as a slow-release reservoir, increasing dry-season base flow from a catchment. On the other hand, a forested catchment may yield less total water than a non-forested one. But even in this case, the useable and non-destructive yield can be enhanced.

Maintenance of Water Quality: Rain falling on a forest is intercepted and filtered through a mass of soil and roots. As a result, water flowing from an undisturbed catchment area is generally high-quality. Disturbance to the catchment and changes in land use can lead to sedimentation and nutrient pollution. This can affect water availability and associated benefits, such as fisheries. The quality of water for human consumption, agricultural use and industrial use also can be affected.

Water Table Regulation: Forests can play an important role in water table regulation. Over time, equilibrium develops between vegetation and the water table. Deforesting a catchment may lead to greater infiltration high in the catchment and rising water tables lower down. This may bring salt water nearer to the surface and affect crops and water quality. Conversely, in other watersheds, water table replenishment may be disrupted. Deforestation can lead to falling water tables if denuded land becomes heavily eroded or compacted and water runs off before it can infiltrate.

Biodiversity Services

Forests provide some of the most biodiversity-rich ecosystems on earth and are believed to provide habitat for an estimated 90 per cent of threatened and endangered species. Forests house myriad examples of genetic diversity within individual species. Similarly, ecosystems within forests adapt to local and landscape-level variations in the environment. Ecosystems become resilient to environmental disturbance and stress by maintaining diversity on all levels. Biodiversity has intrinsic value as well as providing practical benefits. Medicines, plant derivatives and other non-timber products may form a foundation for the livelihoods of forest-dwelling people.

Biodiversity provides the basis for bio-prospecting for new medicines. Agricultural systems depend on pollination and gain resilience from naturally occurring biodiversity. Aspects of biodiversity, such as soil building and nutrient cycling, have general and ubiquitous importance. In addition, social value, from recreation to spiritual and cultural benefits, is increasingly recognised as fundamental to human health.

The Rationale For Market Approaches

Forests' great social value and their many environmental services are frequently not realised by forest landholders. This is because these benefits often are experienced some distance from where they are generated, and no mechanisms exist for compensating forest owners for their services.

As a result, these off-site services are often described as "externalities" which forest landholders effectively provide for the public free of charge. At the same time, landholders who allow damage to forests, and thus reduce their supply of offsite benefits, are rarely penalised.

Forest conservation adovcates support market approaches because it is thought that capturing the financial value of forest services will promote good stewardship and discourage more degrading uses of forests. Market approaches have gained prominence as frustration has increased with regulatory approaches - often thought to be inefficient, expensive and inequitable.

While markets can be powerful mechanisms for improving human welfare, they are not without risks. Market performance depends on numerous site-specific factors, including existing power relations, degrees of concentration in demand and supply, the supply of information on trading conditions, and the level of transaction costs. With the right conditions, markets can provide a powerful boost to well-being. Where conditions are less favourable, markets may lead to greater degradation, while at the same time reinforcing existing inequities. The emerging challenge is to find the best market tools that, together with the right regulatory framework, will encourage just and efficient forest conservation.

Types of Markets and Payment Mechanisms

By the very nature of their adaptability, forests and their ecosystems vary greatly. In the same way, market mechanisms will vary according to their particular ecological, social and political context. However, there are several basic types of market approaches, organised here according to their level of public involvement: self-organised private deals, open trading schemes and public payment schemes.

Self-organised Private Deals

This approach includes direct, usually closed, transactions between those who benefit from forest services and those who provide them. This includes deals such as voluntary certification and eco-labelling schemes, direct purchases of land and purchases of development rights to land, as well as direct payment schemes between offsite beneficiaries of forest services and landholders responsible for the services. In France, Perrier-Vittel, a company that sells bottled water, pays upstream landowners to use best management practices on their land to ensure that the company has a supply of quality water. Other examples include conservation groups or businesses motivated by corporate conscience or marketing considerations to pay forest holders for conserving biodiversity. Private deals, typically limited in scope and transparency, benefit from clear property rights and enforceable contracts, although clear rights and enforcement mechanisms are not always necessary. In most cases, little other public involvement is warranted.

Open Trading Schemes

This approach is used when a government defines an environmental service commodity to be traded and devises regulations to create demand. In New South Wales, Australia, for instance, the government is piloting proposals for salinity credit trading rooted in broader basin-wide salinity targets. Based on these targets, the government has allocated licenses to dischargers of salinity. The idea is that those wishing to exceed their salinity quota can do so if they purchase salinity credits from those who have taken action to reduce salinity, e.g. by protecting and managing native vegetation. Other examples include tradable development rights pioneered in urban areas of the U.S, the trading of wetland mitigation credits and emerging nutrient trading schemes in some U.S states.

The most prominent example of open trading is the emerging national and international carbon trading market. Rooted in the Kyoto Protocol signed in 1997, carbon trading has evolved from a marginal and largely voluntary exercise to a mainstream mechanism for reaching local and international emission reduction targets. Despite a recent decision by the U.S to renounce its commitment to Kyoto, the treaty has stimulated a number of national and regional trading initiatives. In the August 3, 2001, *Washington Post*, a CO_2 trader was quoted as saying that he believes the CO_2 market could be worth tens of billions of dollars by the end of the decade.

Forests are a key tool for reducing and storing carbon and trading in forest-based carbon offsets is likely to grow. (As part of the political settlement reached in COP6 in Bonn, a number of limits were placed on the use of forest-based carbon offsets in achieving national emission reduction targets.) Any market-based system of trading credits requires a transparent framework, accurate accounting and verification systems.

Public Payment Schemes

This approach is used when a government provides the institutional foundation for a programme and directly invests in it as well. Examples include the U.S. Conservation and Wetland Reserve Programmes, wherein the government pays farmers for managing lands in ways that reduce soil erosion and runoff. In 1998, in response to the Yangtze River floods as well as concern over soil erosion and deforestation, the Chinese government began to plan a Forest Benefit Compensation Fund to be financed by the government and private sector beneficiaries in upper basin areas.

Public payment schemes can be administered by purely public agencies or hybrid partnerships with the civil and private sectors. This approach involves both indirect subsidies and direct payments to forest landowners. Prices paid by governments are often determined by political or budgetary considerations, rather than strict economic evaluation of the environmental benefits involved.

The Process of Developing Markets and Instruments for Services

Developing markets for forest services is, in many senses, similar to developing any new market. However, the process differs in some key aspects. It is similar in that entrepreneurship, local constraints and opportunity will decide the speed and extent to which a market is developed.

Because most forest services are currently treated as free goods, it is perhaps most different in that developing a market often requires converting these freely-accessed goods and services into commodities and property. This is inherently a political process, whereby different stakeholders' rights and responsibilities are questioned, new rules are established, and new entitlements are established. This process occurs in three broad phases (see Figure 1).

In the first phase, the linkages between forest actions and their consequences are gaining attention. In all cases, an entrepreneur operating either in the public or private sector, and operating as an individual or an entity, shows leadership and mobilises action by informing stakeholders of the existing problems and opportunities. This action generates willingness to pay for protection from the problems and provides a basis for interested stakeholders entering into negotiations.

In the second phase, the structure is defined. Supporting rules and processes begin to emerge. Except in purely private deals, drafting regulations requires a political process. The regulations define the service, settle the particular rights and duties of the stakeholders and provide a platform for negotiating payments.

Figure 1

Phases of Development of Markets and Instruments for Forest Services

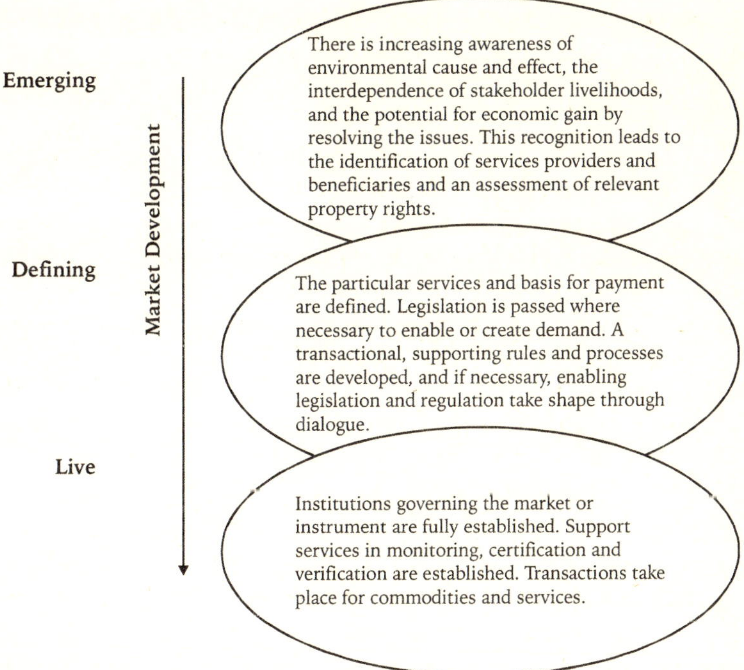

In the final phase, the market becomes live. Transactions take place and money changes hands. Service contracts and agreements are established, along with supporting institutions, such as accounting standards, monitoring and certification mechanisms. The apparent neatness of this scheme is intended for illustrative purposes only. In reality, many stakeholders intervene and interact on various activities within the different phases. Moreover the process is iterative, progressing at different speeds in different contexts, and in some cases involving setbacks.

Key Questions in Developing New Markets

Experience shows that developing new markets and market-based instruments that add financial value to forests is complex. Interested parties must be identified and they must adopt precise roles in transactions. These transactions must be developed by negotiation and supported by rules, contracts and methods of verification. Despite the diversity of contexts and economic opportunities, innovators face many common issues when considering the development of new markets. A preliminary list of these issues follows.

What Environmental Services are Provided?

A key step in market development involves identifying the ecological conditions that provide direct and demonstrable benefits to people. Better management of the forests may improve the quantity, quality or integrity of the existing services already provided—or it may provide new services altogether. Market development can be accelerated if there is a perception that a service is becoming scarce and thus more valuable. This could apply to habitat loss or declining water quality. Action can be driven by the costs of alternatives or the consequences of service failure. Ultimately, the specific service that is marketed will depend on the particular needs of the buyer.

For example, an Australian airline might feel that its public image is served best by funding protection of kangaroo habitat. With respect to watershed protection services, hydropower companies may be interested in controlling sedimentation, while water supply companies may be more interested in reducing nitrogen and phosphorus pollution.

What is the Economic Value of the Environmental Service?

To generate willingness to pay for specific environmental services, it is critical that beneficiaries recognise the value of environmental services for their welfare. Impacts may be direct, e.g. the provision of clean drinking water, or they may be indirect, e.g. the reduction of sedimentation and improved hydropower efficiency leading to cheaper and more regular electricity supplies. A number of methods exist for estimating the economic value of environmental services.

Contingent valuation surveys are an increasingly common method involving questionnaires asking beneficiaries their willingness to pay for the continued delivery of a specific service or their willingness to accept compensation for their loss of the service. Another method involves estimating the cost of replacing the particular service, assuming this is possible.

What is the Cultural, Legal and Regulatory Context?

Developing a new market instrument for a particular service involves a unique set of stakeholders and governance structures. It also must correspond to that local ecosystem. Most markets, with its unique regulatory, fiscal, and legal context, will require substantial creativity, political leadership and willingness by stakeholders to consider new approaches. As knowledge develops in many cases, continued adaptation will also be needed.

What are the Rights and Responsibilities of Stakeholders?

Property rights are particularly important. Societies differ in how they handle the legal and customary rights of stakeholders in forests. These property rights are often insecure, overlapping and contested, and they rarely explicitly address forest services. If rights over services are not previously decided, developing a market will entail assigning or clarifying them. For example, do landowners have a responsibility to protect forest services or a right to be compensated for providing them? Special attention is required to ensure that the less powerful sectors of society do not lose opportunities and access to resources. Market developers must be fully cognisant of existing power relations, vested interests and the implications of their proposals.

Who are the Potential Buyers and Sellers?

The use of market tools to restore, protect or enhance an environmental service would be impossible without sellers able to deliver the service and buyers financially able to pay. After determining ownership or property rights, the next question must be whether that person is willing to sell. Equally important is the existence of funds sufficient to finance regular delivery of the service. In addition, beneficiaries may be unwilling to pay for a service, such as clean water, which they may consider a right and to which they have always had access.

Can the Service be Measured and Monitored?

Services must be defined in order to enable transactions. A service can be defined in terms of a particular commodity, or simply on the basis of assumed land value. For example, carbon credits can be used to offset emissions or biodiversity credits can be used to offset development. Hydrological services can be defined in terms of water quality indicators or stream flow reliability. Depending on the quantity, quality or uniqueness of a forest service, it may be difficult to adequately define a commodity or determine a payment level.

What Support Services are Required to Enable the Market?

In many cases, there is a need for new institutions, ranging from private sector contracts to public entities, to facilitate payment for services. Also, markets require structures for financing, verification, monitoring, accounting and certification. Other necessary structures include business advisory services, planning devices and consultants, independent environmental advisory groups and capacity building. Due to the risks involved in any emerging market, insurance companies and banks can play a critical role by bolstering the security of transactions.

Who Benefits?

Sharing the benefits of market creation is important for equity reasons, but it also is critical to the success or failure of payment systems. Where new markets negatively impact particular stakeholder groups, the stakeholders in question will have an interest in undermining its viability. Depending on who these groups are, and how much power they have, the additional risks introduced by inequitable benefit-sharing are potentially significant and may lead to market failure.

Making Progress and Making Deals

While there are many innovative deals and programmes in the world, trading in environmental services remains a nascent and marginal affair. The players are just beginning to grasp the potential ways in which markets can help protect forest services and improve well-being. Innovative investments and programmes should be pursued by all parties — forest holders looking for compensation, private investors looking to lower costs or reduce risks, community groups seeking to ensure continued supplies of natural capital and governments looking out for the public good. Pursuing this agenda entails gaining knowledge about market approaches, building institutions to facilitate them and making deals – forging ahead with innovative investments and programmes.

Gaining Knowledge

A better understanding of some key dimensions of forest services will facilitate the development of new mechanisms.

- *Biophysical relationships* - It is vital to advance scientific understanding of the biophysical relationships between forest management activities, the flow of services from forests and the resulting impacts off-site. Better data, modeling and analysis will increase confidence and decrease uncertainty about service delivery.

- *Risk management* - It is equally important to understand and develop a range of financial instruments to deal with the uncertainty of these markets. This will most likely entail the creative application of existing instruments such as reinsurance, and guarantees — and the creation of completely new instruments.

- *Property rights definition* - The role of property rights and regulations is another critical area for development and learning. For example, how can markets be constructed to provide additional incentive for conservation without contradicting existing regulations and without providing "perverse incentives" for poor land use? Lessons from currently emerging experiences will no doubt prove helpful to innovators everywhere.

- *Spreading benefits* - The role of equity and participation in markets requires additional study. How can mechanisms achieve the outcomes desired by investors, while also ensuring equitable treatment of relevant stakeholders? What social standards or criteria should be put in place to ensure adequate participation? Are there particular mechanisms that can be used to achieve poverty alleviation as well as conservation outcomes?

- *Comparing options* - Finally, it is critical to understand the different market mechanisms, the conditions in which one might be favoured over another and the success of existing instruments and institutions. Describing innovative experiences and "lessons learned" to business and conservation audiences will improve and accelerate the adoption of market approaches.

Building Institutions

To function efficiently, effectively and equitably, all markets require enabling institutions, such as support services, common auditing procedures and contracts. Because marketing forest services is an embryonic field, enabling institutions are only beginning to be developed. Stakeholders may adapt some of these institutions from models established in other areas, but it also may be necessary to construct some institutions specifically for the forest services market. Three institutions lie at the core of market development — assessment methodologies, registries and certification standards.

- *Assessment methodologies* - Standard measurement tools are necessary because they will ensure transparency and repeatability — essential qualities for market development. For example, Winrock International, an NGO, has been working with a variety of organisations on carbon inventory and monitoring protocols. Similar work for hydrological and biological services is underway by State Forests of New South Wales, Australia. These efforts require more support in order to be fully developed and adopted as credible, standard approaches by market players.

- *Property Rights and Registries* - The value of property rights is largely dependent on the existence of formal and unified registries. Recording ownership of property rights with a single authority is critical for reducing transaction risks. Additionally, a registry contains individually serialised records of scientifically verified and measured environmental services. In addition to guaranteeing ownership, a registry can assure potential buyers that credible measuring and monitoring have taken place in a transparent scientific manner. Registries can assure buyers that no double counting had taken place. By developing documentary records of their achievements and establishing title to such services, owners of forests will be more likely to receive value from these services and less dependent on timber for revenue. In Australia, the Catchment Ecosystem Services Investment Centre is developing steps to assist with brokering environmental services deals. Initial steps include creating a registry and developing criteria for environmental services. In the U.S., the GHG Registry SM has been designed to facilitate the development of a robust GHG trading market. It is modeled on the U.S. EPA's Allowance Tracking System (ATS) for the SO2 (Acid Rain) programme.

- *Certification* - Certification is a voluntary procedure involving an independent third party that evaluates performance using specific criteria. The Forest Stewardship Council, an accrediting organisation, has established an international system to certify forest management using social, environmental and economic criteria. But this system is limited to certifying sustainable management for forest products such as timber, not services. It is urgently important to develop principles and criteria for certifying the management of forest services.

Making Deals

Developing markets means invoking a wide variety of tools and understanding the flexibility of each. Innovators located in areas with weak public institutions may find that self-organised private deals are the most effective. Those in highly regulated environments may find that the additional effort to set up a trading system is more than compensated by dramatically increased efficiency in reaching goals. Where public institutions play an important role, public payment schemes are more likely to work.

There is no substitute for experience, and learning-by-doing is one of the best ways to gain that experience. The existing stock of knowledge has come from those innovators who have forged ahead despite uncertainty and lack of precedent. Business leaders, NGOs and governments should encourage innovation within their own organisations — and in collaboration with other sectors. Those who innovate will be recognised as leaders in the broader global community.

11

Creating Private Property Rights in Wildlife

ROBERT J. SMITH

During man's relatively brief existence on this planet, he has relied on the bounty of its flora and fauna for his existence. He has harvested wildlife for food, clothing, shelter, medicines, beasts of burden, pets, and companionship. Over most of this period, this harvesting and exploitation had little impact on those resources. Human population was very low, and most animal and plant populations were relatively large. Animal and plant communities, populations, and species that became extinct did so from other than human causes. Only in recent centuries has man's exploitation of wildlife begun to have a deleterious effect. This was the result of rapid population growth, more efficient means of capture and kill, and expansion into new continents, especially islands and tropical areas where many species of wildlife had evolved with small, localised populations and without contact with man or his camp followers, such as dogs, cats, and rats. Western exploration and colonisation quickly created serious problems of overharvesting and overexploitation of wildlife and led to a slow development of human-caused extinctions.

However, there is increasing evidence that primitive man also had a profound impact on many species. Humans did not live in the idyllic harmony with nature that has been so rapturously portrayed by the more romantic environmentalists who question the direction of modern life and call for a new environmental ethic. At least some of the large mammals, such as mammoths and mastodons, that roamed the earth during the Pleistocene Epoch and immediately thereafter and whose disappearance has been attributed to natural causes, were forced into extinction by primitive men with primitive tools. They drove herds over cliffs, into swamps, or into box canyons, often setting massive grassland fires to assist in the drive.

It would appear that pre-civilised, or at least preindustrial, man exploited wildlife just as carelessly and as effectively as does modern man. The mythologised American Indians also used wasteful and ecologically unsound methods of hunting and killing. Regarding the Plains Indians' exploitation of the buffalo, Baden, Stroup, and Thurman point out that

> It is generally held that these highly diverse groups shared a common reverence for the land and the interdependencies of nature that provided man his niche...The actual behaviour attributed to these cultures reads like an admonition from Francis of Assisi, the patron saint of the ecology movement.

> Prior to the introduction of the horse, the hunting of bison was uncertain, and relatively unproductive. In the pre-horse period the capture of a buffalo was comparatively rare. Its biomass was highly valued and hence fully utilised.

> In effect, the introduction of the horse, steel tools and later firearms lowered the price of the animal. As the price fell due to technological adaptation, patterns of utilisation changed dramatically. During this period many buffalo were killed by Indians merely for the tongue and the two strips of back strap. By 1840 the Indian had driven the buffalo from portions of its original habitat.[1]

It is not true, however, that the Indians' "wasteful" use of wildlife occurred after the introduction of white man's tools. There is abundant evidence that the Indians engaged in wasteful harvesting methods whenever the opportunity arose. Long before they had horses and rifles, the Plains tribes regularly set vast prairie fires in the late summer and fall in order to stampede buffalo herds over cliffs or bluffs. "A successful drive produced a large number of carcasses, often more than could be used before the meat spoiled....At a large kill much of the meat spoiled before it could be processed."[2]

Fires were so commonly used that millions of acres were blackened each year. The Washo tribe of the Great Basin lived in an extremely hostile environment, barely eking out a precarious existence, yet, during the fall they would drive such enormous numbers of jackrabbits into their nets that after the skins were taken, the meat was left to rot.[3]

Overexploitation of wildlife is not a peculiar characteristic of Western man, nor is it a consequence of some form of the modern economic system or the much maligned "commerce" so frequently condemned by environmentalists. Whenever and wherever there have been incentives to overharvest or deplete wildlife it has taken place, whether by primitive or modern man.

While the environmental movement comprises a diverse amalgamation of different groups, with concerns ranging from visual "pollution" such as clear-cuts, to chemical- and smoke-induced damage to human health, the glue that binds the groups is wildlife conservation and preservation. Even here there is a wide and growing chasm among organisations. The more "conservative" groups are interested in wildlife management, such as increasing the numbers of commonly hunted species of fish and game. The "middle-of-the-roaders" are interested in developing sustained-yield management programmes for species where it is clear that wise management and international cooperation can achieve better results than ending all harvesting. The "liberals" push for an end to the exploitation of most species and a complete ban on all trade in most threatened wildlife. Finally, there are the animal rights groups, the "radicals," who value the rights of animals to life and liberty at least as highly as human rights.

The growth, influence, and public support of these organisations has increased rapidly over the past two decades. Since 1970, conservationists have painted a dismal picture of an increasing struggle for survival of wildlife, with one species after another being pushed to the brink of extinction. There is no doubt that human-caused extinctions are occurring at an increasing rate and generating momentum for the environmental movement.

Norman Myers treats the problem of disappearing species in considerable detail and emphasises the accelerated rate of extinction:

> At least 90 per cent of all species that have existed have disappeared. But almost all of them have gone under by virtue of natural processes. Only in the recent past, perhaps from around 50,000 years ago, has man exerted much influence. As a primitive hunter, man probably proved himself capable of eliminating species, *albeit* as a relatively rare occurrence. From the year A.D. 1600, however, he became able, through advancing technology, to overhunt animals to extinction in just a few years, and to disrupt extensive environments just as rapidly. Between

the years 1600 and 1900, man eliminated around seventy-five known species.... Since 1900 man has eliminated around another seventy-five known species....The rate from the year 1600 to 1900, roughly one species every 4 years, and the rate during most of the present century, about one species per year, are to be compared with a rate of possibly one per 1000 years during the "great dying" of the dinosaurs.[4]

Since 1960, human population growth and worldwide development have greatly accelerated the extinction rate, which may have reached 1,000 species per year by 1975. Myers projects that by the late 1980s it may reach one species per hour.[5] This is, indeed, a dismal picture, and it is important to recognise that it is not merely small and local populations of wildlife that have suffered. Many animal species that have disappeared or have been drastically reduced were at one time found in truly enormous numbers.

The passenger pigeon, which was native to North America, was once probably the most numerous species of bird on earth. At its peak, its population may have numbered around 3 billion. Its migrating flocks darkened the skies over towns and cities and sounded like an approaching tornado, yet they were extinct by 1914, mainly because of massive market-hunting. A similar fate befell the great auk, a large, flightless seabird that nested in vast numbers on islands in the North Atlantic. It was exterminated by whalers and fishermen who slaughtered it for food, eggs, feathers, and oil.

Many other species that survived overexploitation, although often only barely so, were slaughtered in equally staggering numbers. The Spanish explorers described the buffalo herds as a limitless "brown sea." At one time, 75 million roamed the western plains, but by 1895 there were little more than 800 left, most in captivity or on private ranches. Vast flocks of ducks, geese, prairie chickens, and shorebirds were decimated by market hunters to provide relatively inexpensive meat for the large cities. Naturalist Frank Graham, Jr., points out:

> In 1873 Chicago markets bought 600,000 prairie chickens at $3.25 a dozen. Frank M. Chapman, the ornithologist, recalled as a boy in the 1870s the glut of prairie chickens in the butcher shops....By the end of the century he had to travel to the sand hills of Nebraska to find them in any numbers.[6]

It is obvious, however, that not all natural resources or wildlife have disappeared or even been seriously depleted. Environmentalists, journalists, and writers draw our attention to the most shocking cases. But there are many species that are more common today than they were at any previous time. Many plant and animal species exist in large numbers today that were not present in North America before the arrival of the white man. Furthermore, certain animals and plants are thriving under some specific ownership and management conditions but vanishing under other conditions. It is extremely important to examine these cases in order to understand why overexploitation of some resources and wildlife takes place and why other living or renewable resources are managed on a self-sustaining basis.

Why are some species disappearing and others thriving? First, we can examine what is disappearing and what is not. Apparently, few environmentalists have taken the time to do this in their haste to catalogue extinct and vanishing species. It is true that the prairie chicken nearly vanished - the heath hen is extinct, the Atwater's greater prairie chicken has been reduced to endangered species status, and the rest are uncommon, localised, and greatly reduced in numbers- but what about other chickens? Why is the Atwater's greater prairie chicken on the endangered species list but not the Rhode Island Red, the Leghorn, or the Barred Rock? These chickens are not even native American birds. They came to this continent with the first European settlers, and the small flocks at the settlement in Jamestown, Virginia, in 1607 became the basis for a broiler industry that produces 3 billion birds a year.[7]

Similar questions can be asked: Why was the American buffalo nearly exterminated but not the Hereford, the Angus, or the Jersey cow? Why are salmon and trout habitually overfished in the nation's lakes, rivers, and streams, often to the point of endangering the species, while the same species thrive in fish farms and privately owned lakes and ponds? Why do cattle and sheep ranchers overgraze the public lands but maintain lush pastures on their own property? Why are rare birds and mammals taken from the wild in a manner that often harms them and depletes the population, but carefully raised and nurtured in aviaries, game ranches, and hunting preserves? Which would be picked at the optimum ripeness, blackberries along a roadside or blackberries in a farmer's garden? In all of these cases, it is clear that the problem of overexploitation or overharvesting is a result of the resource's being under public rather than private ownership.

The difference in their management is a direct result of two totally different forms of property rights and ownership: public, communal, or common property vs. private property. Wherever we have public ownership we find overuse, waste, and extinction; but private ownership results in sustained-yield use and preservation. Although it may be philosophically or emotionally pleasing to environmentalists to persist in maintaining that wildlife, the oceans, and natural resources belong to mankind, the inevitable result of such thinking is the opposite of what they desire.

Harold Demsetz defines communal ownership as

> A right which can be exercised by all members of the community....The community denies...to individual citizens the right to interfere with any person's exercise of communally-owned rights. Private ownership implies that the community recognises the right of the owner to exclude others from exercising the owner's private rights....

> Suppose that land is communally owned.... If a person seeks to maximise the value of his communal rights, he will tend to overhunt and overwork the land because some of the costs of his doing so are borne by others. The stock of game and the richness of the soil will be diminished too quickly....

> If a single person owns the land, he will attempt to maximise its present value by taking into account alternative future time streams of benefits and costs and selecting that one which he believes will maximise the present value of his privately-owned land rights...It is very difficult to see how the existing communal owners can reach an agreement that takes account of these costs.[8]

It is important to recognise that this distinction between the destructive overuse of common property resources and careful sustained-yield use of private property resources does not merely apply to a comparison of wild with domesticated populations of plants and animals.

Common property problems involving wildlife have been especially prevalent in America, and they continue to be extremely vexing precisely because of American wildlife law. The President's Council on Environmental Quality has stressed that "under U.S. law, native wildlife belongs to the people; it is not private property, even on private land."[9] Victor B. Scheffer points out that

> Wildlife in the United States is the property of the people...No animal may be reduced to private ownership except by permission of the state. Wild animals do not belong to the owner of the land upon which the animals live, though the owner can post his land against entry and thus restrain the public from using its property.[10]

In Europe, native wildlife often belongs to private landowners or is managed under a combination of private and public property. Some European countries have fewer problems of overexploitation of wildlife, regardless of population pressures and economic and political systems. In other words, we find precisely what economic analysis has predicted about the treatment of common property and private property wildlife resources.

The salmon fishery provides a nearly perfect example of the differences between private and common property management. Salmon are anadromous fish. They hatch in the clear, shallow waters of the upper reaches of rivers, go downstream to the sea where they grow to maturity, and then return upstream to spawn another generation in the same rivers where they were hatched. Management of a fishery should be a relatively low-cost operation because the only requirements are to maintain a high-quality spawning environment and to prevent overfishing. The fish don't need to be fed because they grow to maturity in the sea and return as a highly valuable source of protein.

Yet most of the salmon in the United States have either disappeared from their ancestral rivers or are being rapidly depleted precisely as a result of their being treated as part of the common heritage. As a common property resource, they belong to everyone, can be caught by everyone, and essentially belong to no one. They run a gantlet of competing users at every stage of their migration. They are sought at sea by commercial fishing boats, off the river mouths by sports fishermen, in the rivers by netting operations, and upstream by rod fishermen and Indians, all of whom are interested in taking all the salmon they can before others do. In addition, foreign fishing fleets take all the salmon they can beyond the 200-mile limit. Furthermore, the rivers are also common property; dams and logging operations often make them impassable, and pollution makes them uninhabitable.

Because no one owns the salmon, each user is pitted against all other users, and the result has been a rapid depletion of the stock. As each group catches fewer fish, they turn to the government for special legislation that will limit, control, and regulate the catch and the fishing techniques used by their competitors. As a result, fishing techniques have become increasingly inefficient and costly. But this still does not prevent overfishing.

Douglass C. North and Roger LeRoy Miller have commented on the alarming decline of the Bristol Bay salmon fishery in Alaska.

> In an attempt to reverse the trend, regulations over the season, fishing hours, boats, and gear increased in complexity.
>
> Naturally, the results in real life are more like a nightmare than a fairy tale. Fishermen are poor because they are forced to use inefficient equipment and to fish only a small fraction of the time, and of course there are far too many of them. The consumer pays a much higher price for red salmon than would be necessary if efficient methods were used. Despite the ever-growing intertwining bonds of regulations, the preservation of the salmon run is still not assured.
>
> The root of the problem lies in the current non-ownership arrangement. It is not in the interests of any individual fisherman to concern himself with perpetuation of the salmon run. Quite the contrary: It is rather in his interests to catch as many fish as he can during any one season.[11]

The consequence of treating the salmon as common property has been to destroy the salmon resource and to waste economic resources through regulations. John Baden describes the results:

> The regulations are such that it is increasingly more difficult and more expensive to obtain salmon. . . . The state and federal regulations are now such that salmon....have a negative social value, if one includes the cost of government management of the industry. Given current institutions, society would be better off if the salmon disappeared; it costs society more to get a salmon than it is worth for food.[12]

The environmental literature has reported at length on the sad plight of the salmon fisheries; but many authors appear either to be unaware of common property theory or, worse, to ignore it because it clashes with their philosophical view that wildlife should belong to everyone. So we continue to read their attacks on profit-seekers, free enterprise, big business, and a consumptionist lifestyle. And we continue to watch the salmon disappear.

Fortunately, salmon are a highly desirable fish, and American entrepreneurs are attempting to remove them from the common property trap, even though they generally run afoul of our wildlife laws, which act to prevent private property solutions. Obviously, under a common property system, there is no incentive for any user to restock the fishery by creating a private hatchery because everyone else would merely receive a cost-free benefit from his investment. But attempts at salmon mariculture or farming are being made where private owners can avoid the problem of having their fish migrate along common property rivers.

One such effort is being made by Bay Center Mariculture Company in Washington State. Until recently, raising salmon in ponds in order to market them for profit was not permitted in Washington, although taxpayer-supported state fish hatcheries were used to help alleviate overexploitation. Bay Center reports:

It has been discovered in recent years that salmon go through their early development and growth much faster in brackish water than in the fresh water we are accustomed to expect them in during their early life. We take advantage of this by both sea-farming salmon and raising them to market size in controlled ponds. Because of the specificity of salmon to return to the spot of their birth, it is possible to release them at an early age and allow them to mature in the sea and return to the hatchery site a year earlier than they would if raised in nature in freshwater streams. The expected return is a very small per centage, but still very profitable.[13]

In 1977 Weyerhauser initiated a similar system in southern Oregon, where the young salmon will be raised in ponds at the head of a bay and then released into the bay, whence they will go out to sea. They will not be able to prevent fishermen from catching the returning salmon at the mouth of the bay, but Weyerhauser expects to make a profit if one per cent of the adult salmon return to the company's fish ladders. In the wild, approximately 2 to 5 per cent of the salmon survive to return to their spawning grounds.

Such programmes not only solve wildlife conservation problems at private rather than at taxpayer expense, but they reduce the take on the wild common property populations and, as a positive externality, provide food for the wild food chain.[14]

Outside the United States we find a strikingly different situation. In Iceland and in some northern European countries, the salmon fishery is in much healthier shape because the rights to the salmon or the salmon rivers are privately owned. Some of the finest stretches of rivers are owned or leased by individuals, groups of fishermen, or fishing lodges, and the salmon are not overfished. It is in the economic self-interest of the resource owners to conserve the salmon. Limits are effectively placed on the number of fish that can be caught, enough fish are released to maintain a healthy population, and the owners carefully protect their streams and see that agricultural and grazing activities do not adversely affect the quality of the water.

In Iceland private property rights have also been extended to the common eider, a large sea-dwelling duck of the North Atlantic. The eider supplies meat, eggs, and especially eiderdown (a brownish down under the breast feathers of the female that she plucks for her nest to insulate the eggs) for the farmers and local populations. As early as 1281, the civil and ecclesiastical codes stated that the eiders "belong to the occupiers of the lands where they occurred." The farmers have protected the eider nesting colonies for centuries and actually farm them. Robin W. Doughty writes:

> Skuli Magnusson, pioneer agriculturalist and industrialist, and others, who protected and farmed the nesting places of wild eiders for down and eggs, promoted the concept of farming. In the 1770s, Magnusson protected a very large colony on the island of Videy, where reportedly he gathered and cleaned about 90 pounds of down from his "favourite" birds. On a visit to the same place in 1810, Sir George MacKenzie noted that eider ducks were "assembled in great numbers to nestle," and that severe penalties were imposed on persons killing them.[15]

The number of such farming operations grew during the nineteenth century and peaked during the 1920s, when there were more than 250 farms. Because many farmers have moved into towns, the number of farms has declined, currently totaling about 200, and eiderdown production is now about half of its peak figures. One result of this decline has been an increase in predation on the eiders by gulls, ravens, and feral mink and fox. Even though the number of wild predators has increased, the eider are still carefully protected on the existing farms, where property owners shoot and poison predators.

Iceland's management of the wild eider as a private property resource has been a great success. The private eider farms have benefited both the property owners and the eider population. The farmers have protected the birds from overexploitation, from poachers, and from natural predators. They have also created artificial nesting sites in which the female will nest. The combined provision of protected nesting areas and artificial nesting sites has served to maintain a thriving population.

The individual self-interest of private property owners has caused them to protect and carefully manage a valuable wildlife resource. Because private property rights were extended to the wild eider, the harvesters of eiderdown and eider eggs were allowed to manage the resource in a nonconsumptive and nondestructive manner. If they had killed the eider to obtain the down, they would have been killing the fabled golden goose. Instead, they harvested the resource on a sustained-yield basis and realised a far greater return.

If the eider had been treated as a common property resource, the only way the Icelanders could have captured any economic value from the resource would have been to take all they could before other users did the same. It would not have been profitable to wait for the eider to line their nests with down; someone else might have collected it first. The rational course of action for each user would have been to kill the eider and immediately appropriate all of the down. It also would not have been in anyone's self-interest to invest in conservation programmes, such as nest site construction or predator control. All of the other down collectors would have benefited from the actions of the conservationist. We would have seen the creation of a new fable: "Who killed the eider that provided the golden down?"

It is also instructive to compare wildlife ownership and management in Great Britain with that in the United States. Throughout much of British history, the right to harvest wildlife belonged to owners of the land.[16] For centuries landowners in England were entitled by law to the wildlife on their lands. This was supported by "qualification statutes," which determined the amount of land or income that was necessary in order to have the right to own and harvest game. The landowner owned the game, and he had the right to what was supported by his land. It was recognised that the economic incentives engendered by the right to own wildlife often led the landowners not to use their land for agricultural purposes that would be harmful to wildlife. To the extent that wildlife was valued as a source of economic gain, or even of hunting pleasure, wildlife would be protected, and it would be hunted on a sustained-yield basis to ensure the preservation of breeding stock and the continuous replenishment of the population. Landowners even employed gamekeepers who managed the game, protecting it against predators, preserving habitat, and guarding against poachers.[17]

The evidence indicates that there have been far fewer problems of vanishing wildlife in Britain than in the United States. We can still witness the benefits of the private ownership system in the management of Britain's most famous game bird, the red grouse. The opening of grouse season each August is reverently referred to as the "Glorious Twelfth." So highly prized and desired are the grouse that restaurants throughout the London area compete to see which can deliver the first meal of grouse. Commercial airlines, private jets, and even helicopters are used in the race, and the resulting roast grouse may cost $85.

The grouse are clearly a valuable resource, and they provide a substantial income to the landowners of the great estates in Scotland, where wealthy hunters gather to participate in the hunts. Reservations for the shoots are required well in advance, and social status and wealth may be necessary in order to hunt on the best and most productive lands. Landowners obviously have incentives to carefully manage the grouse on a sustained-yield basis, to keep the habitat in prime condition, to prevent the heather from being burned, to control predators, and to keep out poachers. As a result, the birds are not overexploited, in spite of heavy shooting and even though commercial marketing is allowed. Compare the results of this system with the overexploitation of the American prairie chicken.

Unfortunately, because so much of the ownership of land in Britain, especially the great estates, was based on government grants of privilege, the landowner's property right to wildlife increasingly came under attack as a special privilege accorded to the wealthy. In the eighteenth century, the great egalitarian and democratic debates over property rights in land and wildlife resulted in America's rejection of the English system. Private property rights in wildlife and game were viewed as part of an undemocratic system emanating from government grants to the ruling classes. But in rejecting them, Americans threw out the baby with the bath water. The English system was attacked because it supposedly benefited the rich at the expense of the poor, and it was held that wildlife in America should belong to all. From the earliest time American law adopted a policy of "free taking," which recognised everyone's right to take game. Only the law of trespass kept anyone from entering another's property, and its efficacy was severely restricted by the requirement that the land be posted against hunting if the owner wanted exclusive use of it.

Thomas A. Lund essentially approves of the egalitarian approach to wildlife law. Nevertheless, he states:

> Early American law sought to dispatch the antidemocratic vices of its legacy while at the same time to preserve the system's virtues...Following these seemingly harmless improvements to the English system, American wildlife populations fell as if afflicted by a plague. While the incursion of new settlers into habitat played a role in this decline, subsequent proof that wildlife can coexist and even thrive alongside agricultural development shows that the spread of primitive farming and ranching does not bear principal responsibility. Instead the affliction upon wildlife must be attributed to a surprising source: those democratic policies that were injected into wildlife law. Their unforeseen impact was to undercut totally the bases upon which English wildlife law had been so effective.[18]

Another especially illustrative example of private property rights in wildlife appears in the Montagnais Indians of Quebec and Labrador.[19] The Montagnais dwelled in the forests of the Labrador Peninsula, hunting such fur-bearing animals as caribou, deer, and beaver. They treated wildlife as a common property resource, with everyone sharing in the bounty of the hunt. Because game was plentiful and the Indian population was relatively low, the common property resource system was able to work. "The externality was clearly present. Hunting could be practiced freely and was carried on without assessing its impact on other hunters. But these external effects were of such small significance that it did not pay anyone to take them into account."[20]

However, with the arrival of the French fur traders in the 1600s, the demand for beaver began to rise rapidly. As the value of the furs rose, there was a corresponding increase in the exploitation of the resource. Increasing use of the common property resource would have led to overexploitation of the beaver. However,

> With the beaver increasing in value, scarcity, depletion, and localised extinction could be predicted under the existing system of property

rights. But unlike the buffalo, virtually condemned to extinction as common property, the beaver were protected by the evolution of private property rights among the hunters. By the early to mid-18th century, the transition to private hunting grounds was almost complete and the Montagnais were managing the beaver on a sustained-yield basis.[21]

It was a highly sophisticated system. The Montagnais blazed trees with their family crests to delineate their hunting grounds, practiced retaliation against poachers and trespassers, developed a seasonal allotment system, and marked beaver houses.

Animal resources were husbanded. Sometimes conservation practices were carried on extensively. Family hunting territories were divided into quarters. Each year the family hunted in a different quarter in rotation, leaving a tract in the centre as a sort of bank, not to be hunted over unless forced to do so by a shortage in the regular tract.[22]

This remarkably advanced system lasted for over a century and certainly served to prevent the extinction of the beaver. Unfortunately, more whites entered the region and began to treat the beaver as a common property resource, trapping them themselves rather than trading with the Indians, and the beaver began to disappear. Finally, the Indians were forced to abandon their private property system and joined the whites in a rapid overexploitation of the beaver. Baden, Stroup, and Thurman sum up this sorry return to a common property system:

With the significant intrusion of the white trapper in the 19th century, the Indians property rights were violated. Because the Indian could not exclude the white trapper from the benefits of conservation both joined in trapping out the beaver.... In essence, the Indians lost their ability to enforce property rights and rationally stopped practicing resource conservation.[23]

Another example of how private ownership can successfully preserve wildlife is found on game ranches, hunting preserves, safari parks, and animal and bird farms. Many of these private ventures, especially the game ranches, were established to generate profits from private hunting. Consequently, there has been a tremendous outcry from environmentalists and conservationists because the animals are raised for profit and some of them are killed. Yet, if emotional responses can be put aside, it seems clear that these game ranches produce many positive results. Many of the animals they stock are rapidly disappearing in their native countries because of pressures resulting from a rapidly expanding human population. Native habitats are disappearing through the encroachment of agriculture, cattle grazing, timber harvesting, and desertification arising from overexploitation of common property water resources, overgrazing of grasslands, and overutilisation of brush, scrub, and trees for firewood and shelter. So serious are these problems and so insoluble under a common property system that there is little hope of saving many species of wildlife in the developing countries. Indeed, some of the more spectacular and most sought-after big-game mammals may now have healthier and more stable populations on some of the game ranches than in their native countries.

As human population growth accelerates throughout the developing world, as annual incomes hover at the subsistence level, and as rising costs of petroleum and agricultural fertilisers and chemicals continue to compound the misery, fewer and fewer internal resources will be available to preserve wildlife. In spite of the noble intentions of India and some African and South American nations to preserve their vanishing wildlife, there are few signs of any real accomplishments. Impoverished people have little patience with elephants rampaging through their meagre crops or with lions, cheetahs, and leopards preying on their livestock. Poachers take a terrible toll, for food as well as for ivory and spotted cat skins that bring high prices on the black market. And there is ready evidence that government officials in many of these countries, while publicly showing great concern for wildlife, are profiting handsomely from the illegal trade. If these common property resources are going to be depleted anyway, what incentives do they have to act otherwise?

Many environmentalists bemoan the fact that the once free-roaming animal herds of the African continent are now kept in captivity for the benefit of American hunters and safari park visitors. But free-roaming is a relative concept. These animals are certainly free-roaming within the boundaries of the game ranches, and many of these ranches are enormous. Furthermore, as the African plains are increasingly delimited with hostile political boundaries and with warring armies and starving populations, there seems little point in mounting campaigns against so-called immoral game ranches and preserves. The growth of agriculture and cattle ranching in these countries is also restricting the free-roaming nature of the wildlife herds.

If the profits gained by giving hunters access to exotic game can provide the economic incentive for these landowners to manage the animals on a sustained-yield basis, some species will be saved. The same holds for the profits to be derived from visitors to game parks and preserves. In fact, the protection provided at some of the parks, preserves, and gardens has actually produced a glut of some animals. There have been well-publicised efforts by some preserves to return their surplus animals to Africa. Lions from America have even been taken to Africa to appear in movies that were filmed there. While we read of zoological parks attempting to discover reversible birth control techniques in order to control their tiger populations, we continue to read about the never-ending difficulties of preserving the remaining tigers in the wild. Perhaps we should judge all of these activities by their achievements rather than by their motives, for it may turn out that in the future the developing countries will be restocked with their native fauna from specimens now thriving on game ranches and preserves.

It is important not to cloud the issue of common vs. private property resources with philosophical judgments regarding competing economic systems or with attacks on a free-market system or profit-seeking activities. The books on disappearing wildlife abound with stories of captive breeding of birds and mammals and the successful preservation of species that have either become extremely rare or have disappeared entirely in the wild.

Among the many mammals in this category are Père David's deer, which does not exist in the wild and was "discovered" living in the royal zoological park in Peking. It has been saved from extinction by breeding in zoos and private parks. The European bison, or wisent, was reduced to three animals in 1927 and now survives in preserves in Poland. There are similar stories for the Asiatic lion, Prsewalski's Mongolian wild horse, the Arabian oryx, the wild ass or onager, and the Bactrian camel.

The same is true for many members of the parrot family. The familiar budgerigar, or budgie, is commonly kept as a pet in the United States and is bred in enormous numbers by thousands of breeders. Practically the entire trade is supplied by captive-bred birds. This demonstrates another conservation aspect of extending private property rights to wildlife, as captive breeding can supply the market demand for the birds and reduce or eliminate the demand on wild populations.

In all of these examples it is clear that the single most important element in wildlife survival was their removal from common property ownership. Under private property ownership, others were prevented from exploiting the resource, and there were incentives for the owners to preserve them. Furthermore, these incentives were not solely motivated by the possibility of economic gain. With the exception of game ranches, economic gain has seldom been the primary motivation behind most captive breeding projects. Many of these examples were fostered for the pleasure of owning and breeding attractive or rare wildlife, as well as for more "altruistic" reasons, such as a deep commitment to the preservation of vanishing wildlife. Private ownership includes not only hunting preserves, commercial bird breeders, parrot jungles, and safari parks, it also includes wildlife sanctuaries, World Wildlife Fund preserves, and a multitude of private, nonprofit conservation and preservation projects.

The problems of environmental degradation, overexploitation of natural resources, and depletion of wildlife all derive from their existence as common property resources. Wherever we find an approach to the extension of private property rights in these areas, we find superior results. Wherever we have exclusive private ownership, whether it is organised around a profit-seeking or nonprofit undertaking, there are incentives for the private owners to preserve the resource. Self-interest drives the private property owners to careful management and protection of their resource.

It is important not to fall into the trap of believing that the different results arising from these two forms of resource management can be changed through education or persuasion. The methods of using or exploiting the resources are inherent in the incentives that are necessarily a part of each system. The overuse of common property resources and the preservation of private property resources are both examples of rational behaviour by resource users. It is not a case of irrational vs. rational behaviour. In both cases we are witnessing rational behaviour, for resource users are acting in the only manner available to them to obtain any economic or psychological value from the resource.

It has nothing to do with the need for a new environmental ethic. Asking people to revere resources and wildlife won't bring about the peaceable kingdom when the only way a person can survive is to use up the resource before someone else does. Adopting a property system that directs and channels man's innate self-interest into behaviour that preserves natural resources and wildlife will cause people to act as *if* they were motivated by a new conservationist ethic.

Any resource held in common - whether land, air, the upper atmosphere and outer space, the oceans, lakes, streams, outdoor recreational resources, fisheries, wildlife, or game - can be used simultaneously by more than one individual or group for more than one purpose with many of the multiple uses conflicting. No one has exclusive rights to the resource, nor can any one prevent others from using it for either the same or any non-compatible use. By its very nature a common property resource is owned by everyone and owned by no one. Since everyone uses it there is overuse, waste, and extinction. No one has an incentive to maintain or preserve it. The only way any of the users can capture any value, economic or otherwise, is to exploit the resource as rapidly as possible before someone else does.

But private ownership allows the owner to capture the full capital value of the resource, and self-interest and economic incentive drive the owner to maintain its long-term capital value. The owner of the resource wants to enjoy the benefits of the resource today, tomorrow, and ten years from now, and therefore he will attempt to manage it on a sustained-yield basis.

Given the nature of man and the motivating forces of self-interest and economic incentive, we can see why the buffalo nearly vanished, but not the Hereford; why the Atwater's greater prairie chicken is endangered, but not the red grouse; why the common salmon fisheries of the United States are overfished, but not the private salmon streams of Europe.

It should be equally clear that the analysis applies to broader environmental issues. Many of the most beautiful national parks are suffering from severe overuse and a near destruction of their recreational values, but most private parks are maintained in far better condition. The public grazing lands have been repeatedly over-grazed, while lush private grazing lands are maintained by private ranchers. National forests are carelessly logged and overharvested, but private forests are carefully managed and cut on a sustained-yield basis, and costly nursery tree farms have been developed. In addition, the basic concept of self-interest explains why people don't litter their own yards but do litter public parks and streets, and why people don't dump old refrigerators and tires in their own farm ponds or swimming pools, but repeatedly dump them in the unowned streams, rivers, and swamps.

Perhaps the most important treatment of the common property syndrome was that of the noted biologist and environmentalist, Garrett Hardin:

> Picture a pasture open to all. It is to be expected that each herdsman will try to keep as many cattle as possible on the commons. Such an arrangement may work reasonably satisfactorily for centuries because tribal wars, poaching, and disease keep the numbers of both man and beast well below the carrying capacity of the land. Finally, however, comes the day of reckoning, that is, the day when the long-desired goal of social stability becomes a reality. At this point, the inherent logic of the commons remorselessly generates tragedy. As a rational being, each herdsman seeks to maximise his gain. Explicitly or implicitly, more or less consciously, he asks, "What is the utility *to me* of adding one more animal to my herd?" The utility has one negative and one positive component.
>
> 1. The positive component is a function of the increment of one animal. Since the herdsman receives all the proceeds from the sale of the additional animal, the positive utility is nearly + 1.
>
> 2. The negative component is a function of the additional overgrazing created by one more animal. Since, however, the effects of overgrazing are shared by all the herdsmen, the negative utility for any particular decision-making herdsman is only a fraction of -1.
>
> Adding together the component partial utilities, the rational herdsman concludes that the only sensible course for him to pursue is to add another animal to his herd. And another, and another...But this is the conclusion reached by each and every rational herdsman sharing a commons. Therein is the tragedy. Each man is locked into a system that

compels him to increase his herd without limit-in a world that is limited. Ruin is the destination toward which all men rush, each pursuing his own best interest in a society that believes in the freedom of the commons. Freedom in a commons brings ruin to all.[24]

Unfortunately, the environmentalists have either overlooked the economic analysis of common property resources or they have deliberately ignored it because of the profound difficulties it raises for a philosophy that is opposed to private ownership of natural resources and wildlife. Certainly, the former is true, but there is also evidence of the latter, since some of the more scholarly treatments of wildlife problems refer rather fleetingly to the problems of common property resources and to Hardin's analysis. But they mainly refer to Hardin when they are following his proposals for limiting human population growth to preserve the natural world. Meanwhile, the logic of the commons continues to generate a remorseless loss of the world's wildlife.

Where, then, do most environmentalists place the blame for our environmental problems? Most environmental literature and treatments by the public media consistently repeat attacks on man's greed, self-interest, consumerism, piggishness, the profit system, the market economy, and Western political, legal, economic, religious, and property institutions and beliefs. *Newsweek* points to "entrepreneurial greed." The *New York Times* mentions "commercial exploitation." Michael Satchell writes, "The road ahead seems to lead to more extinctions on the altar of consumerism.[25]

This level of thinking is especially prevalent in popular magazines published by the major wildlife, conservation, and environmental organisations, ranging from the Audubon Society to the World Wildlife Fund. A typical example, written by the editor, John Strohm, appeared in a recent issue of *International Wildlife*. He asked why there were growing problems of over-fishing, over-cutting, over-grazing, over-ploughing, and disappearing species of plants and animals. Following a discussion of too many people, too many wasteful demands, and the haves vs. the have-nots, he wrote: "We must eliminate waste...and adopt a more frugal lifestyle. You cut a tree, you plant a tree. You catch a fish, you leave some behind so you can fish tomorrow."[26]

But even the most serious, scholarly books on wildlife conservation fail to rise much above this level. Norman Myers, in *The*

Sinking Ark, writes about how we like hamburgers, our consumerist lifestyle, our desire to be consumers and fat cats, and our love of inexpensive foreign beef.[27] David Ehrenfeld, in *Conserving Life on Earth,* finds the problem to be economic success and increases in consumption. He calls for the disassociation of progress from growth, blames the pet trade, attacks private land ownership as unplanned, and points out that one of his students had discovered that people believe that China is the only country that has successfully coped with environmental problems. Ehrenfeld calls for the creation of a radically altered economic system that would produce labour-intensive and non-polluting goods, such as guitars, rather than goods that are capital-intensive and highly polluting, such as snowmobiles.[28]

These examples make one wonder whether the intellectual leaders of the environmental movement are more interested in a visionary society patterned after Rousseau and Thoreau than in grappling with the difficult problems of finding a method of conserving life on earth.

Two major themes repeatedly appear in the environmental literature as the source of all our problems with environmental preservation and conservation: commerce and the free-enterprise or free-market economic system. We find emotional references to the role of commerce in the overexploitation of wildlife. Victor Scheffer has written:

> By market hunting I mean the trapping and clubbing of fur-bearers, the shooting of animals for the pet-food industry, the live-capturing of rabbits for coursing, the capturing of hawks for sale to falconers, and similar pursuits of wild birds and mammals for commercial ends. The Friends of the Earth organisation has already taken the position that the use of wild animal products as objects of commerce should be discouraged.[29]

In *Time of the Turtle,* Jack Rudloe describes a conversation with Dr. Archie Carr, widely recognised as the world's foremost sea turtle scientist:

> Inevitably a discussion on the morality of eating this heavenly tasting, but nearly extinct creature arose. Was it wrong to eat it when we knew the endangered status of the green turtle?
>
> "That touches on a very deep and fundamental problem" said Archie in a defensive tone. "If we had gone out and bought it, paid cash for it, and encouraged its commercial sale, then we would have been wrong. More than anything else, the commercialisation of turtle meat

and products has pushed the species to extinction. It isn't the Indian eating a few turtles on the beach for subsistence. At least five thousand turtles each year are being slaughtered in Nicaragua right up the coast from us. Thousands used to be shipped into the Keys for soup. And last year thirty thousand pounds of meat came in from Colombia, Mexico, and even from the Middle East to the United States. It isn't coastal people sharing the meat among themselves in the village that's the problem, it's that mass marketing that's going on in the world trade that is."[30]

However, it seems totally inappropriate to refer to the harvesting of unowned wild populations as commerce. Environmentalists and scientists make exceptions for wildlife harvesting by native populations and Indians, whether it is for food, clothing, medicine, ornamentation, religious rites and ceremonies, or for pets. Yet, when European exploration and settlement began to affect wildlife populations by providing food, clothing, and ornamentation for growing settlements, it was referred to as commerce. It is somehow suggested that exchanging wildlife for money is somehow immoral.

The overharvesting of the earth's wildlife can no more be labelled commerce than the primitive gathering of wild fruits, vegetables, and grains can be called agriculture - even if those products are subsequently sold in the marketplace. Jacques-Yves Cousteau has written: "Unlike livestock rearing and fish farming, fishing and hunting are really sheer pillaging of the environment and quickly lead to the destruction of the very resources they seek to cull."[31]

The disappearance of wildlife has nothing to do with commerce or a lack of reverence for wildlife. The more rapid disappearance of common property wildlife during the past century is due to the fact that much larger human populations are using it, more efficient means of capture and kill are employed, and a larger number of uses have been found for the resources. Furthermore, many of the most populous species of wildlife had been reduced to a severely depleted state long before the development of modern business and commerce.

The second common theme used by most environmentalists is the private enterprise system and the free market. Myers has written at length about the evils of the system:

But within the American system, with its emphasis on private profit to be derived from private property, common property of society's heritage gets short shrift; farmers' interests are allowed precedence over the nation's needs.

The guts of the issue are that private interests do not necessarily run parallel with public interests. In certain circumstances, private interests undermine public interests. This runs counter to much conventional wisdom concerning a free-market economy, as exemplified by Adam Smith's dictum that, through pursuing his own interests, an individual is "led by an invisible hand to promote the public interest."[32]

In the concluding section of Myers's book, he develops his strategy for conservation:

> Much decline of species is due to the deficiencies and failures of the market-place system, which sometimes favours the here-and-now needs of private individuals to the detriment of long-term needs of the community in general. Moreover, the market-place tends to ignore the value of resources without a price tag.... Governments should step in to take account of the fact that the open market-place does not cater for all economic needs of society, and especially that the market-place ignores and even depletes the common heritage Species.[33]

Ehrenfeld argues from a nearly identical position:

> The greatest barrier to the implementation of a strong and unified conservation policy is the difficulty of protecting those parts of the public domain that have traditionally been exploited by private interests. Both Keynes and later Hardin have explained that there is no logically consistent reason why an individual, corporation, or nation acting in self-interest should voluntarily abstain from exploiting the public domain even when the result will be certain destruction of the valuable features of that domain. Hardin, in fact, goes somewhat farther in "The Tragedy of the Commons" by asserting that it is usually damaging to private interests (at least in the short run) to act for the collective good as far as the commons are concerned.
>
> If private interests cannot be expected to protect the public domain, then external regulation by public agencies, governments, or international authorities is needed. If that regulation is effective, the commons will be managed to provide the maximum *sustained* yield of natural products, which in turn will ultimately maximise the sum total of the profits for the various interests that rely on the commons. This concept, simple in theory, is difficult to put into practice.[34]

It is difficult to decide how to deal with Myers's and Ehrenfeld's critiques of free enterprise and the tragedy of the commons. One can understand their philosophical objection to private ownership of property and to the free-market economy, yet it is difficult to deal with their analyses of the tragedy of the commons. Nowhere does Hardin state that the tragedy of the commons is the result of free enterprise, the profit system, or the existence of private property.[35] Since common property is the antithesis of private property, there is simply no way in which private property can be the cause of the tragedy. Hardin clearly stresses that it is by treating a resource as common property that we become locked in its inexorable destruction. The conservationist-minded common property resource user has no more incentive to restrain his overexploitation of the resource than the most shortsighted and greedy common property resource user has. That is precisely why such a system remorselessly generates tragedy.

Regarding the whaling industry in the ocean commons, Hardin mentions the Japanese and the Russians.[36] But the ocean's common property resources are being overexploited by the ships of all nations, regardless of economic and political systems. It is certainly true that ships sailing out from the United States are seeking to take all the fish they can, but ships from socialist and communist nations are following a similar course. Indeed, because individual profit-and-loss calculations play such a small role in the economic activities of socialist nations, there is less incentive for their fleets to desist in their overexploitation when it becomes economically unprofitable than there is for a private American fishing boat. One can only conclude that writers like Myers and Ehrenfeld have either completely failed to grasp the concept of the commons or they have deliberately misrepresented it in order to support their philosophical positions. It is true that Hardin is somewhat ambiguous in recommending a solution to the tragedy of the commons. He points out that there is no technological solution and that a political solution must be found that will create some form of ownership for the earth's resources. While he does not explicitly recommend private property rights, he appears to lean in that direction in his discussion of parks as commons:

> What shall we do? We have several options. We might sell them off as private property. We might keep them as public property, but allocate the right to enter them. The allocation might be on the basis of wealth, by the use of an auction system. It might be on the basis of merit, as defined by some agreed-upon standards. It might be by lottery. Or it

might be on a first-come, first-served basis, administered to long queues. These, I think, are all the reasonable possibilities. They are all objectionable. But we must choose-or acquiesce in the destruction of the commons that we call our National Parks.[37]

At least in certain areas Hardin views the creation of private property rights as a solution: "The tragedy of the commons as a food basket is averted by private property, or something formally like it."[38]

Unfortunately Hardin swings in the other direction in his treatment of the ocean commons:

Faced with the tragedy of the commons, we can have only one rational response: change the system. To what? Basically, there are only two possibilities: free enterprise and socialism. Free enterprise in the oceans would require some sort of fences, real or figurative. It is doubtful if we can create territories in the ocean by fencing. If not, we must-if we have the will to do it-adopt the other alternative and socialise the oceans: create an international agency *with teeth*. Such an agency must issue not recommendations but directives; and enforce them.[39]

It is possible that in perceiving the difficulties involved in solving the intricate boundary problems Hardin seems to acquiesce to some form of state or socialist ownership and control of ocean resources. Hardin does stress that our experience with the League of Nations and the United Nations gives us little hope that a system of state ownership would succeed. Indeed, the decade-long debates at the Law of the Seas conferences have repeatedly been mired over the establishment of suprastate authority over the manganese nodules carpeting vast areas of the ocean bottom - a far more tractable problem than managing the highly fugitive wildlife resources of the oceans. However, the problem of "fencing" the seas may not be as difficult to solve as Hardin imagined. North and Miller have written:

Notice that, until recently, it would have been immeasurably difficult, or even impossible, to maintain and enforce private rights in the ocean, but the invention of modern electronic sensing equipment has now made the policing of large bodies of water relatively cheap and easy. Through the centuries it has often become feasible for common property to give way to private property precisely because technology has made possible the enforcement of private rights (exclusivity). We are not saying that making the oceans into private property is "good." We are saying that doing so would lead to more output and fewer ecological disasters....[40]

The problems of overexploitation and extinction of wildlife appear to derive consistently from their being treated as a common property resource. Example after example bears this out. It is also predicted by the economic analysis of common property resources. Both the economic analysis of common property resources and Hardin's treatment of the tragedy of the commons suggest that the only way to avoid the tragedy of the commons in natural resources and wildlife is to end the common property system by creating a system of private property rights.

Private property rights have worked successfully in a broad array of cases to preserve wildlife and resolve the tragedy of the commons. Experience and the logical implications of common property resource theory suggest that private property rights are far superior to state or public property rights partly because of the unambiguous exclusivity of private property rights and the difficult problem of preventing too many from using the public domain under a system of state ownership. Furthermore, private property owners have a direct and immediate incentive not to mismanage their own property, while government owners or managers do not have the same incentives, nor are there many incentives that prevent all of the public from overusing the resources held in the public domain.

It seems that Hardin's proposal that resolution of the tragedy of the commons comes down to a choice between private ownership or government ownership is insufficient. State ownership appears to be little more than a more regulated commons. We witness the same overuse and destruction of the public domain as we do in the purest commons.

The proper path toward resolving the vexing issues of wildlife conservation lies in removing wildlife from common property resource treatment and creating private property rights. This entails an outright rejection of the concept that wildlife should be viewed as the common heritage of all mankind. It also poses a direct challenge to the basic philosophical beliefs of many environmentalists. But if we are to resolve the tragedy of the commons and preserve our natural resources and wildlife, we must create a new paradigm for the environmental movement: private property rights in natural resources and wildlife.

Notes

1. John Baden, Richard L. Stroup, and Walter Thurman, "Good Intentions and Self-interest: Lessons from the American Indian", in *Earth Day Reconsidered*, ed. John Baden (Washington, D.C.: The Heritage Foundation, 1980), pp. 7-9.

2. Francis Haines, *The Buffalo* (New York: Thomas Y. Crowell, 1970), p. 20.

3. James F. Downs, The Two Worlds of the Washo: An Indian Tribe of California and Nevada (New York: Holt, Rinehart and Winston, 1966).

4. Norman Myers, *The Sinking Ark* (Oxford: Pergamon Press, 1979), p. 4.

The distinction between natural and human-caused extinction seems to be a philosophically loaded one. Once we make human beings some special type of causative agent apart from the rest of nature, it is very easy to begin to single out specific groups of humans as "good" or "bad," e.g., conservation organisations are good and multinational timber companies are bad. What does it matter to the species whether its extinction results from natural or human causes?-it is just as dead.

5. *Ibid.*, pp. 4-5.

6. Frank Graham, Jr., *Man's Dominion: The Story of Conservation in America* (NewYork: M. Evans and Co., 1971), pp. 23-24.

7. Robert E. Cook, Harvey L. Bumgardner, and William E. Shaklee, "How Chicken on Sunday Became an Anyday Treat," in *The 1975 Yearbook of Agriculture: That We May Eat* (Washington, D.C.: Government Printing Office, 1975), p. 125.

8. Harold Demsetz, "Toward a Theory of Property Rights," *American Economic Review* 57 (May 1967): 354-56.

9. Council on Environmental Quality, *The Sixth Annual Report of the Council on Environmental Quality* (Washington, D.C.: Government Printing Office, 1975), p. 257.

10. Victor B. Scheffer, *A Voice for Wildlife: A Call for a New Ethic in Conservation* (New York: Charles Scribner's Sons, 1974), p. 138.

11. Douglass C. North and Roger LeRoy Miller, *The Economics of Public Issues*, 2nd ed. (New York: Harper & Row, 1973), pp. 109-11.

12. John Baden, *Persona Grata: An Interview with John Baden*, World Research Ink (November-December 1979), p. 22.

13. Bay Center Mariculture Company brochure, Bay City (Willapa Bay), Washington

14. Another interesting experiment is being conducted by Maine Sea Farms, where the common property problem has been eliminated by raising the salmon in a fenced seawater cove. They are fed a special diet of ground shrimp and crab, achieve a nearly 50 per cent survival rate, and are marketed twenty months later as 9- to 12-ounce "yearlings" in the luxury fresh fish markets of the Northeast.

15. Robin W. Doughty, *Farming Iceland's Seafowl: The Eider Duck*, Sea Frontiers (November-December 1979), p. 346.

16. For an important discussion of British and American wildlife laws, see Thomas A. Lund, American Wildlife Law (Berkeley and Los Angeles: University of California Press, 1980).

18. Lund, *American Wildlife Law*, p. 103.

19. This remarkable development of private property rights in land and especially in wildlife by an aboriginal people was first treated by anthropologists Frank Speck and Eleanor Leacock. More recently it has been subjected to economic analysis by Harold Demsetz and by John Baden, Richard Stroup, and Walter Thurman. See Frank G. Speck, "The Basis of American Indian Ownership of Land," Old Penn Weekly Review (January 16, 1915); idem, "A Report on Tribal Boundaries and Hunting Areas of the Malechite Indians of New Brunswick," American Anthropologist 48 (1946); Eleanor B. Leacock, "The Montagnais 'Hunting Territory' and the Fur Trade," American Anthropologist 56, no. 5, pt.2, memoir no. 78 (October 1954); Demsetz, "Toward a Theory of Property Rights"; and Baden, Stroup, and Thurman, "Good Intentions and Self-Interest."

20. Demsetz, *Toward a Theory of Property Rights*, pp. 351-52.

21. Baden, Stroup, and Thurman, *Good Intentions and Self-Interest*, p. 10.

22. Demsetz, *Toward a Theory of Property Rights*, p. 353.

23. Baden, Stroup, and Thurman, *Good Intentions and Self-Interest*, pp. 10-11.

24. Garrett Hardin, "The Tragedy of the Commons," *Science 162* (December 13, 1968): 1244.

25. Michael Satchell, "How the Smuggling Trade is Wiping Out Wildlife," *Parade* (December 3,1978), p. 25.

26. John Strohm, "An Open Letter to Colleen," *International Wildlife* September-October 1978), p. 19.

27. Myers, *The Sinking Ark*.

28. David W. Ehrenfeld, *Conserving Life on Earth* (New York: Oxford University Press, 1972).

29. Scheffer, *A Voice for Wildlife*, p. 210.

30. Jack Rudloe, *Time of the Turtle* (New York: Penguin Books, 1980), pp. 251-52.

31. Jacques-Yves Cousteau, "The Endangered Sea," *Vision* (July-August 1974), p. 53.

32. Myers, *The Sinking Ark*, pp. 88-89.

33. *Ibid.*, pp. 235-36.

34. Ehrenfeld, *Conserving Life on Earth*, p. 322.

35. Discussing pollution of the commons, Hardin does say, "We are locked into a system of 'fouling our own nest,' so long as we behave as independent, rational free enterprisers." See Hardin, "Tragedy of the Commons," p. 1245. However, it is evident that socialist and communist nations have been even less successful in solving pollution problems. See Robert J. Smith, "The Environment under Socialism," Policy Review 2 (Fall 1977): 113-18, for a brief review of the Soviet experience with environmental degradation.

36. Garrett Hardin, *Exploring New Ethics for Survival: The Voyage of the Spaceship Beagle* (New York: The Viking Press, 1972), p. 121.

37. Hardin, *The Tragedy of the Commons*, p. 1245.

38. *Ibid.*

39. Hardin, *Exploring New Ethics*, p. 121.

40. North and Miller, *The Economics of Public Issues*, p. 112.

12

How to Protect Kenya's People and Wildlife

JAMES SHIKWATI

Direct income from wildlife tourism contributes about 5 per cent of Kenya's gross national product, accounting for just over a tenth of national wage employment and over a third of the nation's annual foreign exchange earnings (Emerton 1999). Yet the people who sacrifice to protect the source of this income—wildlife—receive little value from it.

All the wildlife in Kenya is owned or controlled by the government. Due to financial constraints and the conservation laws inherited at independence, the international conservation community has indirectly taken over this resource. The Kenyan government's over-reliance on aid (an estimated $150 million from the international conservation community goes to Kenya Wildlife Service) has made it insensitive to the people's plight. In addition, whites and Asians hold senior positions in the conservation organisations. Most of the expert conservationists are whites from the West (Bonner 1993). Most of the safari companies and camps are owned by multinationals - Africans are secretaries, cooks, and drivers.

Popular notions suggest that Kenyan natives are keen on exterminating wildlife. Yet conservation and consumptive utilisation were part of the African culture prior to colonialism. Africans co-existed with wildlife and would only kill them for defence and when they wanted to use their skins and meat. The Kiswahili word for wild animals is "Wanyama," from "Nyama," the Kiswahili word for meat. The Africans did not put animals into parks for protection; they protected themselves from the animals by putting up thorn fences or digging trenches around their homesteads.

After the scramble for Africa that saw imperialists control the continent, it became apparent to the foreign occupying forces that the pristine environment in Africa was being destroyed. The destruction was largely due to the white hunters and white occupation of the fertile African lands. The remaining land, either poor in fertility or prone to infestation by the tsetse fly and malaria-bearing mosquitoes, experienced conflict between humans and wildlife. That conflict continues.

Wildlife has invaded farmlands, destroyed crops, and killed people. In June 2002, in the Voi region, hungry lions killed fifty-four sheep. This led an angry Voi member of parliament to threaten to mobilise the community to kill one elephant for every shamba (farm) destroyed by animals. He cited the ineffectiveness of the Kenya Wildlife Service and called for it's disbanding.

The livestock and crop losses caused by wildlife impact heavily on individual ranchers, pastoralists, and arable agriculturalists. The conflict between humans and animals is intense where forested parks border farmlands; this includes the Imenti, Nyeri, Trans Mara and Kwale districts. Other incidents occur where rangeland has pockets of agriculture such as Kimana, Leroghi and Taita districts.

On average, more than 15 people are killed by wild animals each year, with the highest number recorded at 55 people in 1992. According to the Kenya Wildlife Service, elephants cause 75 per cent of human deaths from wildlife. The government offers 30,000 Kenya shillings ($389) as compensation for each person killed. The bureaucracy involved to get the compensation may take more than 10 years. This has made the locals rightly conclude that the government values wildlife more than people.

One study estimates that the net cost to the Kenyan economy from maintaining nearly 61,000 square kilometers (23,552 square miles) of land under protected areas is US $203 million. This is some 2.3 per cent of gross domestic product, equivalent to supporting 4.2 million Kenyans (Emerton 1999).

During the antipoaching war between 1988 and 1999, when the government issued "shoot to kill" orders against poachers, many of those shot were poor rural folks. Instead of enlisting them in the fight against poachers, the government resorted to burning villages near the parks, as in Kora. The local communities were caught in the cross-fire between conservationists and wealthy government officials sponsoring poachers to get trophies to sell to their European and Asian accomplices.

The Maasai Mara National Reserve received US $26 million from tourism in 1988. Only 1 per cent went to the local Maasai. Tourism firms received 45 per cent, hotels 35 per cent, shops 5 per cent, taxes 5 per cent, Narok Council and wages 5 per cent respectively, according to the Intermediate Technology Development Group, a non-profit organisation. Only 2 per cent of tourism industry profits go to the local people in Kenya, says this group. And the bulk of this tiny percentage goes to local leaders and those with capital and know-how to exploit the tourist market.

As Aldrich-Moodie accurately observes, the poor populations of the world must make a living from their natural surroundings (Aldrich- Moodie and Kwong 1997). Otherwise they will have little incentive to preserve these surroundings, including the wildlife that inhabits them. Only people who do not make a living in the vicinity of the wildlife reserves have the luxury of questioning whether or not human beings have the right to control wild animals.

According to Richard Stroup and John Baden (1998, 39), if property rights to a resource are not fully defined and enforceable, those who put a relatively low value on its use may use the resource without compensating anyone else. This is the case in Kenya. The government and western conservationists have ignored the local communities in their quest to manage the wildlife resource-even though the World Commission on Environment and Development of the United Nations in 1982 called for recognition of local communities' traditional rights to the land and resources they use to sustain their way of life.

When everyone owns wildlife—that is, when government owns it—no one will take care of it. The public or non-profit decision maker who cannot personally gain from more efficient utilisation of the wildlife resource will not be keen to minimise wastage.

What is urgently needed locally is the strengthening of the institutions of justice to ensure the rule of law and the devolution of property rights to the local communities. At the international level, it is important to have the western world listen to the plight of the people around the wildlife conservation areas. The former director of Kenya Wildlife Service, David Western, observed correctly that if villagers living around a park made money from wildlife, the park would in effect become the villagers' bank and the wild animals in the park their assets. This would provide a powerful incentive against poaching. People are not likely to rob their own bank.

The farming communities should be exposed to farming methods that increase yield per hectare in order to reduce competition for space with the wildlife. They should also be allowed to co-own wildlife around their farms and/or be entitled to shares in the parks, now owned by the government, around their farms. Landowners whose land the animals occupy outside official protected areas should be included in the Kenya Wildlife Service board of management. This representation will help check disputes between the government agency and the locals.

Another approach would be to decentralise the Kenya Wildlife Service into regional committees that are managed by elected representatives from the ranches, farmers, trust land, and government land representatives. An efficient licensing procedure for supplying locally caught game to restaurants would provide an incentive for local participation in conservation and encourage employment.

Locals should be well informed on wildlife and conservation issues. A network should be put in place to facilitate communication between the African Kenyans, the white Kenyans and external wildlife experts to stop the trend of outsiders dictating issues locally. A study is urgently needed to clarify issues on possible wildlife ownership and local responsibility for wildlife.

There is absolutely nothing immoral in having people own wildlife. It is immoral to have them trampled to death and their crops destroyed with no gain in sight. It is illogical to have people drown in poverty when they can profitably gain from wildlife.

References

Aldrich-Moodie, B., and Jo Kwong (1997), *Environmental Education*. London: Institute of Economic Affairs.

Bonner, Raymond (1993), *At the Hand of Man; Peril and Hope for Africa's Wildlife*. New York: Alfred A. Knopf.

Emerton, Lucy (1999), "Community Conservation Research in Africa: Principles and Comparative Practice". Paper No. 9. Manchester, UK: Institute for Development Policy and Management, University of Manchester.

Stroup, Richard, and John Baden. (1998), "Property Rights and Natural Resource Management". in The Federal Judges' Desk Reference to Environmental Economics, ed. John A. Baden. San Francisco: Pacific Research Institute, 35–65.

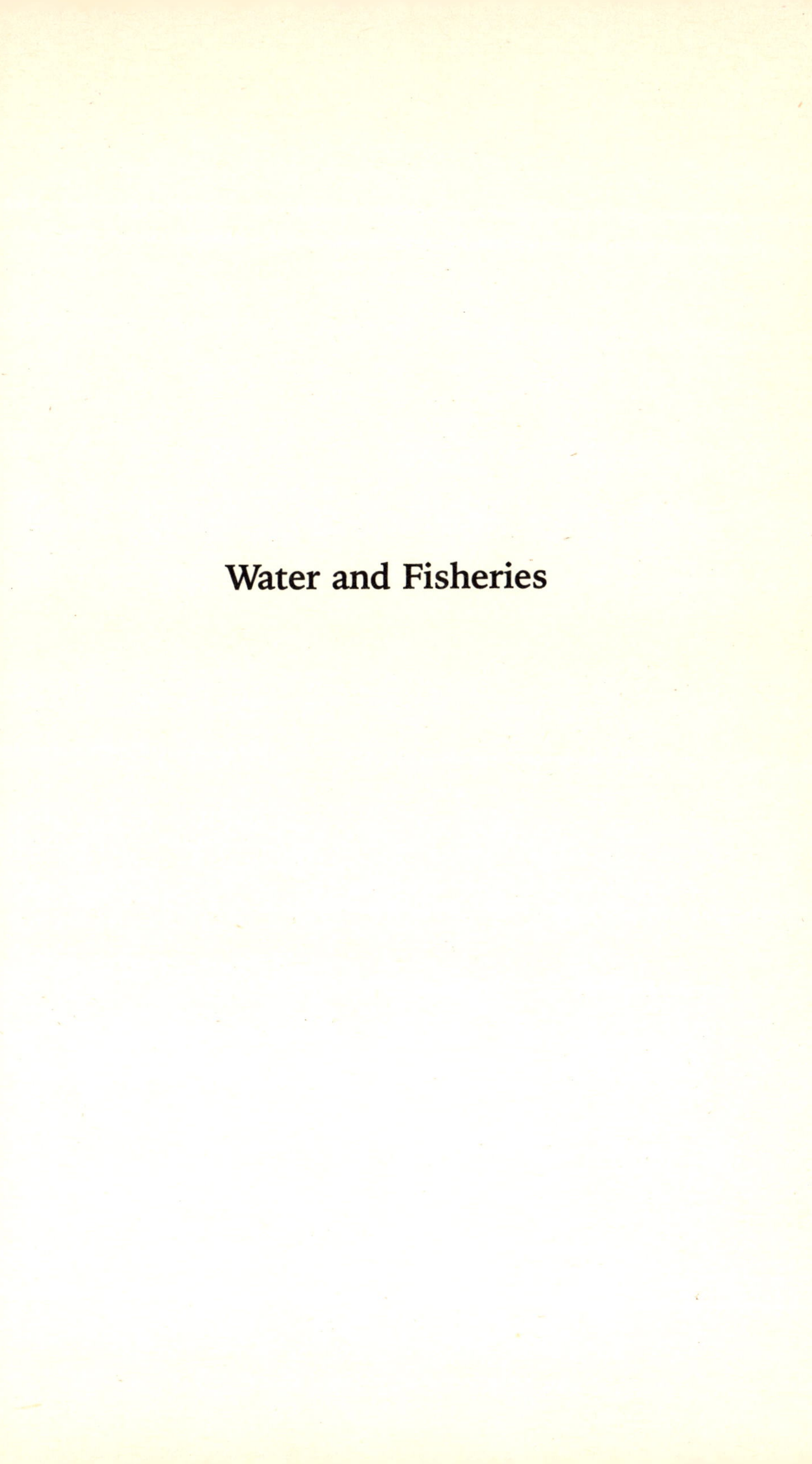

Water and Fisheries

13

Managing Water Resources: Communities and Markets

PARTH J. SHAH

AMBRISH MEHTA

India has over the last 50 years spent $50 billion on developing water resources and another $7.5 billion on drinking water, with little to show for the money (Devraj 2002). Apart from big dams and irrigation systems, the government has encouraged the digging of millions of tubewells and borewells energised by electric and diesel-driven pumps that now provide half of the country's irrigation. Still, 120 million people in India do not have access to safe drinking water, and 21 per cent of all communicable diseases in this country are water related.

Doomsday scenarios that give the impression that there is too little water and that demand will rise exponentially in the future are far from the truth. India is one of the wettest countries of the world, receiving about 4000 billion cubic metres (BCM) of rainfall every year. India currently uses barely a third of potentially available water supplies (FAO 2002).

Nature has endowed India with plenty of water, why then, does it not reach those who need it?

Water Scarcity: The Tragedy of the Commons

> What is common to many is taken least care of, for all men have greater regard for what is their own than what they possess in common with others.
>
> *— Aristotle*

According to biologist Garret Hardin's famous formulation, environmental problems are the result of the 'tragedy of the commons'. Underlying any environmental problem is a resource that is commonly or collectively or publicly owned. A common resource with an unrestricted access gives rise to problems such as depleting fish populations, overuse of water, clear-cutting of forests, poaching of animals. What is the resource that is commonly owned in each of these environmental problems? Fish, water, forests, and wild animals. Among all these resources, water is the most open-access commons.

Historically, access to a common resource—village pastures for cattle grazing, forests for fruits and fuelwood, wild animals for hunting, river water for agricultural use—was controlled by norms and customs, either articulate or inarticulate. Ever-increasing demand for these resources due to growing population, accelerating economic development, and improving technologies began to put pressure on the informal norms and customs that controlled use of these resources. Unfortunately instead of building on the informal arrangements that had worked well, a completely new method was adopted. The state took over the ownership and management of common resources. The process of state control began under the British but continued unabated after Independence.

The solution is to put these resources back in the hands of the people—convert the informal arrangements that had worked before into formal and legally enforceable rules and contracts.

Management and ownership of the commons must be transferred to local communities and user groups. Simultaneously, the state must withdraw from day-to-day management and focus on the overall policy and legal framework within which people will own and manage their common resources.

The single change of legal ownership and genuine participation of people would address many of anomalies of the current state management. Ownership brings responsibility. State subsidies, which largely went to the richer groups, will continually decline, as people will take charge of service provision; full cost recovery will reduce overuse and misuse; rights-based allocation decisions will become more immune from political calculus.

The government of India's recent policy, National Water Policy 2002, acknowledges the changed realities and emphasises a new institutional set-up for managing water resources. The basic principles of this policy are: 1) water should be treated as an economic good instead of a free service; 2) the approach should be demand driven and not supply driven, 3) the government should function as a facilitator and not service provider, and 4) users should be fully responsible for operation and maintenance of the services. These policy directives are a right step, but only the first step.

Community Water Rights: A Fair and Sustainable Solution

Riparian Rights

Because of its transitory nature, the rules governing the use of water have evolved in different ways than those governing other resources. The British common law jurist, William Blackstone, asserted that "water is a moving, wandering thing, and must of necessity continue common by the law of nature; so that I can only have a temporary, transient, usufructuary property therein" (quoted in Webb, 1931). The riparian doctrine that evolved under English common law gave all landowners and other residents living on the banks of a natural stream or river a right to an undiminished quantity and quality of water. This right is restricted to riparian owners, that is, to those whose lands border a stream or river. Non-riparian owners cannot possess this right. Moreover the right is correlative–it must be shared coequally with the other riparian owners. The doctrine prohibits alteration of the stream. Instead it requires that a stream be allowed to flow in its natural watercourse as it was accustomed to flow undiminished, unobstructed and unchanged in quality. The right is for the use of water and there is no ownership in the corpus of the water.

It is clear that the riparian doctrine works well in humid areas where the rainfall is abundant and consumption is confined mainly to domestic and livestock use and diversion of water for irrigation is unnecessary. Under these conditions riparian owners can exercise their usufructuary rights without harming downstream owners.

This doctrine of riparian rights is still prevalent in many parts of the world including the eastern states of USA. In India too it exists in law. But in practice it is often not followed and what we find is a mixture of state monopoly and a "free for all" situation.

Prior Appropriation Rights

The riparian rights doctrine seems to be ill-suited to arid areas where rainfall is less. Here the use requires diversion of water in quantities greater than necessary for domestic and livestock consumption and this would impair the downstream riparians. Moreover, the availability of water varies from season to season and year to year. This would mean that not everyone's demands can be satisfied in low water years. Hence an alternative system of water rights evolved in the arid western regions of the United States during the second half of nineteenth century.

The first major rush for settlement in west came from the miners who flocked to these areas in great numbers in search of gold. The rules and customs developed by the miners for the mining claims also played a major role in the evolution of this doctrine. Since the miners were in fact squatters on public domain, they applied the law of the public domain, first in time, first in right. After the days of the pan and shovel gave way to ditches and sluice boxes, questions of right to use the streams arose. And the miners applied the same rules to water as they had to the land. He who diverted water first had the prior right to it to the extent of his diversion for use on both riparian and non-riparian lands. To perfect the right, ditches had to be dug with diligence and water applied to beneficial use. It was not to be wasted. And as with mining claims, the right ceased when the use ceased. Here was the genesis of a new property right in water.

The same rules were later applied to diversion for irrigation as initial farmers were all disappointed miners who had failed to strike gold and decided to settle and start agriculture. These rules represented a marked departure from riparian doctrine in that they: 1) granted to the first appropriator an exclusive right to the water and conditioned other rights upon the prior rights of those who came before, 2) permitted diversion of water to non-riparian lands, 3) limited the amount of water that could be claimed to that which could be put to beneficial use, and 4) allowed for transfer and exchange of rights.

Initially one had to simply take and use water to establish the claim. Later, this was formalised by posting a notice of intent to take water and recording such notice in the county of local jurisdiction. Legislation and court decisions started recognising this method, known as prior appropriation, which eventually was adopted into a formalised framework of a water rights allocation system. The courts have recognised the possessory interest in water use, although not in the corpus of the water. The different western states each have developed administrative regulations and procedures for appropriating water under the major use categories. As it involves recording of who diverted water when and how much, this system requires an elaborate and sophisticated system for defining, recording and enforcing of rights. All western states have established such a centralised administrative system under the supervision of the state engineer for the same.

Many countries now have elaborate systems of clearly articulated and centrally administered water rights for all diversions from surface waters based either on the historical use or limited duration licensing arrangements. Eventually we shall also need to create such systems for water rights on all surface and groundwater resources. But this in itself is a huge undertaking, requiring detailed studies and appropriately designed laws and institutions for creating and administrating the water rights system, and would require years to accomplish. A beginning can be made by first establishing rights on project waters.

Project Allocations: Towards Legal and Long-term Contractual Arrangement

In recent years, many countries across the world—Sri Lanka, Nepal, Philippines, Indonesia, Mexico, and Turkey are some of the well-known ones—have initiated programmes to transfer some or all responsibilities for management of irrigation and related water services to local organisations of farmers. The scope of the responsibilities transferred, physical conditions of the services, and institutional elements of the transfers vary from country to country.

On the other hand farmer developed and managed local irrigation and agricultural drainage infrastructure services have been in existence for decades, if not centuries in many developed countries. Most of the new projects built by the government in these countries kept these farmer-developed systems intact and transferred the management of newly developed services to farmer organisations. In the United States, for instance, local Water Service Entities (WSEs) of farmers were formed to finance, construct, and operate systems serving up to 100,000 hectares, independent of the usual government agencies. Local WSEs, without government assistance, built systems serving over 80 per cent of the lands irrigated in California today–the vast majority completed in the late 1800s. In many European countries, in areas prone to inundation by storm water, people organised and dealt with flood control in the same manner.

The story is not restricted to the developed countries with European traditions. Farmer-owned and managed irrigation systems were in existence for several hundred to two thousand years in Nepal, Sri Lanka, China, and also in tank irrigation in south India. They served over 70 per cent of irrigated land in Nepal, still dominate in Bali, and at pre-independence served most land area in south India and Sri Lanka.

All these schemes across different cultures and nations share and exhibit a consistent and common set of institutional principles basic to sustainable, fair, and affordable use. They provide crucial lessons for the evaluation of the ongoing transfer programmes. Water rights among farmers in sub-areas receiving water under the schemes and between different schemes on the same stream are established by priority and are honoured by the farmers. Farmers own equity in the facilities and meet all operations and maintenance (O&M) costs and, through representatives, manage the enterprise themselves. Farmers usually share O&M tasks at the immediate neighbourhood (tertiary or higher) level, but often employ operators who route water and assure distribution in the primary and secondary facilities, following the agreed upon rules. Operation and maintenance costs are shared in proportion to service. The maintenance objective is to sustain a reliable system into the future. Often the entities provide more than one service; irrigation with agricultural drainage being common. Government provided little assistance in the past and, only in rare instances provides technical advice now.

These lessons from the well-established and proven principles should form the basis of the all transfer programmes in India.

The success of any programme aimed at transferring control of water to user communities is dependent on the establishment of well-defined, proportional property rights.

Clearly, in countries such as India that do not have a historically established water rights framework, riparian or prior appropriation rights are infeasible, and the state itself will have to create such a system.

In the meantime, a plausible alternative is to use the current system of 'project allocations' made by the government, which already serve as quasi-rights. Indeed, in most developing countries public authorities decide how to allocate water using guidelines or laws establishing priorities and also specify the uses to which water can be put. In this context, the most basic categories of water users are farmers, industries, rural households and urban households. The water rights themselves are obtained by the users without paying any charge.

The public allocation of water among different user groups can be defined in various ways. The allocations may be volumetric, i.e. a share of the water available in a reservoir/lake or a share of the flow from a stream or dammed river. Alternatively, they may be defined in terms of shifts or hours of availability. In practice usually a combination of these methods is used.

Water Delivery Services – Taking Water to the User

In India, project authorities generally enter into long term contracts with municipal corporations and other government agencies for supply of fixed quantities of water for various purposes. The major categories of project water use are rural (domestic use and irrigation) and urban (domestic use and industrial use).

Thus if the currently used system of public allocations is firmed up and formalised, then it can serve as a system of property rights over surface and ground waters. These quantitative allocations will ultimately need to be converted to proportional allocations of water rights for irrigation, domestic (rural and urban) or industrial use. Also a framework is required where the rights can be traded among various user groups as their needs change over time.

Introducing a system of tradable water rights allows a price and an opportunity cost to be assigned to the value of the water right. Open market forces will lead to the most efficient allocation of these water rights.

It is important to note that a water right, whether purchased or publicly allocated free of charge, does *not* mean that the holder obtains free access to the water supply network.

"Many confuse the water charge with the price of water rights. Under a tradable water rights regime, the water charge should equal the operations and maintenance cost of the infrastructure, whereas the price of water rights would be the market price for the ... right to use the water" (Holden and Thobani 1997).

Once the project allocations are decided, Water User Associations should be given the responsibility of operating and managing the delivery systems, either on their own or by hiring a private agency. Implementing this should not pose much of a problem since the physical delivery infrastructure (i.e. pipelines, treatment facilities etc) is already in place in most areas.

This idea of WUAs taking charge of local water delivery services is not new in India, especially in rural regions. In fact traditional systems of farmer managed irrigation schemes (for instance tank irrigation in South India and 'phad' in Maharashtra) have begun to be formalised with the introduction of Participatory Irrigation Management (PIM) schemes.

Similarly, there are many cases of rural domestic water delivery systems being managed by local WUAs. There is no reason why such models cannot be replicated in urban areas and for industrial water users.

Rural Domestic Water Supply

Government projects that aim to restore community participation in water management (such as the recently launched *Swajaldhara* scheme) focus on constructing water supply infrastructure without ensuring that there is someone with the capacity to manage and operate it over the long term. The urgency to provide safe drinking water most often leads to bypassing of existing local institutions of resource management.

As a result, even in rural areas where the delivery installations are already in place, the standard of service is extremely low, and the installed facilities deteriorate rapidly.

Domestic water supply in rural areas is best managed by the local community. In fact, India's unique *Panchayati raj* system has great untapped potential for providing civic amenities.

Tamil Nadu

The Water and Sanitation Project of the Tamil Nadu government (supported by DANIDA) emphasises the establishment of village based management systems (VMSs) for water supply and sanitation in 300 villages spread over two districts. The core institution is the Village Water and Sanitation Committee, (VWSC), a statutory standing committee of the *panchayats*. It represents the village community in the project and comprises all the elected members of the *panchayats*, representatives from user groups, NGOs, health organisations, and traditional *gram sabha* leaders.

Bodies such as the Block Water and Sanitation Committee and the District Water and Sanitation Committee assist the VWSC through legal, administrative, and technical support, training, auditing, and monitoring. The legal basis for collection of user fees is provided by the Tamil Nadu Government Order of 1998, which specifies the modalities of setting up connections and collection of payments for the same.

The project provides inputs to villagers in efficient water management, and contributes towards building new water supply installations and repairing existing facilities. Villagers are trained in the building, operation, and maintenance of water installations. Every village *panchayat* gets 50 man-days of training and capacity building. The costs are usually paid back within a year. This model attempts to do away with obsolete government orders and regulations that hinder the smooth functioning of the VWSCs and sound use of financial resources.

Olavanna *Gram Panchayat*, Kerala

This is an example of how the government can move from a provider to a facilitator. This was perhaps the first initiative in rural India where the user communities were meeting the full capital and operation and maintenance costs of their drinking water scheme.

The Olavanna *gram panchayat* in Kozhikode district had drinking water scarcity.

The three rivers flowing through the *panchayat* are saline. The *Gram Panchayat* commissioned the first piped water scheme in 1987 in Vettuvedankunnu ward. Later five neighbouring families in the hamlet of Kambili-Paramba pooled resources and installed a small one HP pump.

Encouraged by this initiative and supported by the *panchayat* president, 54 other households of Kambili-Paramba got together in 1989, and with a contribution of Rs. 4,500 each, formed a registered co-operative society to provide drinking water for their own needs. This was a piped water supply scheme consisting of an intake well, pump set, over-head tank and distribution system. Since 1991, several such private societies have been formed and similar small piped water supply schemes commissioned. Today, there are 26 such private co-operative societies operating in the *gram panchayat* and six more societies are in the process of constructing their schemes. The *panchayat* has successfully shifted its role from being a provider to a facilitator and it has performed the regulatory function to sustain and encourage this novel project.

The *panchayat* does not provide any funds to these societies as capital costs or for the operation and maintenance (O&M). It also needs to be noted that not a single private scheme has failed till date.

The process of initiating a private piped water supply scheme is as follows. After enlisting all households who wish to benefit from a piped water supply scheme, the beneficiaries get together, draft their by-laws and register their co-operative society. This process is facilitated by the *gram panchayat*, which in turn supports a group of individuals who are willing to mobilise the beneficiaries and take the responsibility of running the project in an open and democratic manner. Members of the society are asked to pay their membership fees, which vary from Rs. 4,500 to Rs. 12,500 per household. The amount differs across societies because the costs of individual schemes vary. Land is purchased for the open well and for the overhead storage tank. The location of the well is decided by consensus.

Once the scheme is ready, water is available 24 hours a day, except in the summer months of April and May. During this period, water supply is reduced by mutual agreement among the beneficiaries, to about 10 hours a day. As the users are also the managers, self-regulation brings in an element of responsibility and ensures that there are no unnecessary complaints.

The General Body (GB) of each society, consisting of all users (averaging around 50 households per scheme) elects an Executive Committee (EC) of 7-11 members, including a President, Vice-President, Secretary, and Treasurer. The EC is elected annually and runs the day-to-day affairs of the society. The EC has to obtain the permission of the GB to utilise the society's funds, including the payment for the construction of the water supply scheme. Each year, a General Body Meeting (GBM) is held to scrutinise the accounts, discuss the annual report, and elect the EC for the following year. Transparency in everyday functioning is a critical factor contributing to the sustainability of the society. Each Society has prepared detailed by-laws for efficient functioning.

Households who wish to join the scheme after it has been commissioned have to pay twice the initial membership fees to offset the initial risk taken and efforts made by the members who started the society. The beneficiaries pay all the O&M costs, including the cost of hiring a pump operator and energy costs. The society makes sure that there is an operating surplus for future repairs and maintenance. In order to assist the poorer families, the societies accept their contribution in instalments. In some societies, the poor are given an opportunity to earn wages during the construction of the scheme that partly funds their contribution.

All members of a society must permit other members to lay pipelines through their property.

However, this must be done without causing any damage to the property. All members must take individual connections from the main line to their houses, at their own expense.

Water must not be used for irrigation under any circumstances. Storing water for irrigation, if detected, invites penal action by the EC. However, a show cause notice must be issued to the member concerned before initiating any action. If a member sells his house, the water connection is also transferred. These sales must be intimated in writing to the EC. In no case will the EC pay back the initial contribution of the original member. The purchaser will automatically become a member of the society.

Each household is allowed 400 litres per day. Water meters are installed to check consumption, and excess consumption attracts a penalty. This is indeed a remarkable achievement, for not only can the societies collect the full O&M costs, but they can also impose such high penalties for excess usage. This is in stark contrast to the experience of the water boards where meters are tampered, the O&M costs not paid, and where it would be impossible to regulate and restrict the use of water at critical times. (DFID 1999)

Pazhakulam, Kerala

Here, more than 1150 springs have already been developed with the active participation of the community. It is interesting to note that more than 90 per cent of the spring users are from the below poverty line category.

Even though the springs are located in remote areas, the project was able to bring down the cost. The cost is considerably less than other water schemes implemented by PWD, Irrigation Department, Rural Development Department and other government organisations.

This is mainly due to the adoption of simple technology (gravity flow), use of local materials and effective supervision by the water committees and NGO staff. This project has demonstrated that with active involvement of the beneficiary group and locally available materials per capita cost can be considerably reduced. In more than half of the springs (630), the per capita cost was below Rs. 200. Only in 22 springs has the cost exceeded more than Rs. 500.

Generally in the gravity scheme the annual O&M costs approximately Rs. 90 per family, where as the pumping scheme cost more than double. Each user family is expected to contribute Rs. 5 every month towards the O&M cost. An O&M account is operated at the local post office or bank depending on accessibility. Water quality testing (bacteriological and chemical) has been carried out on a quarterly basis and the results are shared with the water committees and the community. The results are also discussed in the training sessions (Kurup 2003).

Rural Irrigation

Participatory Irrigation Management

Participatory irrigation management in the context of India refers to the transfer of irrigation management and responsibility from state government agencies (mainly the irrigation departments) to Water Users Associations or farmers organisations. It implies the involvement of irrigation users in all aspects and all levels of irrigation management including the initial planning and design of new irrigation projects or improvements, as well as the construction, supervision, and financing, decision rules, operation, maintenance, monitoring, and evaluation of the system.

Over the last two decades various initiatives have attempted to introduce participatory forms of irrigation management. But many of these efforts have been half hearted, with no serious effort on the part of the various state governments to really transfer systems or responsibilities to farmer organisations. A few WUAs have been formed below the minor outlets in some of the major canal systems and the sporadic efforts of NGOs remained localised. But over the last few years various state governments have woken up to the fact that irrigation systems are crumbling, and PIM is an effective way to sustain these systems.

WUAs in Ozar, Nasik

In July 2001 the Government of Maharashtra issued a resolution notifying that irrigation water was to be supplied only to farmers' WUAs. However this plan to transform the state's irrigation systems through farmer participation remains largely on paper, with a few exceptions.

Over the past 10 years, more than 15 WUAs have been formed along the right bank of the Waghad dam in Maharashtra. Water is released from the dam a fixed number of times, and the WUAs work to ensure equitable distribution, with special attention to farmers located at the furthest reaches of the canal. A network of 'outlet committees' composed of farmers and WUA canal inspectors ensure fair distribution.

The WUAs have come together to form a federation that interacts regularly with the Maharashtra government's Irrigation Department, to ensure each association gets its fair share of water allocation.

Pricing: Earlier, under the Irrigation Department management, water prices were based on a crop area basis, where a farmer is supposed to use water till his irrigation needs are fulfilled, and then pass it on to the next user. But this system caused inefficient and wasteful irrigation since "a farmer with 1 acre can take 10 hours to irrigate his lands and another with one acre can take 5 hours, and both would pay the same cess…" (Bose 2003) Now, under the WUA managed system, prices are based on a fixed volume of water, not on land or type of crop irrigated. According to WUA records, while earlier 150 cusecs of water would irrigate 150 hectares of land, many farmers now irrigate 200 hectares with the same amount.

Also, the Irrigation Department is able to recover over 90 per cent of costs from the WUAs whereas before it managed only 25 per cent recovery from individual farmers. Rising water efficiency has also resulted in higher crop and land productivity. Since the management is local, the system is transparent, and there is no water theft. Farmers do not have to resort to bribing authorities for water any more. There have been cases of WUAs getting together to videotape evidence of suspected water theft, and getting guilty officials suspended. There are also encouraging signs of Department officials dealing directly with farmers and promoting formation of WUAs.

While it is now acknowledged that WUAs have the potential to solve many irrigation sector problems such as poor management, corruption and inequity in water distribution, governments have done little to spread awareness among farmers. Vested interests of irrigation department officials, procedural formalities, and lack of funds with the Irrigation Department have come in the way of speedy implementation in most other parts of the country. (Bose 2003)

Andhra Pradesh

The Andhra Pradesh Farmers' Management and Irrigation Systems Act, 1997 was enacted to form farmers' organisations with a view to involve farmers in irrigation management and ultimately transfer the management to farmers. Elections were conducted in June 1997 and 10,292 Water Users' Associations (WUAs) were constituted along with 174 Distributary Committees.

Out of Rs. 200 per acre collected towards water tax, Rs. 100 would be adjusted to the government account to meet the cost of infrastructure. Out of the remaining 50 per cent i.e., Rs. 100 per acre, Rs. 10 will be adjusted to the gram panchayats and the remaining Rs. 90 per acre would be adjusted amongst Water Users' Associations/ Distributory Committees/ Project Committees. Initial indications are that there has been a significant increase in the state's irrigation area (510,000 ha or 10 per cent): over 73 per cent of WUA presidents report an increase in irrigated area (Blackburn *et al.* 2000 and Pangare 2001).

Urban Water Supply

While there are many success stories of rural communities actively participating in operation and management of water delivery services, the same cannot be said of the urban areas in India.

In many of the urban water privatisation cases in the rest of the world, the outcome has been far from ideal – water privatisation has become synonymous with steep rate increases, health crises, water riots and general social turmoil. Lists of 'water privatisation fiascos' have been drawn up, concluding that privatising water supply is not a viable option for large cities.

In this section we consider some examples of city water privatisation, the ensuing problems, and suggest why these experiments might have gone wrong. We then propose a workable alternative, which tries to introduce competition in the water delivery market by focusing on ward-level contracts.

Manila

In 1997, following the World Bank's advice, the Philippines government privatised the Metropolitan Waterworks and Sewerage System (MWSS) of Manila city. The aim was to reduce cost of water, improve services and increase the number of connections.

Manila was divided into two zones, (ostensibly to encourage competition), and contracts were awarded to two oligarch families, each of whom had major international water companies as partners. MWSS granted the rights to operate and expand water and sewerage service to Manila Water (co-owned by Bechtel and the Ayala family) and Maynilad Water (co-owned by Ondeo/Suez and the Lopez family).

The contracts were in the form of 25-year lease agreements, and guarantees were made regarding stable water rates (at 4.6 pesos) till

2007. It was also promised that there would be 100 per cent infrastructure coverage by 2007, and US$ 7.5 billion worth of new investments over 25 years. It was estimated that the city would thus save US$ 4 billion over the 25 years. In 1997, the 25-year lease agreement in Manila was the biggest water privatisation in the world.

However, a couple of years into the agreement, the privatisation experiment began to unravel. Consumers' and other activists' groups began protesting about rapidly increasing water rates, non compliance with expansion and service targets, worker lay-offs, the non-democratic and non-transparent nature of the privatisation process, and the weak regulatory framework.

In December 2002, after five years of controversy, Maynilad Water decided to terminate their water contract in Manila when the regulatory commission rejected an additional rate increase to 27 pesos. (Approval had been granted for six previous rate increases and countless other contractual obligations had been re-negotiated away since the contract was signed.)

When Maynilad decided to exit, control of the waterworks reverted to MWSS. Maynilad claimed that the city had not met its obligations and has brought the dispute to the International Chamber of Commerce. Maynilad is seeking US$303 million in compensation from the government. In addition, MWSS will now have to take on $530 million in loan payments to creditors. Ultimately, it is the residents of Manila who will have to bear these costs.

Coochabamba, Bolivia

On advice of the World Bank, Cochabamba, a city of 800,000 in Bolivia, put its water system up for auction in 1999. Only one bidder showed up. The company, called Aguas del Tunari, a consortium led by London-based International Water Limited (IWL), which is jointly owned by the Italian utility Edison and the US-based Bechtel Enterprise Holdings. The agreement guaranteed the company an average profit of 16 per cent per year over the 40-year life of the contract.

In a few months after securing the concession, in November 1999, the company increased water rates so much so that people finally came out on the streets to oppose them. Ultimately the government decided to cancel the contract and in April 2000 the company left the country. It however filed suit against the Bolivian government asking for $25 million in compensation. The case is being heard in Washington DC in an arbitration court run by the World Bank.

Interestingly there are various versions of how much the water rate was increased by the private company. They range from 10 per cent to 300 per cent. Even after almost three years (Aguas left in April 2000), it is hard to determine what exactly happened in Cochabamba.

According to Jim Shultz of the Democracy Centre in Cochabamba, who also won a prize for his reporting of the protests, "Our own water bill, for example, leapt from $12 per month in December to nearly $30 in January. Similar increases hit almost everyone we know. By US standards that may not be much, but (it was) for the many Bolivian families who often earn as little as $100 per month." (The Democracy Centre 2002)

Betchel's own reply:

Aguas del Tunari increased the water supply by 30 per cent during its first two months of operation and persuaded the government to reverse the rate structure, so that those who used the least water would pay the least per unit. It was the government, however, that set the rates. It was also the government that insisted that those rates be increased to cover not only operating costs, but also years of accumulated utility debt as well as certain unnecessary capital projects.

It is important to understand the difference between water rates (the unit rate paid for water) and water bills, which depend on the amount of water actually used. For the poorest people in Cochabamba rates went up little, barely 10 per cent. This is in contrast to the figures of 200 or 300 per cent that some have claimed. Unfortunately, water bills sometimes went up a lot more than rates. That is because as Aguas del Tunari improved service, increasing the hours of water service and the pressure at which it was delivered, people used a lot more water. Unfortunately, a campaign to inform residents of the changes and improvements to the service failed to prepare them for the shock of higher bills (The Democracy Centre 2002).

What went wrong in these and other so called privatisation 'fiascos'? It does not take long to see that in all these cases, a public sector monopoly was simply being replaced by a private sector monopoly. It is unfettered *competition*, not merely privatisation in itself that reduces costs, increases efficiency, and improves services.

When the government takes it upon itself to transfer public utilities to private entities, inevitably, a monopoly (or a duopoly as in the case of Manila) results, with the associated efficiency losses and lack of transparency. Water users, not the government, should be able to decide who will operate and mange their supply systems.

Also long term contracts such as the ones attempted in Manila defeat the very idea of competition. A 25-year contract in effect results in a 25-year monopoly, and the local community does not have the option of opting out in favour of better offers.

Another key issue here is the difference between private *utilities* and private *operators*.

The argument that "privateers" will raise water charges to boost profits may apply in the case of privatised utilities, since they may, with the regulators approval, set their own rates. Indeed, this is what happened both in Manila and Coochabamba. However this does not follow in the case of private operators since:

- The municipality (or, as we propose, Water Users Associations) retains control over rate setting.

- No city would outsource water delivery services to a private firm unless this results in reduced cost or provision of these services, and savings for the city exchequer and the ratepayers. (Tsybine and Evans 2003)

In a situation of multiple private operators, any one of them can remain in business only by offering a competitive rate.

Selling an entire water utility to a private company may, as opponents point out, result in a loss of local control over water resources. However, community control would actually increase in the case of short-term operation and management contracts. The community retains ownership of the system, controls the rates, and has the option of not renewing the contract if the company's performance is inadequate. With a short-term contract, there is more competitive pressure, and this encourages the company to provide high-quality services at the market rate (Tsybine and Evans 2003).

Private provision of water services is most successful where the operation and maintenance contracts are offered not by the central authority, but by the local water users, thus encouraging competition.

Ward Level Water Users Associations

The most commonly proposed method for privatisation of water supply in urban areas is for local governments or municipalities to hire a private company to manage and operate the water delivery systems, i.e., the entire network of pipelines from the supply source to the use points at households, and the associated logistics and maintenance activities.

What is needed is to introduce competition in this area, with several private companies competing among themselves to provide water delivery services to different parts of the city, so that water users benefit from better service and lower, more competitive water rates.

Domestic and industrial water users could form Water Users Associations at the ward or constituency level (The ward is usually the smallest administrative unit in an urban area). These "ward" level Water User Associations (WUAs) would invite bids from private companies to operate and maintain the water delivery infrastructure.

The difference here would be that depending on the locality, the proximity to the supply source, and the nature of demand, different companies would offer services at different rates - leading to competition among them for ward-level contracts.

Various types of contracts are possible, depending on the degree of private sector participation, ranging from Divestiture/Build Own Operate contracts to service or management contracts. It is possible to incorporate mechanisms into these contracts that allow for a more competitive market to develop. The very fact that different private companies compete amongst themselves for the numerous ward level contracts ensures a degree of competition. Besides this, there are several other devices to introduce competition in the contracts:

- *Inset Appointments*: Simply put, this means that larger consumers groups are allowed to appoint a company other than the one contracted by the local WUA.

- *Common carriage*: This occurs when one service provider shares the use of another's assets, such as its pipe network or treatment works (The telecom and electricity service providers already use this).

- • Removing the monopoly of water companies in making connections to the main source - customers must be given the right to do so as well.

- • *Recognising the market for different 'grades' of water*: The water supply to the city does not have to be only of 'drinking' quality – Depending on the purpose of use, different types of water can be supplied: Piped water for non-drinking use (washing, gardening, coolers), bottled water for drinking purposes, recycled water for industrial use, reclaimed (or treated wastewater) for irrigating greenbelts, parks, golf courses, etc. This kind of "product differentiation" is sure to encourage competition and reduce costs.

- • Dual supply systems to deliver these differing grades of water, while possibly impractical for existing domestic users, would work well in case of large industrial sites or newly developing residential areas.

Getting the right amount of water to the right place, at the right time, requires localised solutions that put control over water resources in the hands of those who use them.

The principles recommended here are old wisdom: user rights, and user ownership and responsibility for managing collective services and common resources. This wisdom somehow died in recent years—historians would debate the reasons—but now is the right time to recognise it. The modern disciplines of new public management, new resource economics, public choice, and new institutional economics provide further support to the old wisdom. New technologies have made it possible for people to acquire necessary information and take prudent decisions. All the ingredients for sustained development and wise use of natural resources are present; we need only the courage and foresight to bring them together.

References

Anderson, T.L. (ed) (1983), *Water Rights: Scarce Resource Allocation, Bureaucracy and the Environment* San Francisco: Pacific Institute for Public Policy Research, USA.

Anderson, T.L. & P. Snyder (1997), *Water Markets* Washington DC: Cato Institute, USA.

Bate, Roger (August 2002), *Water – Can Property Rights and Markets Replace Conflict?*

Sustainable Development: Promoting Progress or Perpetuating Poverty, edited by Julian Morris, London: Profile Books, Chapter 15.

Becker, N. (1995), "Value of moving from central planning to a market system: lessons from the Israeli Water system" *Agricultural Economics*, 12; pp. 11–21.

Blackburn, James, R. Chambers, J. Gaventa (2000), *Mainstreaming participation in development*, OED Working Paper Series No. 10, Washington.

Bose, Ruksan (2003), *Watering Down a Success Story*. Down to Earth, May 31. Centre for Science and Environment. Delhi

Campbell, Melanie (1997), "A Simulation of Kansas Water Rights: Background Information on Kansas Water Rights," Topika, Kansas.

Chowdhary, Supriya Roy (2002), "The Cauvery Dispute," *The Hindu*, October 3.

De Alessi, Michael, *Fishing for Solutions*, IEA Studies on the Environment No. 11, pp. 14-24.

Department for International Development (DFID). (1999) Water and Sanitation Programme, *Villagers treat water as an economic good: Olavanna, Kerala*. Write Media. New Delhi.

Devraj Ranjit (2002), *Water policy flows past community rights*, New Delhi: Inter press service news agency, April 5.

Finnegan, William (2002), "Leasing the Rain", *Frontline World* and *Now with Bill movers*, June.

Government of India (2002), *National Water Policy Report*, April.

Food and Agriculture Organisation of the United Nations: Water Resources, Development and Management Service (2002), *AQUASTAT Information System on Water in Agriculture: Review of Water Resource Statistics by Country*. Rome. Available on-line at: http://www.fao.org/waicent/faoinfo/agricult/agl/aglw/aquastat/water_res/index.htm.

Frederiksen, H., J. Berkoff, W. Barber (1993), "Water Resources Management in Asia" *World Bank Technical Paper 212* Washington DC: Asia Technical Department.

Harden, Sarah (2001),"The challenges after winning back our water - The fight against water privatisation in Bolivia," *Upstream Journal*, January.

Holden, and M. Thobani (1996), "Tradable Water Rights: a property rights approach to resolving water shortages and promoting investment," *World Bank Policy Research Working Paper 1627*, Washington, DC.

Kurup K.B. (2003), *Community management in rural water supply through the development of natural springs*, IRC International Water and Sanitation Centre

Livingstone, M.L. (1995), "Designing Water Institutions: Market Failures and Institutional Response" *Water Resources Management*, 9:203–220.

Ostrom, E. (1990), *Governing the Commons* Cambridge: Cambridge University Press

Pangare, Ganesh (2001), *Indian Network on Participatory Management Newsletter* No. 11, New Delhi, August.

Rajiv Gandhi Drinking Water Mission and Water & Sanitation Programme South Asia, "Water no longer a pipe dream in villages," *Jalvaani* Vol 2 No 4.

Repetto, R. (1986), "Skimming the Water: Rent-seeking and the performance of public irrigation systems". *Research Report No. 4* Washington DC: World Resources Institute.

Rosegrant, M.W. and H.P. Binswanger (1994), "Markets in tradable water rights: Potential for efficiency gains in developing country water resource allocation," *World Development*, 22(11): 1613–1625, UK: Elsevier Science Ltd.

Schultz, Jim (2000), "Bolivia's Water War Victory," *Earth Island Journal*, Autumn.

Sebastian, Sunny (Aug 2001), "The Waterman of Rajasthan," *India's National Magazine* from the publishers of *The Hindu*, Volume 18 - Issue 17.

Shah, T. (1991), "Managing Conjunctive Water Use in Canal Commands: Analysis for Mahi Right Bank Canal, Gujarat," from: Meinzen-Dick, R. and Svendsen, M. (Eds.) *Future Directions for Indian Irrigation* Washington DC: International Food Policy Research Institution.

Shiva, Vandana (2003), "The World Bank, WTO and Corporate Control over Water", *Navdanya*, p 17.

Shiva, V. and R. Holla, A. Jafri, K. Jalees (2003), "Corporate hijack of water", *Navdanya*, p 32.

Singh, Nivirkar, and Alan Richards (2001), "Inter State water disputes in India: Institutions and Policies," October.

Starr, Kayla M. (2002), "The Blue Gold Rush," *Sentinent Times*, December.

The Democracy Center (2002), "Bechtel vs. Bolivia Riley Bechtel's response."

——————. (2000), "The water rate hikes by Aguas del Tunari: The real numbers."

Thobani, M. (1998), "Meeting water needs in developing countries: Resolving issues in establishing tradable water rights" in Easter, K W, *et al.*

Tsybine, A. and Evans, D. (2003), "Revisiting the Public Interest in Private Water". *Policy Brief No. 22*. Reason Foundation. Los Angeles.

World Bank (2000), "India's water resources management," *World Bank Newsletter*, New Delhi, July 6.

14

Water: Can Property Rights and Markets Replace Conflict?

ROGER BATE

Introduction

According to the UN, 2.7 billion people will face severe water shortages by 2025 and it warns that there may be wars over water. Already, many disputes in the Middle East concern water. Former President Anwar Sadat of Egypt claimed in 1979 that 'the only matter that could take Egypt to war again is water'.[1] Where peace is achieved in a region, there is often much emphasis on the resolution of water conflicts.[2] Hydrology and politics are strongly interlinked.

The availability of fresh water varies widely from country to country. In 1995, Canada had over 105,000 cubic metres of fresh water per person; Tunisia had only 500, Algeria 625 and South Africa 1,400. In semi-arid and arid regions, quantitative water scarcity is a serious concern. Precipitation is typically low and unevenly distributed, while evapotranspiration is high. Water scarcity is thus related to climate and is exacerbated by misuse, inefficiency, increasing populations and economic development. In these regions, pollution often constitutes an additional problem, but even if the pollution problem were resolved, water scarcity would remain.

In other parts of the world, where rainfall is abundant and evapotranspiration low, water scarcity still exists but is more typically a matter of quality than quantity. In North America, agricultural pollution alone is estimated to cost $9 billion every year.[3]

However, water quality is an even bigger problem in low-income countries. According to the World Bank, poor quality water is responsible for around 3 million deaths a year in low-income countries, mostly from diseases such as dysentery, diarrhoea, hepatitis and cholera.

Industrial pollution is also a major problem in low-income countries.[4] When a commodity is scarce, conflicts often arise over its use. This is especially true of commodities that are essential for human survival, such as water. Governance in general can be seen as an attempt to resolve such conflicts through the construction of allocation schemes.

Methods of water allocation range from highly centralised governmental control,[5] to communal allocation with internal self-regulation,[6] to private markets enforced by communal reputation and ostracism and/or national law.

This chapter discusses alternative allocation mechanisms for water and evaluates their relative success in ensuring stable, high quality supplies at low cost to society. It also provides insights into what schemes might most successfully be employed in the future, especially considering the risk of international conflict.

Water Allocation by Central Government

In most semi-arid countries, water has been allocated by centralised government administration. The track record of such schemes has not been impressive. Despite growing water scarcity and the high costs of hydraulic infrastructure, "water is typically underpriced and used wastefully, the infrastructure is frequently poorly conceived, built and operated, and delivery is often unreliable."[7] At the same time, there are high fiscal costs stemming from the "construction of hydraulic infrastructure; from the institutional bureaucracy to support the design and execution of the projects and to set and collect water tariffs; and from the cost of operating and maintaining the system".[8]

Too little emphasis has been placed on the planning and design of water delivery projects. A major reason for this is the perverse incentives resulting from political allocation. Indeed, irrigation projects, which are the largest water users in developing countries, have changed little since Helen Ingram found in 1973 that economic studies had been used to "clothe politically desirable projects in the figleaf of economic respectability."[9]

In a similar vein, a comprehensive report for the World Bank in 1996 concludes that "Many large multipurpose hydraulic projects (irrigation, hydropower, flood control, urban use, etc.) were undertaken on political rather than economic grounds. Costs tend to be high because of: inappropriate design, stemming in part from poor studies done prior to start-up; long gestation periods resulting from funding shortfalls due to changing government priorities and poor capital programming and budgeting; few managerial incentives to control costs; and reported corruption that typically involves kickbacks from construction companies."[10]

An earlier World Bank report highlights some egregious examples of water mismanagement.[11] For instance, between 1974 and 1993 the Peruvian Government invested $3.4 billion (in constant 1993 dollars) in nine coastal multipurpose projects. These projects achieved only 6.6 per cent of the planned expansion in supply and none of the planned hydroelectric generation capacity. The primary justification for the projects was irrigation and, while the estimated cost per hectare ranged from $10,000 to $56,000, at completion irrigated land in the area was selling for about $3,000 per hectare. Similar statistics can be found elsewhere. Sri Lanka invested 6 per cent of its GDP in a single project, the Mahaweli Development Programme, which was vastly inefficient and led to severe social tensions.[12]

Economic Inefficiency of Administered Allocation

The main reason for such waste is that water is often purposely sold at a very low price or no price at all. Authorities in many arid areas in both developing and developed countries ration water to urban, industrial enterprise while subsidising farmers to grow low-value, water-intensive crops. The effect of under-priced water is that farmers use inefficient irrigation technologies to produce uneconomic goods at the expense of lucrative alternative economic activities. The opportunity costs of this misallocation can be vast.[13]

Furthermore, government control of water has often favoured the relatively wealthy and has not been effective at ensuring access to water for the poor.[14] The poor are often excluded from piped municipal water and must resort to expensive private water truckers to meet their daily consumption requirements. In its review of water vending, the World Bank showed that the unit cost of water supplied from water trucks to poor communities was between four and one hundred times more expensive than piped supplies.[15] In addition, the poor can rarely afford proper sewage arrangements. Similarly, government control has often failed to address environmental problems.[16] For example, 10 per cent of the Pakistani irrigation system faces chronic salinity problems.[17]

Historically, administrative approaches to water allocation have tended to favour large-scale investments over water conservation. There are few rewards for administrators who make painstaking improvements in water efficiencies via better pricing policies. In addition, the glamour of large projects and the attendant publicity and power they bring provide strong incentives to both administrators and the primary agricultural users.[18] Meanwhile, administrators are often captured by interest groups. Attempts to set prices that reflect the true cost of water provision are typically unpopular.[19]

But the incentive to conserve water is weakened without appropriate prices. A key feature of successful allocation mechanisms is the financial connection between suppliers and users of irrigation systems. Elinor Ostrom explains that allocation mechanisms which build in the eventual users' interests work well because the users naturally want to monitor resource use among themselves – at a very low cost to the system. When the users' interests are not satisfied, expensive auditing systems become necessary, but are rarely supplied.[20]

Discussing the failure of political allocation schemes in Africa, Robert Bates argues that "inefficiencies persist *because* they are politically useful; economic inefficiencies afford governments means of retaining political power."[21] Bates explains that when a government artificially lowers the price of water, supply from private sources is reduced, users demand more and the result is excess demand. One result of this process is that other agricultural inputs, particularly land, become more valuable, and the administratively-created shortage creates an economic premium for those who own such assets. As the market cannot match supply and demand without prices, the existing supply has to be rationed – usually by those who operate the political market.

"Public programmes which distribute farm credit, tractor-hire services, seeds and fertilisers and which bestow access to government-managed irrigation schemes and public land, thus become instruments of political organisation in the countryside of Africa."[22]

Furthermore, central control of irrigation schemes gives local users no incentive to maintain the engineering facilities, which inevitably deteriorate. The resulting fall in reliability of water supply stimulates greater government investment and central control, and farmers become even more reluctant to pay.[23]

Consultants, engineering firms and construction companies know this political process well, and understand the mechanism whereby the allocation of funds is politically negotiated. Feasibility studies for new investments rarely warn that the construction of storage dams or the upgrading of canals may be financially unattractive.[24]

When centralised administrative solutions fail – for example because supply augmentation is unsuccessful or simply leads to increased use – governments look for alternatives. The two most popular schemes are centrally administered quotas and prices.

Pricing

In principle, pricing enables public authorities set user fees for water at levels that reflect the opportunity cost of provision, thereby inducing water conservation and making more water available to higher value uses.[25] In practice, no central government has set water prices in this way and pricing is primarily used to recover costs of water delivery.[26] [27] Nevertheless, where demand is sensitive to pricing it can have a considerable influence on the long run quantity demanded. Some countries with low water capacity per person have higher tariffs, such as Tunisia, but others, such as South Africa, do not. Countries with naturally high water levels have varied tariffs – in other words, scarcity of water is not a strong determinant of higher tariffs.[28]

Farmers are the main users of water in semi-arid areas and oppose raising water prices because it increases their variable costs. Their assets also depreciate as higher water prices reduce the value of their land. Farmers' resistance to increased fees is therefore grounded in their incentives. If the fees charged for water on some projects were to be raised (and enforced) sufficiently to recover both the recurrent and capital costs, many farmers would be better off not irrigating. Robert Repetto found this to be the case for five of the six countries he analysed in Asia.[29]

However, a common problem for pricing schemes is that increases in fees usually accrue to the central treasury and will only rarely be re-invested in better institutional facilities, unless bribes are paid. Farmers thus often have good reason to oppose price rises.

Occasionally, governments will allow local associations to keep water fees for reinvestment and farmer opposition to fees is lowered.[30] Repetto found in only one country out of six (the Philippines) did the fees collected cover the operating and maintenance costs of the systems (this was probably because in the Philippines revenues were hypothecated); in no country were the capital costs covered.[31]

In several of the countries surveyed, Repetto found that the amount of bribes and fees paid to private tube-well operators demonstrated the farmers' willingness to pay far more than the subsidised price for reliably available water. Ironically, it is poor farmers who suffer most from the under-pricing of water. Low prices inflate demand, so authorities have to impose limits on water use. When this happens, the politically insignificant poor are excluded from receiving supply, while better-off, politically relevant farmers obtain water at highly subsidised prices.[32]

Quotas

Quotas can constrain demand for water (if they are set lower than the quantity demanded).[33] Indeed, if demand is relatively unresponsive to changes in price (as is likely to be the case in places where water is extremely scarce)[34], quotas are more successful at constraining demand than water pricing.[35] This may be important if limiting water use to a specific maximum is vital to ensure sufficient water for ecosystems or subsistence users. Similarly, if demand for water increases due to economic development, quotas limit supply to a specific, ostensibly sustainable amount. Compared to water pricing, however, fixed, un-tradable quotas provide less of an incentive to increase the efficiency of water usage.

In addition to the above factors, quotas provide official recognition of access rights and can be given for free. Further, if supply is augmented through dam development and irrigation .canals, farmers are more likely to escape paying for developments from which they alone benefit under a quota scheme.[36] It is thus not surprising that of the two schemes, farmers in less developed countries seem to prefer water quotas over pricing regimes[37], and quotas have been the favoured solution to restraining demand in LDCs.[38]

Quota systems have proven insufficient to constrain demand. Often new quotas given to farmers (either incumbents or new arrivals) following supply augmentation eventually increase pressure on supply. Nevertheless, it is now the norm in LDCs to use quotas, coupled with low levels of pricing, to limit demand.[39]

Water Markets

Quotas and prices, separately or in combination, have been criticised for not allowing sufficient flexibility in allocation.[40] Inflexible allocation usually leads to inequity and inefficiency, because as prices remain low, the value of quotas increases, with latent demand unfulfilled, which leads to economic stagnation, increased power to quota holders and even illegal exchanges of quotas.

There are exceptions: where there is the administrative capability regularly to calculate shadow prices and change water usage, as in countries such as France, pricing and quota strategies have been relatively flexible. But most developing countries do not have such administrative capacity. Scholars therefore suggest a third option, water markets, where quotas are legally traded among users.

Market allocation requires well-defined and freely exchangeable property rights. Unfortunately, transaction costs, which include the costs of creating, monitoring, and enforcing such rights, are often large compared with the value of the water, making it difficult to establish market systems. Problems of externalities, inequity, and devolution of power to local users also make markets politically unpopular.[41]

Furthermore, water use is legally linked to land in most countries of the world, making the legal creation of tradable rights to water technically complex. Meanwhile, at least one solution to this problem, tradable quotas, tends to result in hostility.[42]

Yet the flexibility and efficiency that come from the decentralised knowledge held in water markets has made them popular in many developing countries.

Tradeable water rights allow the price of water to reflect the value of its alternative uses, which creates incentives to put it to more productive use. For example, if farmers were able to sell their water rights at freely negotiated prices, some might sell surplus water to a neighbouring farm where it has a higher value. Often, farmers can generate a surplus by using more efficient irrigation techniques or by switching to less water-intensive crops. In addition, buyers of water rights are likely to conserve water more efficiently. Tradeable water rights can often shift water to higher value uses more cheaply and equitably than alternatives such as building hydraulic infrastructure, confiscating water from farmers, or raising water charges sufficiently to force farmers to conserve water. Although the conveyance infrastructure to transfer traded water must be built if it does not exist already, the cost of doing so may be less than that of generating new water rights.[43]

Trading water rights is particularly beneficial for small farmers, who are often excluded from the political process and have therefore been most vulnerable to reductions in their water allocation over time, and have few other sources of collateral. Poor farmers are assured of sufficient water by their tradeable rights, which also help to reduce the abuses of administrative allocation. As rights are divisible, farmers can mortgage some part of their water rights for small loans, rather than their entire farm holdings.[44]

Although the experience in some countries is that small farmers are generally bought out completely when they trade,[45] the more efficient ones that remain are able to expand by purchasing water use rights.[46] Finally, when comparing the transactions costs of water markets with alternative allocation mechanisms, the 'hidden' transactions costs of the latter must also be taken into account. There are very high costs due to private rent-seeking by the managers of publicly administered irrigation systems. Allocation through markets in well-defined property rights in water will economise on rent-seeking costs.[47]

Case Studies

So far the discussion has mostly focussed on the problems of government control and the theoretical alternative of water markets. We now turn to a brief discussion of the practical implications of water trading, looking at various schemes that exist around the world, both formal and informal.

Informal Markets

Failure of governments to allocate water to the satisfaction of users has led to informal, but often illegal, markets in water transfers in many countries of the world. The Pakistan Water and Power Development Authority found active water trading on 70 per cent of the water courses it studied.[48] Although trades were not officially sanctioned, it was found that where water had been traded agricultural incomes had increased by 40 per cent due to greater control by farmers over their water supply.[49]

Similarly in India, over half the area irrigated by tubewells belonged to farmers who bought water.[50] The estimated gains from trading water in the whole of India were $1.38 billion per year. Yet the only policy statements and governmental actions on water markets in India have been to discourage them because they were illegally using electricity for pumping.[51] Markets in India and Pakistan, and others like them, helped to resolve short run water shortages. Scarcity drove the institutions to make the most efficient use of the available water.[52] But because informal water markets are not supported by existing laws, contracts are not enforceable. Agreements are only struck by users (usually farmers) who know and trust each other well. The lack of legal title also limits transactions to spot sales of water for brief periods of time, and never permanently, and so trades do not fully capture the benefits of an organised market in secure tradeable rights. Unfortunately in many informal water markets (such as those in India and Pakistan) groundwater is depleted because of concerns that markets may be curtailed at any time. Nevertheless, informal markets can occasionally provide the benefits expected from legal markets, such as maintenance of canal delivery systems, because the canals transport purchased water.[53]

Chile

Chile has the most developed and one of the oldest regimes of water-use trading in the world.[54] All water rights were expropriated by the Chilean Government in 1966 and transferred to the State. But Chile's 1981 Water Code re-established tradeable water rights where existing water users (farms, industries, municipalities and power utilities) were granted rights to surface and groundwater.[55]

New or unallocated water rights were auctioned. Except for a few restrictions, the allocated rights could be transferred or sold to anyone for any purpose at freely negotiated prices. The agricultural sector accounts for 89 per cent of the estimated 300,000 owners of water rights.[56] Administration is highly decentralised with the monitoring, distribution and enforcement of water rights carried out by water user associations at the local level – river basin, underground aquifer (for groundwater), primary canal and secondary or tertiary canal. Except for a few large dams and their associated main canals, all hydraulic infrastructure is owned and operated by water users themselves.

Water users enjoy flexibility and control over water rights. In the arid areas north of Santiago there have been many mutually beneficial sales and leases of water, resulting in a voluntary transfer of water to more productive uses. By contrast, in the higher rainfall area south of Santiago, there have been few trades, since the transaction costs of registering the rights and conveying water is greater than the gains from transferring the water.[57]

Chile's transfer of water to more productive uses was carried out voluntarily and without having to raise water charges. In fact, water charges fell following the introduction of tradeable water rights. The decrease occurred because the Chilean Government facilitated the transfer to user groups of the responsibility for carrying out operations and maintenance activities and for setting water tariffs. User groups were able to conduct these activities at a much lower cost than government. Despite the lower water charges, the opportunity to sell water reduces waste.[58]

Perhaps the greatest benefit of the trading approach has been that demands for environmentally destructive dam building have been dropped. The city of La Serena in Chile is able to meet its rapidly growing demand for water by purchasing water rights from farmers at a lower cost, rather than contributing to the construction of a dam. Farmers received an acceptable price for the water and were induced to use more efficient irrigation techniques.[59] Similarly, when Chile's main water company, Empresa Municipal de Obreros Sanitarios (EMOS), realised that it could no longer obtain free water rights, it invested in a programme to significantly reduce physical water losses.

Remarkably, Chile's sustained annual growth of 6 per cent in agriculture during the 1980s was managed without public investments in new hydraulic infrastructure. While this was due in part to heavy investment in water infrastructure in previous decades, the tradeable water rights regime enabled new water uses, including the rapid expansion of fruit production.[60]

Chile has also been successful in increasing access of the poor to potable water, as 99 per cent of urban residents and 94 per cent of rural residents are supplied typically for 24 hours a day. This contrasts sharply with comparable rates of coverage of 63 per cent and 27 per cent in Chile in 1970 and with developing countries elsewhere in the world.[61] While this was due to several factors, such as ensuring that regulated water tariffs reflect the true cost of water, allowing competition among water companies (Santiago alone has seven private companies), and subsidising water consumption for those with low incomes, the ability of water companies to buy water from farmers played a significant role.

Chile still has some problems in water use, particularly with quality. The obligations on holders of non-consumptive rights to release water for public consumption in times of shortage were not clearly defined, so dilution for effluent was occasionally low. This reduced water quality and led to conflict between the recently-privatised hydropower companies and farmers. Some shortcomings in the law have also enabled one hydropower company to obtain huge blocks of non-consumptive rights without charge. Despite these problems, Chile has fewer conflicts and makes better use of its water than its neighbours.[62] [63]

Mexico

In the past few years, the Mexican agricultural sector, and the economy as a whole, have become more market-oriented, and policy makers have increased security of water rights. According to the 1992 Water Law, and its 1994 regulations, users may convert their existing precarious water rights into more secure tradeable 'concessions' with a maturity of between 5 and 50 years (most are about 30 years), to ensure security of tenure.[64]

However, the rights are not as secure as in Chile. Under the Mexican Constitution, all water belongs to the nation, and the Water Law also mentions the possibility of forfeiture of water rights for the public interest if water is not being used efficiently, or if it has not been used for three years. By 1995, 85 per cent by volume of available water in Mexico had been allocated as water use rights and there was widespread leasing and selling of both surface and groundwater rights.[65]

Water trades had been common before 1994, but, as in India and Pakistan, these trades were limited, informal and illegal. Apparently, the authorities tolerated the trades but did nothing to monitor any externalities from them, so, unfortunately, aquifers were drained. The new law recognised and encouraged trade (on either a permanent or temporary basis), which reduced total consumption and alleviated externalities (for example, slowing aquifer depletion).[66]

Most of the recent trades involve farmers selling to industries, water companies or more efficient farms, thereby encouraging investment in more productive activities. Water trading has also allowed unprofitable farmers to reduce their farming debts and to work as labourers on more efficient farms or to seek alternative employment. Although it may not be an ideal solution, having a tradeable asset in water rights gives an inefficient farmer some flexibility.

The Western United States

Shortage of water in the Western United States led to a system of property rights to water based on the prior appropriation doctrine;[67] whoever first diverted water and established beneficial use, obtained primary rights. Successive claimants could only obtain rights that were contingent upon those with prior rights having received their allocations.

Although water rights regimes vary widely between states, their common characteristic is that water use cannot be changed without authorisation of state water authorities. Obtaining authorisation to change water use is often a lengthy and costly business, requiring consent from the relevant governing body after public hearings.

Perhaps the most extreme example of restricting transfer between uses occurs in California. The agricultural sector makes up only 4 per cent of GDP of the state; yet it receives about 44 per cent of the water. Environmental use is allotted 44 per cent, while urban and industrial users receive only 11 per cent. Agricultural water rights vary widely, from cheap, inherited sources to highly subsidised ones. The anomalies that these restrictions cause are extreme; water is so cheap to some users (as low as $2.50 per acre-foot) that rice is cultivated in the desert, while some municipalities have built desalinisation plants to supplement their supplies of water at a cost of $2,000 per acre-foot.[68]

Even worse are the perverse incentives in conserving water. Farmers are forced to operate under a 'use it or lose it' rule, while in towns, rationing during periods of drought is based on family use during periods of plentiful water, which encourages profligacy. Reform is often discussed but assigning to farmers the ability to simply sell their rights would give them millions of dollars in windfall gains on top of the large subsidies that they already receive – a politically unpopular result. On the other hand, farmers fear that their allocation will be reduced over time and with no compensation. Therefore, even in a country with well-developed institutions, there is poor administration of water allocation. The solutions used often defy logic and waste resources, and reform is slow.

A notable contrast to the various restricted water right regimes which exist in the western United States is provided by the Big Thompson water trading scheme of 310,000 acre-feet of water in Colorado.[69] This scheme, which brings headwaters from the Colorado River through a tunnel in the Rocky Mountains to north-eastern Colorado, was partially funded by subscribers in return for use rights. Soon after the scheme was fully operational it became apparent that water demand varied significantly between users and areas within the district. The Northern Colorado Water Conservancy therefore established a system that allowed permanent water right trades. Trades had to demonstrate 'beneficial use' and no sales were allowed to areas outside the District. A central registry records ownership and transfers. The system has become so refined that a simple postcard is used to notify the Conservancy of a transfer. An extremely sophisticated market has evolved for this water and many types of contracts are used, from straight transfers to the purchase and sale of options to water. Within the Conservancy District all the complex infrastructure is in private hands. The Conservancy's role is to record transactions and to check that there is no cheating. The system appears to be operating efficiently within the water District, with supplies going to their highest valued use, although there is undoubtedly an opportunity cost to owners of water rights in not being able to sell their water outside the District.

Australia

Sturgess and Wright studied water transfers along the Murray-Darling River Basin, which stretches 2,530km from the Snowy Mountains of eastern Australia to its mouth in South Australia.[70] In the worst drought years of 1987-88 there were 687 transfers, totalling 340 million cubic metres, with gains estimated at $17 million. In better years, a great deal of trading still took place but the total gains were lower: in 1988-89, the corresponding figures were 280 transfers, 85 million cubic metres and $5.6 million; in 1990-91 the figures were 435 transfers, 120 million cubic metres and $10 million. The researchers concluded that "if benefits of this scale can be obtained by a system of water transfers circumscribed by regional barriers, the benefits that would flow from redefinition of water property rights to allow the free transfer of water between regions.... would be greater still."[71]

Spain

Maass and Anderson examined the centuries old water market of the farmers of Alicante in Spain.[72] The irregularity of water supply in the region led the Alicante farmers to build the Tibi Dam in the late sixteenth century, which became one of the most admired hydraulic works in Europe. Although this temporarily improved supply, it encouraged population growth and demand increased again. A flexible system of allocation developed to improve water use. Over the years, flexibility was achieved when the use of water was split from ownership of land and permanent and temporary transfers of water became legal. Although the water rights were based on allotted irrigation time from a canal, the rights were translated into volumetric units. By the 1970s the market was sophisticated enough to have temporary transfers, measured in minutes or fractions of minutes per day, up to entire seasons.

For the most part, water prices were freely negotiated and agreed at local taverns by the farmers themselves, although brokers did arrange some transactions. In spite of this very free system, concern was raised by officials and non-trading farmers that trades would lead to speculation and higher prices, and eventually a reduction in farming. These opponents of the system demanded a return to the time when water rights were legally tied to land sales, to ensure that water was always used on local farms. These demands were not acted upon. However, attempts to communalise private rights to water were enshrined in the 1985 Water Law, under which rights considered private in 1985 remain so until 2060, when the right will revert to the State, and no new rights will be allocated. Under the 1985 Law, water use will be maintained by the issuance of licenses, the right remaining with the State.[73]

Other Examples

Water markets exist in other countries, including parts of North Africa,[74] South Africa, and Brazil.[75, 76] These markets have similarities with other markets discussed above and will not therefore be given detailed consideration here. Other markets also probably exist, but are yet to be documented, and many other countries are contemplating trading, for obvious reasons. In all the countries discussed in this chapter, allocative improvements have been made from trading and in some cases, such as Chile and India, significant economic gains have been made.

Summary and Conclusions

Governments have sought to control water allocation in most countries of the world, often giving water to politically important users, such as farmers. The usual response to the inevitable insatiable demand for free water has been construction of supply augmentation facilities, such as dams. As demand has continued to rise, governments have attempted to reduce water use by setting quotas and then imposing a tariff on the quota.

For the most part, governments have resisted handing over allocative control to the market. Often political restrictions to trading have reduced the possibilities for better allocation. Many of these restrictions have emanated from concerns that over-depletion of water resources and environmental damage might occur. Concerns about concentration of power in the water market are also common but the literature is not clear on either point. Negative externalities (such as from pollution) have increased under certain trading conditions (usually informal markets), and reduced under others. Where formal markets have been permitted, overall environmental costs have been reduced, even where trading has reduced dilution for pollution.[77]

On balance, water trading countries in arid-areas have a better than average environmental performance.[78] There have been problems in Chile with concentration of power, but this was because of a mistake in allocating (non-consumptive) use rights to power companies. The companies (sometimes inadvertently) lowered the possibility of consumptive use trades by reducing flows in specific locations, although they never actually reduced net flow in the river taken as a whole. As the first to establish comprehensive modern water markets, Chile was bound to make mistakes, but now it and other countries have learnt lessons and are resolving problems through regulation of non-consumptive use rights (ensuring that non-consumptive users return water to the same location from which they withdrew it, to ensure no 'dry' spots).

Furthermore, much of the literature shows that existing allocation systems favour special interests, such as water bureaucracies. These interests differ from those that may dominate in a market, such as major farm estates or power companies. Water, like gas, electricity and telecommunication utilities, is probably best allocated by private provision with government regulation.[79]

This combination brings the flexibility and dispersed knowledge of the market, while leaving sufficient power in government hands to set maximum abstraction limits and control abuses of monopoly power. The theory and practice of water markets, as outlined in this chapter, provide a potential solution to inflexible allocations of water. As water scarcity increases around the globe, it is surely time to see more widespread use of such flexible solutions, not only because of their greater efficiency but also as a means of reducing conflict. The Six Day War was fought partially over water. Israel refuses to leave the Golan Heights or West Bank in part because doing so would reduce the headwaters they control, putting thousands of Israelis at significant risk. Tensions are rising in Egypt about Ethiopian plans to expand water abstraction from the Blue Nile – Egypt draws over 85 per cent of its fresh water from the Nile and any reduction in flow would harm its massive farming enterprises. Will it go to war over water, as Sadat suggested it might in 1979?

Similar problems are occurring elsewhere on the globe. In the past decade, armed conflicts have been fought over water in Bangladesh, Tajikistan, Malaysia, Yugoslavia, Angola, East Timor, Namibia, Botswana, Zambia, Equador and Peru.[80] Of course, water markets and good property rights systems would not by themselves resolve these tensions. But by improving efficiency of water use they would lower effective demand and hence the need to increase abstraction – the cause of the tension. Water markets in India and Pakistan stopped fighting among farmers and engendered cooperation. Markets replaced conflict. Water markets will not settle historic border grievances such as the disputed territory of Kashmir, nor will they stop terrorism driven by religious fanaticism. But water markets can stop water wars. Just as Chile and Mexico have adopted markets to allocate water, so too can Israel and other parts of the world where tensions run high over dihydrogen monoxide.

Notes

1. Tadros, N. (1996), "Shrinking Water Resources: The National Security Issue of This Century", *North West Journal of International Law and Business*, 17: 1091.

2. The Israel-Jordan peace treaty, signed on 26 October 1994, devoted much attention to water allocation decision-making.

3. Dinar, A. and Xepapadeas, A. (1998) "Regulating water quantity and quality in irrigation" *Journal of Environmental Management* 54:273–289.

4. Pearce, F. (1992) *The Dammed: Rivers, Dams, and the Coming World Water Crisis*, London: The Bodley Head

5. Becker, N. (1995) "Value of moving from central planning to a market system: lessons from the Israeli Water system" *Agricultural Economics*, 12; 11–21

6. Ostrom, E. (1990) *Governing the Commons* Cambridge: Cambridge University Press

7. Holden, and Thobani, M. (1996) "Tradable Water Rights: A Property Rights Approach to Resolving Water Shortages and Promoting Investment". *World Bank Policy Research Working Paper 1627*, Washington, DC, USA.

8. Ibid

9. Ingram, H. M. (1973) "The Political Economy Of Regional Water Institutions" *American Journal of Agricultural Economics*, 55.

10. Holden, and Thobani, M. (1996)

11. World Bank (1995a) "Water Policy and Water Markets." Selected Papers and proceedings for the *World Bank's Ninth Annual Irrigation and Drainage Seminar*, Annapolis, Maryland, December 8-10, 1992.

12. Frederiksen, H., Berkoff, J. Barber, W. (1993) "Water Resources Management in Asia" *World Bank Technical Paper 212*. Asia Technical Department. Washington DC.

13. Holden, and Thobani, M. (1996)

14. Kemper, K. E. (1996) *The Cost of Free Water: Water resources allocation and use in the Curu Valley, Ceará, Northeast Brazil* Linköping Studies in Arts and Science; 137. Linköping University, Sweden.

15. The median cost was 12-fold greater.

16. Kemper, K. E. (1996)

17. Frederiksen, H., Berkoff, J. Barber, W. (1993) "Water Resources Management in Asia" *World Bank Technical Paper 212.* Asia Technical Department. Washington DC.

18. Kemper, K. E. (1996)

19. Anderson, T.L. & Snyder, P. (1997) *Water Markets,* Washington DC, Cato Institute, USA ; Holden, and Thobani, M. (1996)

20. Ostrom, E. (1990) *Governing the Commons,* Cambridge: Cambridge University Press

21. Bates, R.H. (1983) *Essays on the Political Economy of Rural Africa,* Cambridge: Cambridge University Press.

22. Ibid

23. Ibid

24. Shanks. B. (1981) "Dams and Disasters: The Social Cost Of Water Development Policies" In: Baden, J. and Stroup, R. (eds.) *Bureaucracy V. Environment: The Environmental Costs of Bureaucratic Governance.* Ann Arbor: The University of Michigan Press.

25. Thobani, M. (1998) "Meeting Water Needs in Developing Countries: Resolving Issues In Establishing Tradable Water Rights" In: Easter, K. W., Rosegrant, M. W. and Dinar, A.(eds.) (1998) *Markets for Water: Potential and Performance.* Kluwer Academic Publishers: Massachusetts, USA.

26. Note that a policy of pricing water to cover the full cost of building and managing the infrastructure (the long-run marginal cost) is not optimal if the infrastructure was ill-conceived and built at high cost. If full cost pricing could be enforced, most farmers, who typically account for the bulk of water use, would find irrigation farming unprofitable

27. Water prices can also be used to reflect external costs, and this approach has been applied in some locations. See: Howitt, R. E. (1998) "Spot prices, options prices and water markets: an analysis of emerging markets in California" in Easter, K.W. *et al.* (1998).

28. Dinar, A. and Subramanian, A. (eds.) (1997). "Water Pricing Experiences: An international perspective" *World Bank Technical Paper No. 386.* Washington DC

29. Repetto, R. (1986) *Skimming the Water: Rent-seeking and the performance of public irrigation systems.* Research Report No. 4 Washington DC: World Resources Institute.

30. Bagadion, B. U. (1988) "The evolution of the policy context: An historical overview" In: Korten, F. F. and Siy Jr., R. Y. (eds.) *Transforming a Bureaucracy: the Experience of the Philippine National Irrigation Administration* West Hartford, Connecticut: Kumarian Press.

31. Repetto, R. (1986) *Skimming the Water: Rent-seeking and the performance of public irrigation systems.* Research Report No. 4 Washington DC: World Resources Institute.

32. Thobani, M. (1998) "Meeting water needs in developing countries: Resolving issues in establishing tradable water rights" in Easter, K. W., *et al* (1998)

33. Quotas are quantitative limits on water use and can be based on time, volume, or can be use-specific (such as per crop).

34. Technically, the price elasticity of water demand is equal or close to zero.

35. Baumol, W. J. and Oates, W .E. (1988) *The Theory of Environmental Policy,* Cambridge: Cambridge University Press

36. Bate, R., Tren, R. and Mooney, L. (1999) "An econometric and institutional economic analysis of water use in the Crocodile River catchment, Mpumalanga Province, South Africa" *Water Research Commission Project K5/855,* Pretoria.

37. Repetto (1986).

38. Easter, K. W., Rosegrant, M. W. and Dinar, A.(eds.) (1998) *Markets for Water: Potential and performance* Kluwer Academic Publishers: Massachusetts, USA.

39. Deacon, E. (1997), Compiled Papers, unpublished.

40. Anderson, T. L. & Snyder, P. (1997) *Water Markets* Washington DC, Cato Institute, USA Backeberg, G. R. (1997), "Water institutions, markets and decentralised resource management" *Agrekon* 36 (4):350-384.

41. Strosser, P. (1997) *Analysing alternative policy instruments for the irrigation sector* Wageningen Agricultural University, The Netherlands.

42. Quotas may in fact change this legal linkage by exchanging a legal concept of 'reasonable use' of water with a stipulation of an exact amount of water. But this legal change is rarely obvious since water is still used on the same riparian land. However, trading in quotas gives a much stronger impression of alienation of water use from land. So, while establishing quotas is the key change to the legal status of the rights, hostility to quotas only occurs when they are exchanged.

43. Provision of conveyance infrastructure is almost certainly a job government will underwrite, if it does not fund it completely. Private finance is used in many schemes at the moment (IFC, 1998 – Private Sector Bulletin).

44. Holden and Thobani (1996).

45. Holtzhausen, K. (1997) Archive of Papers, unpublished.

46. Thobani, M. (1998) "Meeting water needs in developing countries: Resolving issues in establishing tradable water rights" in Easter, K. W., *et al* (1998)

47. Rosegrant, M. W. & Binswanger, H. P. (1994). "Markets in tradable water rights: Potential for efficiency gains in developing country water resource allocation". *World Development*, 22(11):1613–1625. Elsevier Science Ltd, UK.

48. Pakistan Water and Power Development Authority (1990) "Trading of canal and tubewell water for irrigation purposes" P&I Publications 358 (December).

49. Strosser, P. (1997) *Analysing alternative policy instruments for the irrigation sector* Wageningen Agricultural University, The Netherlands.

50. Shah, T. (1991) "Managing Conjunctive Water Use in Canal Commands: Analysis for Mahi Right Bank Canal, Gujarat." In: Meinzen-Dick, R. and Svendsen, M. (eds.) *Future Directions for Indian Irrigation* International Food Policy Research Institution, Washington DC.

51. Saleth, R. M. (1998) "Water markets in India: Economic and Institutional Aspects" In: Easter, K. W., Rosegrant, M. W. and Dinar, A.(eds.) *Markets for Water: Potential and performance* Kluwer Academic Publishers: Massachusetts, USA.

52. Anderson, T. L. (ed.) (1994) *Continental Water Marketing* Montana: Political Economy Research Center, USA

53. Easter, K. W., Rosegrant, M. W. and Dinar, A.(eds.) (1998) *Markets for Water: Potential and performance* Kluwer Academic Publishers: Massachusetts, USA.

54. Livingstone, M. L. (1995) "Designing Water Institutions: Market Failures and Institutional Response". *Water Resources Management*, 9:203–220.

55. Anderson and Snyder (1997) quote from the Chilean Constitution which translated states that "the rights to private individuals, or enterprises, over water, recognised or established by law, grant their holders the property over them".

56. Ríos, M. and Quiroz, J. (1995) "The market for water rights in Chile: Major issues" *Cuadernos de Economica* 97:291–315.

57. Hearne, R. R. and Easter, K. W. (1997) "The economic and financial gains from water markets in Chile" *Agricultural Economics*, 15:187–199.

58. Rosegrant, M. W. and Gazmuri-Schleyer, R.. (1994) "Reforming Water Allocation Policy Through Markets in Tradable Water Rights: Lessons from Chile, Mexico, and California" Washington DC: International Food and Production Technology Division, EPTD Discussion Paper No. 6.

59. Hearne, R.R. and Easter, K.W. (1997).

60. *Ibid*

61. Rosegrant and Gazmuri (1994).

62. Hearne and Easter (1997).

63. Bauer, C.J. (1997). "Bringing water markets down to earth: The political economy of water rights in Chile, 1976–95" *World Development,* 25(5):639–656. Elsevier Science Ltd, UK.

64. Hearne, R. R. (1998) "Institutional and organisational arrangements for water markets in Chile" in Easter, K. W., *et al* (1998).

65. Holden and Thobani (1996)

66. Hearne, R. R. (1998) "Institutional and organisational arrangements for water markets in Chile" in Easter, K. W., *et al* (1998).

67. Anderson, T. L. & Snyder, P. (1997) *Water Markets* Washington DC, Cato Institute, USA Backeberg, G. R. (1997), "Water institutions, markets and decentralised resource management" *Agrekon* 36 (4):350-384.

68. Holden and Thobani (1996).

69. Kemper, K. E. and Simpson, L. D. (1995) *A Water Market in Practice: The Northern Colorado Water Conservancy District.* Linköping: Department of Water and Environmental Studies, (mimeo) Linköping University, Sweden

70. Sturgess, G. L. and M. Wright (1993) *Water Rights In Rural New South Wales: The Evolution of a Property Rights System,* St. Leonards, Australia: The Centre For Independent Studies

71. Ibid

72. Maas, A. and Anderson. R. (1978) *And the deserts shall rejoice: Conflict, growth and justice in arid environments.* MIT Press, Cambridge, Massachusetts.

73. Anderson and Snyder (1997)

74. Landry, C (1999). "Market transfers of water for environmental protection in the western United States". *Water Policy* 1:457–469. Elsevier Science Ltd, USA.

75. Kemper (1996).

76. They are also allegedly forthcoming in Israel and Peru.

77. Landry (1999).

78. Dinar, A. and Subramanian, A. (eds.) (1997). "Water Pricing Experiences: An international perspective" *World Bank Technical Paper No. 386.* Washington DC

79. Anderson and Snyder (1997).

80. Gleick, P. H. (1993) *Water in Crisis* New York, Oxford: Oxford University Press

15

Sustainable Development
and Marine Fisheries

MICHAEL DE ALESSI

World fishery production today is more than six times what it was half a century ago. Since 1960, the quantity of fish destined for direct human consumption has more than tripled and now stands at around 90 million tonnes per year.[1] Estimates put total world capture fisheries and aquaculture production (which includes all marine fish harvests) at 125 million tonnes for 1999, up from 113 million tonnes in 1995.[2] But this apparently rosy picture belies some worrying facts: The recent increase in production has come primarily from aquaculture; after decades of steady increases, capture fishery numbers now fluctuate around 90 million tonnes per year. More disturbingly, stocks of some important commercial fish are severely depleted and many of those are *not* recovering.

If one considered only the plight of the Atlantic cod, it would be tempting to agree with WWF's 1996 claim that "Without a doubt we have exceeded the limits of the seas".[3] Cod is one of the most fecund fish (an average female produces one million eggs) and has been a staple of many diets for centuries. Once one of the world's richest fishing grounds, cod are so scarce today in New England and Atlantic Canada that they are close to commercial extinction.[4] However, many other fisheries are healthy, and recent evidence indicates that even those that have been depleted may be remarkably resilient.[5]

This chapter considers why some fish stocks have been depleted while others have not. It assesses the various attempts that have been made to improve fisheries management and provides insights into which kinds of institutions lead to sustainable management of stocks.

Sustainable Fishing

The notion of sustainability flows directly from the biological sciences, especially the study of natural populations of animals such as fish and wildlife. For most of the twentieth century, fisheries science focused on determining the 'maximum sustainable yield' (MSY); that is, the largest harvest that could be taken year on year from a specific population of fish. This calculation involved not only estimates of growth and fecundity but also of the size of base population needed to maintain stocks.

Unfortunately, due both to the great uncertainties involved in estimating fish populations and – perhaps more importantly – to political gamesmanship,[6] fish populations have been decimated in the name of MSY. As a result, the concept has fallen out of favour. But while the 'maximum' part has been maligned, 'sustainability' remains the holy grail of fisheries management for everyone from biologists to conservationists to some environmentalists.

Of course, without appropriate definition, sustainability is ambiguous and can be used – as was MSY – to justify good, bad and even ugly policies. For the purposes of this chapter, sustainability is taken to mean *an activity **or** a population of a species that is resilient over time*. It is important to consider sustainability of both populations *and* activities for two reasons. First, especially in the developing world, wildlife and fisheries will not be conserved unless the people who depend on them for food and sustenance also prosper. (Anyone worried about where their next meal will come from will hardly be too concerned about the effects of their catches on the long-term health of a particular fish population.) Second, the concern is with sustainable *development,* so activities that benefit people must be given due consideration.

Conservation and development are often presumed to be diametrically opposed to one another. Having made this presumption, some argue that the environment must be sacrificed for development. Others argue that development must be foregone in order to preserve the environment. But these prescriptions are misguided because the underlying presumption is incorrect. While there are certainly many examples of environmental degradation resulting from development, there are also many examples of environmental improvement and economic development being mutually supportive. Indeed, development and the wealth that it creates are in many respects the environment's best friend.[7]

It is also important to distinguish environmental change from environmental degradation – modern environmentalists tend often to confuse the two, seeing all change as bad.[8] In reality, of course, change is the norm and must be embraced as an inevitable part of the sustainability of a system. So, although we may be concerned about the resiliency of an individual fish species, this must be balanced against the resiliency of the marine ecosystem as a whole, as well as the resiliency of the human activities that depend upon the ecosystem and relevant parts of it.

Decentralising Control Over Resources

There is no single answer as to how best to conserve the ocean's resources. However, experience shows that when people are given the opportunity to conserve marine resources, they generally do so.[9] To give people that opportunity, however, there must be a dramatic shift in the way fisheries are managed, away from many current regimes that all too often encourage depletion of resources and the wastes of time, effort and capital. Resource conservation is not happenstance; it is a rational response to a given situation.

In most countries, the political solution to over-harvesting of resources has been the imposition of regulatory controls on fishermen and other resource users. As the above discussion suggests, these controls have largely failed to stem the over-harvesting of important oceanic and terrestrial species. The problem with such regulations is their failure sufficiently to constrain the incentives traditional resource users have to harvest resources. Thus, limiting the number of days a fisherman may put to sea induces him to invest in more equipment, so that on those few days he is at sea he is able to pull in just as many fish as he did before. Meanwhile, creating 'national parks' from which poor local peasant farmers are excluded, and nationalising the remaining wildlife actually encourages peasants to poach animals – especially if the animals threaten to trample on their crops.

Some nations have, however, demonstrated the promise of an approach, which, at its core, recognises the role of economic incentives in conservation and sustainable development. These nations, notably those in the South Pacific for marine resources[10] and southern Africa for wildlife,[11] have decentralised control over resources – effectively making the users the owners of the resources. Resource users now have a proprietary interest in the resources they rely upon for their livelihood, so they have incentives to ensure that their resource increases in value over time, whilst bringing in a steady annual return. We now consider how this system works in more detail.

Property Rights

Property rights essentially define who has the right to do what with a resource. Economists studying natural resources have demonstrated the fundamental importance of property rights institutions to conservation and sustainable use. The allocation of property rights sets the rules of the game. There are, broadly, three types of allocation rules for resources:

- Open access (no one has any property rights);

- Government ownership;

- Private ownership (property rights are held by individuals or groups).

Any attempt to exert control over resources is an attempt to define property rights. When property rights are not well-defined, or cannot be readily enforced, a situation approximating that of open access pertains. Under open access, scarce natural resources tend to suffer from what Garret Hardin termed 'the tragedy of the commons.'[12] When it is impossible legally to exclude others from utilising a resource, users will tend to behave as though the resource is non-renewable, taking as much as possible as soon as possible, regardless of the impact on the stock.

This does not cause problems when the resource is plentiful and harvests are small,[13] but as the pressure grows, so does the potential for depletion. In a system of open access to a valuable resource with low harvesting costs, there are no rewards for restraint, and then, as Hardin described, "ruin is the destination toward which all men rush."

Property rights encourage particular users to consider the harms and benefits they cause because they determine whether the future effects of their current behaviour (either positive or negative) will be borne by the owner. As economist Harold Demsetz put it, "A primary function of property rights is that of guiding incentives to achieve a greater internalisation of externalities."[14] Thus, as property rights become better defined, resource stewardship becomes more attractive and, equally, owners bear more of the costs of rapacious behaviour. Clearly defined and readily enforceable private property rights to marine resources are rare.

However, those few examples that do exist strongly support the arguments of theorists who have promoted private property rights in the oceans as a means to improve resource management.[15] If a resource is held privately, then the owners have incentives to protect, conserve and husband resources. Formally, a resource is deemed to be privately owned when property rights over the resource are well defined and readily enforceable by an identifiable set of residual claimants. The crucial determinant for whether or not a resource is really privately owned, however, is whether the welfare of those making decisions about its use is tied to the economic consequences of their decisions.[16]

It is the lack of private property rights, not economic development or 'greed' that leads to environmental degradation. The reason that development and environment have often been viewed as diametrically opposed to one another is that private property rights – which would have been a bulwark against environmental degradation – have so often been trampled by the state in the name of economic development.

The Creation and Evolution of Property Rights

Whether private rights develop depends not only on the value of resources and the costs of monitoring them but also on the political costs of creating those rights.[17] The process can be mutually reinforcing; as resources become more valuable, owners invest more in creating, monitoring and enforcing private ownership rights, which in turn make resources more valuable, and so on.

An early example of the development of private property rights concerned the trade in beaver pelts and the Montagne Indians in North America.[18] Prior to the arrival of the settlers, beavers were plentiful and not highly valued by the Montagne, so they did not bother to impose any restrictions on harvesting them. But with the rise of the fur trade, the value of beaver pelts increased rapidly and suddenly the beavers became susceptible to depletion. The Montagne responded by rapidly developing a system for allocating certain areas to specific families who could then benefit from conserving the beaver. As beaver were the only resource valuable enough to warrant this kind of protection there were no other harvest restrictions imposed on these territories. Other people were free to roam across them and were even free to kill and eat the beaver as long as they left the pelt behind. Thus, the trade in beaver pelts was 'sustainably developed'.

Another exemplary case study is the American West at the end of the nineteenth century. Much like the oceans not so long ago, few could imagine depleting its vast resources. But as the West was settled, its water and grassy lands became progressively more scarce and more valuable. Research by economists Terry Anderson and P.J. Hill showed that, as the rights to these resources became more valuable, more effort went into enforcing private property rights, and therefore into innovation and resource conservation.[19]

Defining private property by physical barriers was desirable, but there were too few raw materials, so livestock intermingled and monitoring was difficult. However, frontier entrepreneurs soon developed branding systems to identify individual animals, and cattlemen's associations were formed to standardise and register these brands, allowing cattlemen to define and enforce ownership over a valuable, roaming resource. Then, in the 1870s another innovation came along that radically altered the frontier landscape: barbed wire. Barbed wire was an inexpensive and effective means of marking territory, excluding interlopers, and keeping in livestock. It made it easier to enclose property and exert private ownership, and illustrates how private property rights encourage innovation.

Responses to Depletion

Unfortunately, the most common response to open access and depletion has been government intervention, which has meant that ingenuity and innovation have focused on circumventing restrictions, rather than on conserving or even enhancing resources. For the fisheries, these restrictions are, typically, limits on fishing gear, effort and seasons. Yet, so many variables influence harvest that regulators cannot hope to keep up. As seasons are shortened, fishers might respond with larger nets. As larger nets are restricted, more horsepower may take up the slack, and so on. One of the more extreme examples was the Alaskan halibut fishery, where the primary limitation was the length of the fishing season. As the season shortened, larger boats, larger nets, and technologies such as fish-finding sonar began to appear. Before long, a season that was once months long was down to two days, with no discernible reduction in the total harvests.

The halibut story is an extreme one, but the plot is common around the world. Political battles are inevitably fought over pieces of a pie that never gets bigger. Instead of investing in efforts to enlarge the pie, resources are devoted to attempts to grab a bigger share at some else's expense. Moving resource allocations out of the political arena, however, turns a zero-sum game into a positive one.

While government control may define who has the right to fish, it fails to internalise the effects that harvesters have on the resource, and so it has generally failed to conserve marine resources, let alone help to provide a leg up to those in developing countries who depend on the marine environment for sustenance. Fortunately, however, many subsistence fishing communities are already familiar with one form of private property – common property.

Common Property Rights

Private individual property rights offer the greatest rewards for conservation to their owners, but are also the most costly to define and enforce. Thus, in some instances, private communal property may be optimal, depending on the resource and the costs of monitoring and enforcing rules and excluding outsiders. Private communal property rights may range from nearly open access to a strict system of controls and rules, but essentially they define the rights shared by the members of a group with exclusive access to a resource.[20]

Margaret McKean and Elinor Ostrom provide an explanation for the existence of private communal rights: "Common property regimes are a way of privatising the rights to something without dividing it into pieces ... Historically, common property regimes have evolved in places where the demand on a resource is too great to tolerate open access, so property rights in resources have to be created, but some other factor makes it impossible or undesirable to parcel the resource itself".[21]

An example cited by McKean and Ostrom is a very large, forested area where edible flora and fauna are patchily distributed. Private communal rights may not be easily transferable, but the welfare of either the individual or group is tied directly to the health of the resource, thereby generating incentives for conservation and sustainable use.[22] However, limits on transferability of communal property may lead to problems of transition, as transferability bolsters resiliency in the face of pressure from outsiders. If out-transfers are not possible, pressure from outsiders for access often leads to expropriation, either of the resource itself or of the right of access to it.

In many cases common property regimes are not legally recognised, but as long as they are enforceable they can be workable. In the Maine lobster fishery, for example, lobstermen formed 'harbour gangs' to mark territories and turn away outsiders.[23] As a result, lobstermen in these gangs have higher catches, larger lobsters, and larger incomes than lobstermen who fish outside controlled areas. Some of these common access rights have even now been legally recognised by the state of Maine.

Unfortunately, in most places around the world, not only does the legal system not recognise common property rights, it is often biased against them. In fact, many legal systems favour individual property at the expense of common property. Ostrom notes finally that "When resources that were previously controlled by local participants have been nationalised, state control has usually proved to be less effective and efficient than control by those directly affected, if not disastrous in its consequences."[24] Common property rights arise when parcelling is difficult and/or the return from doing so is low.

Otherwise, as resources grow in value and/or monitoring becomes cheaper, private property rights become increasingly attractive.[25] Common property rights are emphasised in this paper because of their prevalence in developing countries, and because in many places where individual rights to resources are too expensive to enforce, they may afford an opportunity for private property rights to gain a foothold, to the benefit of both people and wildlife.

Examples

Coral Reefs

Coral reefs in the South Pacific suffer widely from destructive fishing practices such as fishing with dynamite or cyanide.[26] However, such practices are often proscribed in places where fishing rights are securely owned, most often by a village, clan or community. Biologist Robert Johannes has studied coral reef conservation throughout the Pacific and found village control over local marine resources to be the surest indicator of reef health.[27]

Reef tenure typically extends from the beach to the outer edge of the reef, sometimes even miles out to sea.[28] These reefs are valuable assets to the community and so are fiercely protected. In Fiji some communities employ fish wardens to watch over the reefs. In Johannes' study of Palauan fishers, he found community-managed fisheries employing closed seasons and areas, abiding by size limits and even imposing quotas to ensure conservation.[29]

The experience in much of the Philippines offers a dramatic contrast. Most of the common property regimes there were destroyed by the Spanish Conquest. Today fishing over the reefs is nearly open access and many reefs are dead or deteriorating. The WWF's Hong Kong office looked into the problem of cyanide fishing and found that reef fisheries in Southeast Asia 'work in a sustainable way only in those few places where the rights to fish a particular reef are clearly established'.[30]

In many of these island communities, secure tenure has also led to initiatives such as giant clam farming, redress for coastal pollution and the development of eco-tourism ventures, none of which would have been possible without a healthy environment and the income derived from well-defined property rights. And because of those secure property rights, development centres on a healthy environment.

Japanese Cooperatives

Another formal communal arrangement exists in Japan, where in many places Fishery Cooperative Associations (FCAs) hold the rights to coastal marine resources and impose strict conservation measures on their members. As a result, coastal marine resources in Japan are generally healthy. Cooperative ownership in Japan is so strong that

FCAs have even been able to block polluting coastal developments by asserting the primacy of their fishing rights. As Kenneth Ruddle and Tomoya Akimichi note, "Because fisheries rights have a legal status equal to land ownership under Japanese law, ... a private developer must ... either purchase all of the fisheries rights ... or compensate for any reduction in the quality of the rights".[31]

These cooperatives are, however, not purely private endeavours, since they receive significant government subsidies (as do most Japanese farmers). But they do demonstrate the emphatic link between exclusive control and the stewardship of marine resources.

Oysters in Maryland and Washington State

Much like the Atlantic cod of New England, the oyster fishery in Chesapeake, Maryland, was once a great industry and oysters were a staple of the local diet. Sadly, despite more than a century of warnings, oyster stocks in the Chesapeake have declined precipitously. In 1891, William Brooks, a scientist and Maryland Oyster Commissioner, declared that "all who are familiar with the subject have long been aware that out present system [of open access] can have only one result – extermination."[32] Brooks recommended the creation of privately owned oyster beds, in order to encourage oyster cultivation and stewardship. But regulation was chosen instead. As stocks continued to decline over time, the Maryland government continued to increase its involvement in the fishery, presenting us with a dramatic case of regulatory failure.

It has been said that Maryland has passed more legislation dealing with oysters than any other issue. Restrictions on technology were (and still are) so severe that the skipjacks plying certain Maryland oyster beds are the only commercial fishers in the United States still powered by sail. In recent years disease has also played a part in the continued decline of the Chesapeake oyster, but even before this new threat, oyster harvests were well below one per cent of what they once were.

In marked contrast to public oyster beds in Maryland, oyster beds in Washington state may be owned 'fee simple'[33] – completely privately, and with a title to prove it, just like a house. As a result, harvests of oysters in Washington state look very different from those in Maryland. Additionally, the oysters are harvested by relatively modern means and the beds are often seeded from high-tech hatcheries financed by the oyster growers. Private rights not only allowed oyster growers to protect their beds from over-harvesting, they allowed their industry to develop; to invest in enhancement projects, to invest in biological research, and even to stop pollution.[34]

Individual Transferable Quotas

Although the benefits and feasibility of private ownership are most readily apparent for sedentary species like oysters, they may also be perfectly applicable to more far-ranging species. Many countries are attempting to improve fisheries management by introducing some limited forms of private ownership into the fisheries, frequently by creating a quasi-property right called an Individual Transferable Quota (ITQ). ITQs grant a right to harvest a certain percentage of a Total Allowable Catch (TAC) of fish in a given year and can be bought or sold. Over time, ITQs may also offer a real opportunity to move towards the private ownership of marine resources. Within the past two decades they have been introduced in New Zealand, Iceland, Australia, the United States and Canada. Some developing countries, most notably Namibia, are also beginning to experiment with such systems.[35] Although they are not really private rights, ITQs can be a tremendous step in the right direction.

In contrast to regulation-based controls, they provide positive conservation incentives for those harvesting resources, because the health of the fishery is capitalised into the value of the quota. In other words, the brighter the prospects for future harvests, the higher the value of ITQs, allowing ITQ owners to gain now from steps they take to ensure the long-term health of the fishery. Some banks are even beginning to accept ITQs as collateral, improving access to the fishery by making loans easier to secure for new entrants.

ITQs in New Zealand

Until the introduction of ITQs, fisheries management in New Zealand followed a familiar pattern. Since 1960 the government had condoned free entry into the fisheries and subsidised development, with predictable results; depletion of fish stocks and over-investment in boats, nets and other technologies. The deplorable state of many inshore fisheries, combined with the importance of fish to the New

Zealand economy, forced a rethink of past policies. The result was the Fisheries Act of 1983, which consolidated previous legislation and set out both to improve resource conservation and to increase economic returns from the fisheries. This led to the creation of tradable quotas for some of the deep-water fisheries and, in 1986, the introduction of ITQs for all significant commercial finfish species with the creation of the Quota Management System (QMS). Today, following numerous improvements, the programme appears to have been tremendously successful. Fish stocks are generally healthy, the fisheries receive no subsidies, capacity in the fishing industry has been reduced voluntarily (more efficient quote owners bought out less efficient ones and retired redundant equipment – especially in the deep-water fisheries), and there has been an increase in investment in scientific research into the fisheries.[36] The New Zealand Ministry of Agriculture's Philip Major described a remarkable transformation in attitudes after the creation of the ITQ system: "It's the first group of fishers I've ever encountered who turned down the chance to take more fish".[37]

It has been suggested that ITQs will result in the consolidation of the industry and the elimination of the small-scale fisher. While there has been some consolidation in New Zealand, especially in capital-intensive deep-water fisheries, the total numbers of vessels, full-time employees, and quota owners all increased from 1986 to 1996.[38] Of course, one reason for this is that limits have been placed on the percentage of the overall quota any one fisher may own, ranging from a limit of 45 per cent in a given area for species such as hoki and orange roughy to 10 per cent for rock lobster. While such limits may be economically inefficient – and for that reason it would be desirable to remove them in the longer term – in the short-term they serve to quell objections from those who fear excessive concentration and make it politically easier to shift from open access to a system of privately owned ITQs.

The New Zealand quota system seems to be moving closer and closer to a real system of privately owned fisheries. In the orange roughy fishery, for example, quota owners in 1991 formed the Exploratory Fishing Company (ORH 3B) Ltd., in large part to fund management science and research.[39] Similarly, the owners of scallop ITQs formed the Challenger Scallop Enhancement Company Ltd, which manages the fishery. Through contracts, the company levies money for research, enhancement (a vigorous reseeding programme), monitoring, and enforcement both of ITQ allocations and daily catch limits.[40] They have even contracted with other fleets and owners to ameliorate multi-species effects of other fisheries (in particular, the dredge oyster fishery and the inshore finfish fishery) on habitat and productivity.[41] Once again, when property rights in marine resources are clearly defined and readily enforceable, development proceeds along a path that is truly sustainable.

Overcoming the Political Nature of ITQs

This evolution from political allocation to private ownership is a very important aspect of some ITQ systems. The New Zealand system is the most notable in this respect and has gone the furthest, but the Icelandic system has also gradually moved in the same direction. In many other places, however, ITQs have been explicitly set up in a manner that prevents them from evolving into stronger rights. For example, the ITQ programme in Alaska (which successfully ameliorated the 48-hour halibut derby mentioned earlier) specifically states that ITQs are not private property rights and that they can be taken away without compensation at any time. Such threats of expropriation undermine the credibility of the ITQs as secure, long-term investments, and strike at the very reason why ITQs have had some measure of success in the first place.

As long as an ITQ system remains politically managed, it will be susceptible to many of the pitfalls discussed earlier, limiting the impetus for innovation and sound resource enhancement. It also discourages the exploration of alternative systems for managing resources – such as setting up a scheme under which rights are created in a particular area, rather than a particular species.

These potential pitfalls highlight the importance of giving careful consideration to the structure of the rights being created before an ITQ system is implemented. The central lesson from New Zealand's experience should always be borne in mind: the more an ITQ resembles a private right, the greater is the likelihood that the system itself will adapt and evolve into a system of real private rights, which commensurately have the strongest possible incentives for conservation and sustainable development. The creation of such secure, private-like rights can even help overcome political opposition from powerful groups with emotive and, in many cases, morally and legally justified claims. In New Zealand, the Maoris claimed a significant portion of the fisheries under the Treaty of Waitangi. This claim was settled amicably and the Maoris now own the largest fishing company in New Zealand.

Aquaculture

While the world fish catch has remained relatively stable in recent years, aquaculture production has grown dramatically. In fact, it is primarily responsible for the 20 million tonne growth in world fish production over the last decade.[42] In 1991, world aquaculture production was approximately 13 million metric tonnes, double what it was seven years before.[43] By 1999, that number had jumped to almost 33 million metric tonnes.[44] The reason for these increases is that aquaculture facilities have allowed entrepreneurs to set up private enclosures that "fence" parts of the sea (or even transport it onto land). A fish not harvested today will be there tomorrow, normal rates of mortality notwithstanding. Private ownership has invigorated entrepreneurs to tinker, experiment, and innovate, and, even more importantly, has encouraged others to innovate as well. In this, the experience of aquaculture mirrors that of the cattle ranches in the American West: it was not landowners in the West who invented barbed wire, but entrepreneurs who sought to develop new markets and products.

Of course, aquaculture is not without its problems. Most aquaculture (approximately two-thirds) occurs near the coast or in shallow estuaries, where pollution from outside sources can cripple the operation. In addition, intensive aquaculture in these areas can produce significant amounts of organic pollution, which can reduce levels of oxygen in the water and increase quick-growing algae harmful to marine life. There is also growing concern over the use of antibiotics in some aquaculture. It is worth noting, however, that pollution and environmental degradation generally occur where property rights have not been appropriately defined and/or are not readily enforceable.

Government subsidies and incentives to expropriate coastal areas for aquaculture often further undermine nearshore conservation efforts. In Thailand, for example, aquaculture is heavily subsidised and in many cases farms are built in areas that were previously managed much more sustainably by a system of customary tenure.[45] In Malaysia, the Land Acquisition Act was amended in 1991 to allow the state to appropriate land for any reason deemed beneficial to economic development, including the construction of fishponds. In Ecuador, bribes, corrupt government partnerships and land expropriation are common because "by law, coastal beaches, salt water marshes, and everything else below the high tide line is a national patrimony."[46] Not only shrimp farms but city slums regularly invade these areas, even in national ecological preserves.[47] Alfredo Quarto, a director of the Mangrove Action Project, has pointed out that the main reason why shrimp farmers choose to clear mangrove forests is that they are usually government owned.[48] In other words, government sanctioned open access and expropriation of common property rights are really to blame for coastal habitat destruction in places like Thailand and Ecuador.

Barriers to Sustainable Development of Marine Fisheries

Legal Recognition

One major reason that Japanese cooperatives have been so successful is that they have been recognised by law, which allows them both to defend their rights in court and develop ways of accommodating out-transfers. Unfortunately, in most places around the world, not only does the legal system not recognise private communal rights, it is often biased against them. McKean and Ostrom note that "Some [common property regimes] may have disappeared naturally as communities opted for other arrangements, particularly in the face of technological and economic change, but in most instances common property regimes seem to have been legislated out of existence".[49]

This was certainly the case in the Pacific Northwest, where Native Americans had developed complicated arrangements, both within and between tribes, to manage their salmon fisheries.[50] They relied heavily on fixed nets and weirs along the riverbank, but were careful to allow plenty of fish to pass in order to maintain the spawning runs and ensure a future supply. According to Robert Higgs, "Indian regulation of the fishery, though varying from tribe to tribe, rested on the enforcement of clearly understood property rights. In some cases these rights rested in the tribe as a whole; in other cases in families or individuals."[51] But as the numbers and power of settlers increased, these property rights were quickly expropriated by force. Legal recognition of communal rights would go a long way towards resolving this problem, but unfortunately, especially in developing countries, expropriation is the norm. This may explain much of the current emphasis that many policymakers place on maintaining small fishing communities and their "cherished way of life". Barring legal recognition, sentiment seems to be the next best alternative. Unfortunately, this may do more harm than good, as it tends to work toward entrenching the *status quo*. Property rights institutions, including communal ones, are constantly evolving, and while some communities may choose to maintain a certain way of life, others may not. Empowering people with the property rights to protect their environment is a powerful tool for sustainable development. On the other hand, legislating stasis is bad policy.

Opposition to private rights is often justified by arguing that fisheries are a 'public' resource and that any move in the direction of privatising them will prevent 'the public' from benefiting from the resource. But such arguments are based on prejudice – not reason. If it is agreed that the greatest public benefit likely to result from maintaining both a healthy resource and a healthy and prosperous populace, the available evidence suggests that, generally speaking, private ownership is superior to open access. Maintaining open access may appeal to egalitarian values but is likely to lead to a depleted 'wasteland'. By contrast, a shift to private ownership is more likely to ensure access to a valuable, plentiful resource. It may reasonably be concluded, therefore, that in the interests of 'the public', marine resources should be privatised.

Conclusions

A shift towards private solutions to marine conservation problems would almost certainly lead to better stewardship of marine resources and, as such, must be considered a necessary part of 'sustainable development'. With marine resources privatised, innovation would no longer be about finding ways to fish more quickly; it would be about protecting those resources.

There is, of course, no single answer as to what sort of private ownership schemes might develop – if given the chance. In fisheries where there is great uncertainty and large catch fluctuations, there would no doubt be more risk sharing and group ownership schemes – from village common property regimes (as occur in the South Pacific) to fishing companies made up of ITQ owners (as in New Zealand). In other places, advanced technologies, many of which already exist, might be used to define and protect private property in the oceans, just as branding and barbed wire did in the frontier American West.[52]

While new technologies can help define and enforce property rights, their availability must not blind one to the overarching importance of institutions. As the discussion of common property regimes above has demonstrated, in many places appropriate institutions already exist, though we may not understand them as such. It is worth remembering the words of the anthropologist John Cordell: "It is one thing to contemplate the inshore sea from land's end as a stranger, to observe an apparently empty, featureless, open accessed expanse of water. The image in a fisherman's mind is something very different. Seascapes are blanketed with history and imbued with names, myths, and legends, and elaborate territories that sometimes become exclusive provinces partitioned with traditional rights and owners much like property on land."[53]

People the world over will continue to try to improve their lot, whether or not the environment suffers. Fortunately, the success of attempts to marry conservation with development in countries from Palau to New Zealand points to the only workable solution. By recognising and encouraging more private property rights and by allowing conservation and commerce peacefully to co-exist, both people and the environment will be better off – now that's what I call sustainable development.

Notes

1. FAO (2000). *The State of World Fisheries and Aquaculture: 2000*, Rome: Food and Agriculture Organisation of the United Nations.

2. *Ibid*

3. Associated Press (1996) press release June 7.

4. Commercial extinction occurs when it is no longer economically viable to catch the remaining fish.

5. Myers, R. A., Barrowman, N.J, Hutchings, J.A., Rosenberg, A.A. (1995): "Population Dynamics of Exploited Fish Stocks at Low Population Levels," *Science, 269*, p. 1106-8.

6.   Brubaker, E. (1999) "Cod Don't Vote", *The Next City*, January.

7.   Goklany, I. (2000): "Richer is More Resilient: Dealing with Climate Change and More Urgent Environmental Problems", in R. Bailey (Ed.), *Earth Report 2000*, McGraw Hill, pp. 155-188.

8.   Botkin, D. (2001) *No Man's Garden: Thoreau and a New Vision for Civilization and Nature*, Island Press, Washington, DC.

9.   See, for example the following:

     Demsetz, H. (1967): "Toward a Theory of Property Rights", *American Economic Review*, Vol. 57, pp. 347-59.

     De Alessi, L. (1980): "The Economics of Property Rights: A Review of the Evidence", *Research in Law and Economics*, Vol. 2, pp. 1-47.

     Johannes, R. (1981): *Words of the Lagoon*, Berkeley: University of California Press.

     Ostrom, E. (1990): *Governing the Commons: The Evolution of Institutions for Collective Action*, Cambridge: Cambridge University Press.

10. See Johannes (1981).

11. Sugg, I., Kreuter, U. (1994): *Elephants and Ivory: Lessons from the Trade Ban*. London: Institute of Economic Affairs.

12. Hardin (Hardin, G, 1968: The Tragedy of the Commons, Science, Vol. 162, pp. 1,243-48.) drew on the earlier work of economists such as Gordon (Gordon, H.S, 1954: "The Economic Theory of a Common-Property Resource: The Fishery", *Journal of Political Economy*, Vol. 62, pp. 124-42.) and Scott (Scott, A, 1955: "The Fishery: The Objectives of Sole Ownership", *Journal of Political Economy*, 63, pp. 63-124.), and, unfortunately, did not initially recognise the exclusivity of some communal institutions, so his use of the word 'commons' has been the source of some confusion. Ostrom (1990), for example, documents numerous examples of successful common property management. The real problem seems to be not with commons *per se* but with resources that are not privately owned – that is either 'open access' or controlled by the state

13. Even under open access, harvests may be small either because costs of extraction are high (e.g. if the only technology available is sail boats and rod and line fishing tackle), or because demand for the resource is low.

14. Demsetz (1967), p. 348.

15. See, for example the following:

     Keen, E. (1983): "Common Property in Fisheries: Is Sole Ownership an Option?" *Marine Policy*, 7, pp. 197-211;

     Scott, A. (1988): "Development of Property in the Fishery," *Marine Resource Economics*, 5, p. 289-311;

     Edwards, S. (1994): "Ownership of Renewable Ocean Resources," *Marine Resource Economics*, 9, pp. 253-73.

16. De Alessi, L. (1980)

17. Libecap, G. (1990): *Contracting for Property Rights*, Cambridge: Cambridge University Press.

18. Demsetz (1967)

19. Anderson, T., Hill, P.J. (1975): "The Evolution of Property Rights: A Study of the American West," *Journal of Law and Economics*, 12, pp. 163-79.

20. Ostrom (1990).

21. McKean, M., Ostrom, E. (1995): "Common Property Regimes in the Forest: Just a Relic from the Past?" *Unasylva*, Vol. 46, No. 180, pp. 3-15.

22. Unfortunately, anthropologists, economists and policymakers often promote either individual or group ownership at the expense of the other, even though the distinction is frequently muddled. Adding to the confusion are the varying definitions that different

(and even often the same) schools of thought apply to terms like "the commons", "common property" and "private property". For example, biologist Garret Hardin used the word "commons" to mean open access, anthropologists often use it to mean a strictly monitored form of group ownership, and economists frequently dismiss the concept entirely under the assumption that only individual ownership institutions are private.

23. Acheson, J.M. (1987): "The Lobster Fiefs Revisited", in McCay and Acheson (Eds.) *The Question of the Commons*, Tucson: University of Arizona Press. pp. 37-65.

24. Ostrom, E. (1997): "Private and Common Property Rights", *Encyclopedia of Law and Economics*. Edward Elgar, London and University of Ghent, Ghent, Belgium. Available at http://allserv.rug.ac.be/~gdegeest/generali.htm.

25. Demsetz (1967).

26. Barber, C.V., Pratt, V. (1997): *Sullied Seas: Strategies for Combating Cyanide Fishing in Southeast Asia and Beyond*, World Resources Institute/International Marine Life Alliance.

27. Johannes, R., and Ripen, M. (1996): "Environmental, economic and social implications of the fishery for live coral reef food fish in Asia and the Western Pacific", *SPC Live Reef Fish Information Bulletin*, March.

28. De Alessi, M. (1997): "Holding out for some local heroes", *New Scientist*, March 8, p.46.

29. Johannes, R. (1981): *Words of the Lagoon*, Berkeley: University of California Press.

30. The Economist (11 May 1996), p. 35.

31. Ruddle, R., Akimichi, T. (1989): "Sea Tenure in Japan and the Southwestern Ryukus" in John Cordell (ed.) *A Sea of Small Boats*, Cambridge, MA: Cultural Survival Press, p. 364.

32. Brooks, W.K. (1996) [1891]: *The Oyster*, Baltimore, MD: Johns Hopkins University Press.

33. Also known as 'freehold', this is the tenancy of fee simple absolute in possession.

34. De Alessi, M. (1996): "Oysters and Willapa Bay," Private Conservation Case Study, Washington, DC: Center for Private Conservation.

35. Iyambo, A. (2000): "Managing Fisheries with Rights in Namibia: A Minister's Perspective," in R. Shotten (Ed.) *Use of property rights in fisheries management*, Rome: FAO Fisheries Technical Paper 404/1.

36. See the following:

    McClurg, T. (1997): "Bureaucratic Management versus Private Property: ITQs in New Zealand after Ten Years," in Laura Jones and Michael Walker (eds.) *Fish or Cut Bait!* Vancouver, BC: The Fraser Institute, pp. 91-105.

    Sharp, B. (1997): "From Regulated Access to Transferable Harvesting Rights: Policy Insights from New Zealand," *Marine Policy*, 21(6), pp. 501-17.

37. Quoted in *The Economist* (1994), p. 24, describing a 1993 decision by the hoki fleet to not fish an extra 50,000 tonnes of fish allocated to them by the government.

38. McClurg, (1997).

39. Orange roughy has received a lot of attention in recent years due to new findings about their age and stock sizes, and the industry now fishes them much more conservatively.

40. Arbuckle, M. (2000): "Fisheries Management under ITQs: Innovations in New Zealand's Southern Scallop Fishery", *Oregon State: IIFET proceedings* (www.orst.edu/Dept/IIFET/papers).

41. *Ibid*

42. FAO (2000).

43. FAO (1993): *Aquaculture Production 1985-1991*, Rome: Food and Agriculture Organisation of the United Nations.

44. FAO (2000).

45. Bailey, C. (1988): "The Social Consequences of Tropical Shrimp Mariculture Development," *Ocean and Shoreline Management*, pp. 31-34.

46. Southgate, D. (1992): "Shrimp Mariculture Development in Ecuador: Some Resource Policy Issues," Working Paper #5 of the *Environment and Natural Resources Policy and Training Project*, University of Wisconsin.

47. Southgate (1992).

48. Weber, M. (1996): "The Fish Harvesters: Farm Raising Salmon and Shrimp Makes Millionaires, and also Creates Dead Seas, *E Magazine*, November/December.

49. McKean and Ostrom (1995), p. 3.

50. Higgs, R. (1982): "Legally Induced Technical Regress in the Washington State Salmon Fishery," *Research in Economic History*, 7, p. 55-86.

51. Higgs (1982), p. 59.

52. See De Alessi, M. (1998): *Fishing for Solutions*, London: Institute of Economic Affairs for a discussion of how advanced technologies may be used to enforce private rights to marine resources.

53. Cordell, J. (1989): *A Sea of Small Boats*, Cambridge, MA: Cultural Survival, Inc.

16

Overfishing: The Icelandic Solution

HANNES H. GISSURARSON

Introduction

According to the environmentalist group Greenpeace, commercial fishing fleets are exceeding the ocean's ecological limits. 'Instead of coming to grips with the need for dramatic cuts, nations argue over who will get how much of what remains of dwindling fish stocks. Meanwhile, the financial captains of the global fishing industry plough full steam ahead on their unsustainable, competitive rush to vacuum the oceans and turn fish into cash' (Greenpeace, 1997). Greenpeace asserts that modern technology is to blame. Here it will be argued, on the contrary, that modern technology has facilitated not only fishing and therefore overfishing, but also the management of the fisheries, or rather their self-management. By lowering transaction costs—costs of identifying harmful effects of economic activities, solving them in market transactions, implementing and enforcing the solutions, and so on—modern technology has made feasible the development of property rights to certain marine resources, in particular fish stocks.

Under a certain set of rules, therefore, individual owners of fishing capital can in market transactions further their private interests at the same time as they work for the public interest. A 'competitive rush' to harvest fish can, under certain circumstances, be not only sustainable, but also profitable. More than that: it can lead to the conservation and even the organised growth and improvement of fish stocks. A practical example, examined in this paper, is the way in which the Icelanders have coped with overfishing. They have developed a comprehensive system of individual transferable quotas, ITQs, in all commercially valuable fish stocks in their territorial waters, enabling them to 'turn fish into cash' without, at the same time, having to 'vacuum the oceans'. This paper describes how the Icelandic ITQ system works; how the total allowable catch, TAC, for each fish stock is set; how ITQs were initially allocated and what restrictions apply to their transfers; how the ITQ system is administered and enforced by government; and how the problem of migratory fish is solved.

The Evolution of the ITQ System

The evolution of the Icelandic ITQ system was a process of gradual discovery and difficult bargaining. Initially, politicians, marine biologists and vessel owners were mainly concerned about the conservation of fish stocks. It was only later that they came to realise the economic problem of unlimited access to a limited resource, the 'tragedy of the commons' (Hardin, 1968). From an economic point of view overfishing is similar to pollution. Where access to a fishing ground is free, the cost of adding one more vessel (or another unit of fishing capital) to the fishing fleet on the ground is not borne solely by the vessel owner. His activity has harmful effects on others. The consequences are over-capitalisation and excessive fishing effort. The fishing fleet is much larger than would be most efficient. As an illustration, sixteen boats may be harvesting a lesser catch than that which eight boats could easily harvest.

There is one big difference, however, between pollution an overfishing. Pollution is visible, whereas the economic costs that owners of fishing capital impose on one another are invisible.

Those costs can be, and have been, demonstrated by economists (Gordon, 1954; Scott, 1955), but vessel owners usually come to realise the problem when it is too late—when fishing is exceeding not only the level of highest return on outlays, but also the maximum sustainable yield. Memories of the collapse of herring in the late 1960s may however have facilitated the acceptance by Icelandic vessel owners of what was in effect the enclosure of fishing grounds. Desperation lessens transaction costs (Libecap, 1989).

Another factor lessening transaction costs is homogeneity. Because Iceland's pelagic fisheries were relatively homogeneous, with similar vessels, the introduction of vessel catch quotas and later ITQs was relatively easy. The bargaining process was much more difficult in the heterogeneous demersal fisheries. Owners of small boats, some of them working part-time, did not think, for example, that they had much in common with owners of large freezer trawlers. Indeed, as we have seen, some small boats are still outside the ITQ system. And vessel owners in villages close to the most fertile fishing grounds also thought that they had different interests from other vessel owners, and their strong opposition delayed the introduction of a comprehensive ITQ system for many years.

The main lesson to be learned from this process is that the introduction of ITQs in a fishery, however necessary it may seem to politicians, marine biologists and economists, is by no means a simple task. There are all kinds of interests which may oppose it. A commons like the fish stocks in Icelandic waters will only be enclosed if the private interests of those utilising the commons are made to coincide with the public interest. It was probably crucial for the evolution of the Icelandic ITQ system that the Association of Fishing Vessel Owners repeatedly took the initiative in the process, and that government worked closely with it (Jonsson, 1990), although it inevitably led critics to say that government was in the thrall of the Association of Fishing Vessel Owners. But a cart without a horse to drive it is of little use.

The really important question is: 'Who Cares Whether the Commons is Privatised?' (Buchanan, 1997). It is difficult to see, for example, how vessel owners in the Icelandic demersal fisheries would have agreed to any other initial allocation of quotas in late 1983 than that which was based on catch history. This was the only way for them to continue utilising the fish stocks without much disruption. In this way they could maintain the value of their investments and human capital whereas it would have become almost worthless if government had auctioned off individual quotas to the highest bidders, as some economists proposed.

In essence, the problem in the Icelandic fisheries was the same as in all fisheries utilising modern technology, and operating under free access to fishing grounds: It was, to return to our illustration, that sixteen boats were harvesting even less than that which eight boats could easily harvest. The task therefore was to reduce the number of boats from sixteen to eight. In theory, this could be accomplished by outbidding the owners of the eight excessive boats, by taxation or in an auction of quotas. But in practice, this would have been difficult, if not impossible. In the Icelandic case, what was done was to assign transferable quotas sufficient for the profitable operation of eight boats, to the owners of sixteen boats.

Over time, the eight boat owners who wanted to continue harvesting fish would have a great incentive to buy quotas from their eight colleagues who for one reason or another wanted to leave the fishery. Thus, people were not outbid; they were bought out.

Table 1

Main Stages in the Evolution of the Icelandic ITQ System

1975: Individual quotas in herring fishery

1979: Quotas in herring fishery transferable

1980: Individual quotas in capelin fishery

1983: Vessel owners recommend individual quotas in demersal fisheries

1984: Individual (mostly) transferable quotas in demersal fisheries. Issued for a year

1985: Effort quota option in demersal fisheries. ITQs issued for a year

1986: Individual quotas in capelin fishery transferable. ITQs issued for two years (1986-87)

1988: Individual transferable quotas in all fisheries. Effort quota option retained. ITQs issued for three years (1988-90)

1990: Fisheries Management Act to apply from 1 January 1991

1991: Comprehensive system of transferable share quotas in all fisheries for all vessels over 6 GRT. Effort quota option removed

1993: Supreme Court decides ITQs be taxed as property

1997: Harvesting outside Iceland's Exclusive Economic Zone (EEZ) mostly made subject to the ITQ system

1998: Quota Exchange; legal restrictions on speculation in and concentration of ITQs

2000: Supreme Court upholds initial allocation of ITQs on basis of catch history

---

The Nature of the ITQ System

Economists analysing the 'tragedy of the commons'—the overutilisation of non-exclusive natural resources—generally agree that the tragedy is caused by the absence of private property rights to those resources. In the costly race to extract value from such resources, whether they are plots of land, oilfields, mines, or fish stocks, the rent which could be derived from them is dissipated. 'The business of everybody is the business of nobody.' It was only with the enclosure of land, for example, that the problem of overgrazing was solved, and cultivation replaced simple extraction.

The Exclusive Economic Zones (EEZs) which fishing nations have established in the 20th century may be regarded as important steps towards the enclosure of marine resources. At first sight, however, private property rights in areas of the sea or in individual specimens of fish do not seem technologically feasible, at least not in deep-sea fisheries; such rights would require techniques of fencing or branding, either non-existent or difficult to develop. ITQs may however go far to solve the fisheries problem (Arnason, 1990), precisely because they have some characteristics of private property rights: they are exclusive which means that only those who hold them may harvest fish; they are individual so that the responsibility for their utilisation is clearly defined and lies with individuals; they are divisible which enables fishing firms freely to decide how much of them to hold at any given time; they are transferable which means that market forces are allowed to select the most efficient fishing firms; and they are permanent, making long-term planning possible.

ITQs are not too difficult to administer or enforce, either, although the political problem of their introduction and initial allocation should not be minimised. Therefore, it is not surprising that ITQs are increasingly being used in world fisheries. Between 5 and 10 per cent of world total catches are presently harvested under some kinds of vessel catch quotas. Iceland and New Zealand are the only two countries to have developed a comprehensive ITQ system although ITQs are also widely used in the Netherlands, Australia and some other countries. Despite some weaknesses, the Icelandic ITQ system does not seem too different from the system described by economists as going far to solve the fisheries problem.

Total Allowable Catch

The two pillars of the Icelandic ITQ system are total allowable catches (TACs), and individual transferable quotas (ITQs). TACs are set annually by the Minister of Fisheries for each of the commercially valuable species of fish in Icelandic waters, on the basis of recommendations from the Marine Research Institute, (MRI). Economic considerations—receiving the maximum return on fishing capital—do not seem to play an important role in the setting of TACs although that may change in the future. In the first few years after the introduction of ITQs in the demersal fisheries, the Minister of Fisheries tended to set somewhat larger TACs than recommended by the MRI, mainly because as a politician he was concerned about adverse effects on the economy by sharp reductions in TACs, especially in the fishing villages scattered around Iceland's coastline. This has gradually changed, especially after 1991. In 1995, government even adopted a special rule about the annual TAC in cod: it is to be set at 25 per cent of the fishable biomass, as estimated by the MRI. Thus, the TAC is determined in and by the annual stock assessment. By applying this rule, marine biologists estimate that the chances of stock collapse go down to less than 1 per cent.

In June 2000, the government revised the rule in order to stabilise the setting of TACs in cod between years. It stipulated that the difference in TACs between years should not exceed 30,000 metric tonnes (MT). The MRI had reported weak classes of cod for harvesting in 1999-2001, with an expected strengthening of the stock thereafter. Therefore the cod TAC was set at 220,000 MT for 2000-2001, compared with 203,000 MT under the old rule. Table 2 reproduces the recommendations during 1984-2000 by the MRI for the TAC in cod, the decision by the Minister of Fisheries, and the actual total catch.

The sharp reductions in TACs of cod in 1994-96 are noteworthy. If the members of the Association of Fishing Vessel Owners had not by then begun to think of themselves as stakeholders in the cod fishery, it is doubtful that such sharp reductions could have been accomplished relatively peacefully in a country as heavily dependent on fishing as Iceland is.

Of the 1999-2000 TAC in cod, almost 35,000 MT were reserved for small boats fishing with handline and longline and some 6,500 MT for other purposes, chiefly to compensate for setbacks in other fisheries. A portion of the TACs in haddock, saithe and catfish was reserved in a similar way. It should be mentioned that 1999-2000 TACs for inshore and deep-sea shrimp were provisional, in line with the recommendations of the MRI and pending further research and stock assessment.

Table 2

*Recommended and Set TACs in Cod
and Total Actual Catches, 1984-2000 (in MT)*

Year	Recommended TAC (MRI)	Allocated TAC (Ministry of Fisheries)	Actual Total Catch
1984	200,000	242,000	281,000
1985	200,000	263,000	323,000
1986	300,000	300,000	365,000
1987	300,000	330,000	390,000
1988	300,000	350,000	376,000
1989	300,000	325,000	354,000
1990	250,000	300,000	333,000
1991	240,000	245,000	245,000
1991-92	250,000	265,000	273,000
1992-93	190,000	205,000	240,000
1993-94	150,000	165,000	196,000
1994-95	130,000	155,000	164,000
1995-96	155,000	155,000	169,000
1996-97	186,000	186,000	201,000
1997-98	218,000	218,000	227,000
1998-99	250,000	250,000	N.A.
1999-00	247,000	250,000	N.A.

Source: Marine Research Institute.

Individual Transferable Quotas

ITQs constitute the other pillar of the Icelandic fisheries system. ITQs are shares in the TAC of a fish stock. They are issued to each vessel for an indefinite period of time, in the demersal fisheries initially, on the basis of catch history in 1981-83. The only vessels partly exempt from the system are boats under 6 MT whose owners have chosen to operate under effort restrictions (a given number of allowable fishing days). They harvest, however, a small proportion of the total demersal catch.

The ITQs are transferable both annually and permanently. A legal distinction is therefore made between two kinds of transferable quotas issued to a vessel: her TAC-share, given in percentages, and her

Annual Catch Entitlement, ACE, given in MT, where the ACE is a simple multiple of the TAC for the fishery, and the vessel's TAC-share. For example, if a deep-sea trawler initially received a 0.1 per cent share of the TAC in cod, and if the TAC in the fishing season 1999-2000 is 250,000 MT, then the owner of that vessel may use it to harvest 250 MT of cod in the given year and expect to harvest 0.1 per cent of the TACs set in coming years. His TAC- share is 0.1 per cent, and his ACE in 1999-2000 is 250 MT. He can do one of three things with his quota: 1) he can himself harvest 250 MT over the 1999-2000 season; 2) while keeping his TAC-share, he can sell his ACE, or a part of it, to the owner of another vessel, that is the right to harvest 250 MT, or a part of it, over the 1999-2000 season; 3) he can sell his TAC-share, that is the right to harvest 0.1 per cent share in the TACs set now and in coming years.

Both the TAC-shares and the ACEs are perfectly divisible. The TAC-shares are also perfectly transferable. There are some restrictions on transfers of ACEs, however, with the objective of stabilising local employment. While ACEs can be freely transferred between vessels under the same ownership or within the same region, their transfers between vessels in different regions have to be approved by the Minister of Fisheries after a review by the regional fishermen's union and local authorities. Since few transfers are blocked, in practice the ACEs can be regarded as freely transferable. Over time most of the ITQs have indeed changed hands: In February 2000 only 19 per cent of the quotas initially allocated in the demersal fisheries were still held by those who originally received them (Morgunbladid, 2000).

Since the Icelandic fisheries are mixed fisheries, vessels are bound to come up with different species of fish on the same fishing trips, haddock as well as cod or redfish, to name a few. The TAC-shares in different fish stocks therefore have to be interchangeable.

But species of fish differ in value: 1 MT of cod is for example worth much more than 1 MT of capelin. Cod is therefore used as the common denominator of the system. The term 'cod equivalent' denotes the relative market value of different species of fish, set by a regulation every year. For each vessel having a quota for several species the total quota may be calculated in cod equivalents. Quota transfers between vessels are also often measured in cod equivalents. In the fishing season from September 1st 1998 to August 31st 1999, the cod equivalent values were, for example, as follows: cod 1.00, haddock 1.05, saithe 0.65, redfish 0.70, plaice 1.20, Greenland halibut 2.15, ocean catfish 0.85, witch 1.20, dab 0.65, long rough dab 0.60, capelin 0.08, herring 0.14, nephrops 8.55, shrimp 1.20 and scallops 0.40. While the ITQs are perfectly divisible, and easily transferable, their use and transfers are restricted in some ways: all transfers of TAC-shares (permanent quotas, in percentages) have to be registered with the Fisheries Directorate.

Most transfers of ACEs (quotas over a season, in MT) have to go through the Quota Exchange. The owner of a vessel will lose his quota, measured in cod equivalence, if his vessel harvests less than 50 per cent of the vessel's total quota in two subsequent years. The net transfer of quota from the vessel in any given year must not exceed 50 per cent of her quota. Moreover, no fishing firm may hold more than a given fraction of quotas in each species of fish.

Administration and Enforcement

Two government agencies, under the direction of the Minister of Fisheries, are mainly concerned with administering and enforcing the ITQ system. The Marine Research Institute, (MRI), investigates the state of fish stocks and makes recommendations about annual TACs in different species of fish to the Ministry of Fisheries. The MRI operates research vessels and collects additional information from skippers. It also undertakes basic research in marine biology.

The MRI has a staff of about 170; approximately one-third of its costs of operation are covered by its own revenues. The Fisheries Directorate (FD), oversees the day-to-day administration of the ITQ system, especially the collection of data on harvesting and landings. It has a regular staff of about 60; approximately half of its budget is covered by its own revenues. In addition, the FD employs observers for fishing in distant waters, outside Iceland's EEZ.

The ITQ system is in effect enforced by controlling landings. All marine catch is required by law to be weighed on officially approved scales at the point of landing. Municipal authorities operate the weighing stations and collect weighing fees from the vessels to cover their costs. The officials of the weighing stations record the landings and verify species compositions. There are 67 ports under such landings control in Iceland, and major foreign export ports are controlled as well. A sophisticated computer system links ports of landings to the FD, enabling the transmission of daily catch data to the FD's computer department. All catch data are transmitted to the FD twice a day and processed for dissemination, by several means, through the FD's Web pages, through monthly publications and by phone to skippers and vessel owners checking their catch status. Status reports are sent to vessel owners regularly and upon request. The FD's Web pages of fisheries data show in detail the catch status of individual vessels, quota transfers between different vessels or in different species, quota shares and landings. A third government agency, The Icelandic Coast Guard, under the direction of the Minister of Justice, and with a staff of about 130, monitors fishing vessels at sea and enforces regional closures, with gunboats, helicopters and aeroplanes. As already mentioned, extensive nursery grounds are permanently closed to fishing vessels, and the spawning grounds of cod are closed for a few weeks in late winter during the spawning period. Moreover, the Minister of Fisheries, on the advice of the MRI, has the right of immediate, temporary closure of areas with excessive juvenile fish. There is also a 12 miles limit for large trawlers in most areas. In addition to the surveillance provided by the FD and the Coast Guard, the Ministry of Fisheries itself employs a group of observers of fishing in the Icelandic waters, some of whom take trips on fishing vessels and some of whom travel between ports of landings. Those observers try to ensure compliance with regulations on mesh size, bycatches, and so on. Mesh size has to be 135 mm or equivalent, for example, and in the shrimp fishery a sorting grid is mandatory to avoid the bycatch of juvenile fish. In the demersal fisheries devices for excluding juveniles are also mandatory in certain areas.

The Ministry of Fisheries itself has an office staff of about 20. The Ministry charges holders of ITQs a low fee for the costs of administering and enforcing the ITQ system, with an upper limit of 0.4 per cent of the estimated catch value. The revenue from the fee is about US$8-9 millions a year, and in addition there is revenue from a fee for fishing permits of about US$2 millions a year.

The total net costs of enforcing and administering the ITQ-system, less than US$30 millions a year, including basic marine biology research and guarding the territorial waters, do not seem huge in comparison to the total catch value in the Icelandic fisheries which is, in the late 1990s, on average about US$800 millions a year. Violations of the Fisheries Management Act and the corresponding Ministry of Fisheries regulations carry heavy penalties, such as fines, expropriation of catch and gear and cancellation of fishing permits. While the Ministry of Fisheries has wide discretionary powers in assessing such penalties and a proven willingness to use them, alleged violators have recourse to the courts if unsatisfied with the Ministry's decisions.

Are the Icelandic ITQs Property Rights?

On land, fencing techniques such as barbed wire have enabled individuals to establish property rights in (that is, to exclude others from the utilisation of) land and other immovable objects, whereas branding techniques have enabled them to establish property rights in (that is, again to exclude others from the utilisation of) animals and other movable objects. Fences can however hardly be erected around different areas of the deep sea (although some kinds of fencing may be possible in inshore fisheries), and it is also difficult to see how individual fish in the sea can be branded (at least cod, herring and other species of fish that the Icelanders harvest).

It may be argued therefore that ITQs are substitutes for property rights based either on fencing or branding. They are not exclusive rights to the utilisation of particular areas of the sea, or of particular fish, but rather exclusive rights to harvest a given share of a given total catch of a species of fish. They are rights of extraction rather than property, comparable to rights to extract a certain quantity of timber from a given forest, or to harvest a certain number of deer from a given colony (Hannesson, 1994). While such rights provide incentives to cut the timber and to catch the deer in the most efficient ways, they may not be sufficiently strong to provide the optimal husbandry of the forest or the deer colony.

Nevertheless, ITQs, as described in the fisheries economics literature, have many of the efficient features of individual property rights. They are exclusive, individual, divisible, transferable and permanent. One important feature is that the permanent ITQs, that is, TAC-shares, are share-rights: they are (transferable) rights to harvest, say, 0.1 per cent of the total allowable catch in a species of fish in the foreseeable future. Holders of such rights have a clear interest in the long-term profitability of the resource. There would be a crucial difference in the behaviour of two groups of quota holders, where the members of one group would each have a permanent quota expressed in a given quantity of fish, for example, 250 MT of cod a year, whereas the members of the other groups would each have a permanent quota expressed in a given share of the total catch, for example, 0.1 per cent of the TAC in cod. The latter group would be concerned not only with minimising harvesting costs, but also with setting the TAC in such a way that the long-term profitability of the fish stock in question would be maximised.

Arguably, ITQs, as described in the fisheries economics literature, come as near to being private property rights as is simply feasible in deep-sea fisheries. But what about the Icelandic ITQs, described here? Those ITQs are certainly individual and divisible. They are also exclusive although their exclusivity is somewhat reduced by the continuing existence of exemptions from the system for some boats, under 6 MT. But it is a minor exemption and sooner or later all small boats will probably be integrated into the ITQ system. The Icelandic ITQs are also mostly transferable: the restrictions on quota transfers are not very important. Nevertheless, they are restrictions.

For the system to be more efficient, most economists would argue, ITQs should not be issued to fishing vessels, but to individuals and firms and they should be freely transferable. No restrictions should be imposed either on the relative or absolute amount each individual firm could hold, as is now the case. The ITQs should also be fully recognised by the law as possible collaterals which they are not at the moment. There should not be conditions on their use to discourage speculation in ITQs. More speculation would facilitate transfers in the ITQ market, hasten the reduction of the fishing fleet and enable quota holders to be more flexible in their operations.

The main problem in the Icelandic fisheries is, however, that the ITQs, even if issued to individual vessels for an indefinite period of time since 1990, are not really permanent and secure. The 1990 Fisheries Management Act, a paragraph was inserted to the effect that no assignment of ITQs by this law could constitute any permanent property rights to such quotas or become the ground for compensation if the quotas were taken from their holders. While it is unlikely that the ITQ system would be abolished, or the quotas taken from their present holders, especially since in early 2000, only 19 per cent of the quotas are still in the hands of those to whom they were initially assigned, the unwillingness of the Icelandic Parliament to take any steps legally to recognise the ITQs as property rights, even if they are taxed as such and to all purposes treated as such, has added to the uncertainty facing their holders.

Concluding Remarks

The emergence of ITQs in the Icelandic fisheries has interesting similarities to the emergence of property rights amongst Indians in Labrador, as analysed by Harold Demsetz (1967). For centuries, before the arrival of Europeans, the Indians had hunted beaver primarily for food and the few furs they needed. Since the beaver stock was a non-exclusive resource, the Indians did not have a vested interest in increasing or maintaining it. However, as their needs were small and the technology primitive the negative effects of beaver hunting were insignificant. When European traders arrived, hunting technology improved, and demand for furs greatly increased. The scale of hunting increased so the harmful effects which each hunter had on others by his hunting became significant.

Consequently, the Indians divided themselves into several bands in order to hunt more efficiently. Each band appropriated pieces of land, roughly similar in quality, for it to hunt exclusively. By the middle of the 18th century, the privately allotted territories were relatively stabilised. Thus, the fur trade had encouraged the husbanding of beavers and the prevention of poaching which such husbanding requires.

Demsetz tells this tale to illustrate his main point about property rights. They emerge when harmful or beneficial effects of economic activity emerge, enabling individuals to take them into account. Consider pollution. If I pollute a river in which you swim, or fish salmon, or from which you get your drinking water, with the consequence that you cannot continue your use of the river, it is typically because neither you nor anyone else owns the river, being able to hold me responsible for my activities. While the pollution I cause harms you, it does not cost me anything.

The solution would seem to be to define property rights to the river, just as the Labrador Indians established property rights in different pieces of land. Sometimes, however, the definition of property rights is not feasible: the costs of establishing them are higher than the gains. Demsetz points out that the Indians of the Southwest plains who came into contact with the European market at the same time as the Labrador Indians, did not establish new property rights in response to increased demand for the animals they hunted and improved hunting technology. The reason was that the animals of the plains, such as the buffalo, were primarily grazing animals wandering over wide areas. The cost of husbanding those animals (fencing or branding) was therefore much higher (at least until the introduction of barbed wire) than the cost of husbanding beavers in Labrador which were confined to relatively small areas.

The pelagic species of fish in the Icelandic waters, herring and capelin, are rather similar to the animals of the Southwest plains described by Demsetz: clearly, any territorial rights to those two fish stocks would not have been feasible. Neither fencing nor branding would have been possible. On the other hand, cod and other demersal fish are similar to beavers in the Labrador forests in that they are relatively territorial. The fishing grounds where those species are found are known and rather well-defined. Unlike branding, fencing would in theory have been possible in the demersal fisheries (and even more in the inshore shrimp fisheries, confined to small and clearly demarcated areas).

The interesting question is then why territorial rights were not established in those stocks. Several answers may be suggested. First, there were hardly any legal precedents or possibilities available to fishing vessel owners or legislators. While non-territorial fishing rights in the form of ITQs had already been tried in the pelagic fisheries, and seen to work, ideas about property rights in areas of the sea would have been dismissed as pure fantasy. Second, demersal fishing grounds are very large in scale, creating possible economic inefficiencies of their own as independent units of operations, while vessel catch quotas are perfectly divisible. Third, fencing each fishing ground off would have been quite costly. Instead, under the ITQ system only the Icelandic EEZ is really fenced off. Moreover, the Icelandic fishing fleet includes many multi-purpose vessels, so it was economical to have a comprehensive quota system within which a vessel might switch from harvesting one species to another without many problems. It is also convenient that the quotas are expressed in terms of cod equivalence so fishing vessels can easily solve the problem of by-catch.

On the whole, the evolution of the Icelandic ITQ system can be interpreted as the practical response to the problem of vessel owners imposing economic costs on one another by excessive fishing effort and over-capitalisation—costs which should not be blamed on them, but rather on the lack of property rights and thus the lack of information about those costs (Coase, 1960). It amounts to the enclosure of the fish stocks in Icelandic waters—an enclosure not yet completed.

References

Arnason, R. (1990), "Minimum Information Management in Fisheries", *Canadian Journal of Economics*, Vol.23, pp.630-53.

———. (1994), "On Catch Discarding in Fisheries", *Marine Resource Economics*, Vol.9, pp.189-208.

———. (1996), "Property Rights as an Organisational Framework in Fisheries: The Cases of Six Fishing Nations", in B. L. Crowley (ed.), *Taking Ownership: Property Rights and Fishery Management on the Atlantic Coast*. Halifax, Nova Scotia: Atlantic Institute for Market Studies, pp.99-144.

Buchanan, J.M. (1959), "Positive Economics, Welfare Economics, and Political Economy", *Journal of Law and Economics*, Vol.2, pp. 124-38.

———. (1997), "Who Cares Whether the Commons Are Privatised?" *Post-Socialist Political Economy. Selected Essays*, Cheltenham: Edward Elgar, pp.160-7.

Coase, R.H. (1960), "The Problem of Social Cost", *Journal of Law and Economics*, Vol.3, pp.1-44.

De Alessi, M. (1998), *Fishing for Solutions*. IEA Studies on the Environment No. 11. London: IEA Environment Unit.

Demsetz, H. (1967), "Toward a Theory of Property Rights", *American Economic Review, Papers and Proceedings*, Vol.57, pp.347-59.

Gissurarson, H.H. (1983), "The Fish War: A Lesson from Iceland", *The Journal of Economic Affairs*, Vol.3, pp.220-3.

———. (1990), *Fiskistofnarnir vid Island: Thjodareign eda rikiseign?* Reykjavik: Stofnun Jons Thorlakssonar.

Gordon, H.S. (1954), "The Economic Theory of a Common Property Resource: The Fishery", *Journal of Political Economy*, Vol.62, pp.124-42.

Greenpeace (1997), at http://www.greenpeace.org/~oceans/index.html.

Gylfason, Thorv. (1990), *Stjorn fiskveida er ekki einkamal utgerdarmanna*, ed. Th. Helgason and O. Jonsson, *Hagsaeld i hufi*, Reykjavik: Haskolautgafan og Sjavarutvegsstofnun Haskolans, pp.120-5.

Hannesson, R. (1994), "Trends in Fishery Management", ed. E.A. Loyayza, *Managing Fishery Resources*, World Bank Discussion Paper No. 217.

Hardin, G. (1968), "The Tragedy of the Commons", *Science*, Vol.62, 13 December, pp. 1,243-8.

Johnson, R.N. (1995), "Implications of Taxing Quota Value in an Individual Transferable Quota Fishery", *Marine Resource Economics*, Vol.10, pp.327-40.

———. (1999), "Rents and Taxes in an ITQ Fishery", in R. Arnason and H.H. Gissurarson (eds.), *Individual Transferable Quotas in Theory and Practice*, pp. 205-13. Reykjavik: University of Iceland Press.

Jonsson, B.B. (1975), "Audlindaskattur, idnthroun og efnahagsleg framtid Islands", *Fjarmalatidindi*, Vol.22, pp.103-22.

Jonsson, H. (1990), "Akvardanataka i sjavarutvegi og stjornun fiskveida", *Samfelagstidindi*, Vol.10, pp.99-141.

Libecap, G.D. (1989), *Contracting for Property Rights*, Cambridge: Cambridge University Press.

Lindal, S. (1998), "Nytjastofnar a Islandsmidum—sameign thjodarinnar", in H.H. Gissurarson *et al.* (eds.), *Afmaelisrit David Oddsson fimmtugur*. Reykjavik: Bokafelagid, pp.781-808.

Major, P. (1999), "The Evolution of ITQs in the New Zealand Fisheries", in R. Arnason and H.H. Gissurarson (eds.), *Individual Transferable Quotas in Theory and Practice*, pp. 81-102. Reykjavik: University of Iceland Press.

Moller, M. (1996), "Fyrirkomulag veidileyfagjalds", *Visbending*, 29 February.

Morgunbladid (2000), "80 per cent kvotans hafa skipt um hendur", 18 March.

Runolfsson, B.T. (1999), "ITQs in Iceland: Their Nature and Performance", in R. Arnason and H.H. Gissurarson (eds.), *Individual Transferable Quotas in Theory and Practice*, pp. 103-140. Reykjavik: University of Iceland Press.

Scott, A. (1955), "The Fishery: The Objectives of Sole Ownership", *Journal of Political Economy*, Vol.63, pp.116-24.

Living with Risk:
The Precautionary Principle

17

Escaping Goblins, Only to be Captured by Wolves?

INDUR M. GOKLANY

"What shall we do, what shall we do!" he cried. "Escaping goblins to be caught by wolves!"

— *J. R. R. Tolkien, The Hobbit*

So what do we do if we come to a fork in the road and one way leads through territory inhabited by goblins and the other through country infested by wolves? Which road should we take? Does it even matter if we take one road and not the other?

Policymakers in the environmental and public health arena often face such dilemmas. DDT (dichlorodiphenyltrichloroethane) is a classic case. On one hand, this much-reviled chemical is a proven, cheap, and effective method of reducing malaria, a disease that annually afflicts 300 million people and claims over a million lives worldwide. Over the years, DDT has saved millions of lives not only in Asia and Africa but also in Europe and the Americas. But DDT has also been implicated in the decline of a number of raptors such as the bald eagle and the peregrine falcon. It has been found in various avian eggshells, in the tissues of fish, and in mothers' milk. Some suspect it plays a role in advancing various human cancers and other disorders.

So what should be the policy toward DDT? Should it be banned because of its effects on birds and its hypothesised adverse public health effects? Should its use be encouraged because of its proven ability to combat malaria, one of nature's dread diseases? Or should different policies prevail in different areas, depending on whether those areas are plagued by malaria or host species that might be threatened by DDT?

Today's environmentalists have increasingly invoked the precautionary principle to solve such policy dilemmas. This principle is, essentially, a restatement of a popular rendition of the Hippocratic oath, namely, "first do no harm." Its advocates would have it become a cornerstone for developing policies related to the environment and public health.

Based on the precautionary principle, many environmentalists have supported a global ban on DDT, arguing that, by endangering various avian species, it would harm the environment and that it might possibly contribute to various health problems in human beings. But might not such a ban itself harm public health and result in a higher death toll by postponing, if not foregoing, the conquest of malaria in several developing countries? How do we resolve this dilemma?

The Precautionary Principle

A popular and reasonably good definition of the precautionary principle can be found in the so-called Wingspread Declaration: "When an activity raises threats of harm to human health or the environment, precautionary measures should be taken even if some cause and effect relationships are not established scientifically. In this context the proponent of the activity, rather than the public, should bear the burden of proof."

Many people—environmentalists and others—interpret this principle to imply that if there are any doubts about the safety of a technology, that technology ought to be severely restricted if not banned, unless it can be proven to be absolutely safe. In this vein, one Greenpeace activist stated, "The modus operandi we would like to see is: 'Do not emit a substance unless you have proof it will do no harm to the environment'". Such "absolutist" interpretations capture well the scepticism with which many environmentalists regard technology in general.

Others, noting that there can never be absolute certainty or absolute safety, argue that it is irrational to apply the precautionary principle to policymaking in a world where resources—fiscal and human—are scarce, and that the principle might be counterproductive because it would reduce technological progress toward risk reduction. Technological progress, the critics argue, is the time-tested method for reducing society's vulnerability to all kinds of adversity. For instance, until this century, the major health risks to humanity included inadequate supplies of food; poor access to safe water and sanitation; and insufficient knowledge of basic hygiene, the germ theory, and infectious and vector-borne diseases. Today, because of technological progress in the last century and a half, those risks have been significantly reduced in the developing—and virtually eliminated in the developed—world. As a result, life expectancy, perhaps the single most critical indicator of human well-being, has more than doubled in that period.

Instead of joining the debate regarding the rationality of the precautionary principle and taking sides, I will assume that it is indeed a viable approach to policymaking. And although one must agree that the precautionary principle is not "entrenched in customary law," variations of it can be found in at least 14 international environmental declarations, agreements, and conventions.

Origins of the Principle

While the eminent legal scholar Frank Cross claims that the phrase "precautionary principle" was coined by German bureaucrats in 1965, other scholars claim that it derives from the 1970s and the German articulation of Vorsorgeprinzip, which can be translated as the "precaution" or "foresight" principle. But its roots go deeper. They extend at least to the very first mother who admonished her child that it is "better to be safe than sorry".

The spirit of the precautionary principle can be found in several U.S. laws enacted before the principle acquired international fame, if not its name. The Delaney Clause, which was included in Section 409 of the U.S. Federal Food, Drug and Cosmetic Act of 1958, for instance, essentially outlawed any food additive that was found to induce cancer in real life or in laboratory tests on animals, regardless of the magnitude of the dose. This is perhaps the apotheosis of the absolutist version of the precautionary principle. It can also be argued that the United State's 1970 Clean Air Act effectively operationalised the absolutist version of the precautionary principle. The Act required not only that primary (that is, public health—related) National Ambient Air Quality Standards be established without consideration of social or economic costs, but also that all states meet the standards by a certain date, regardless of the difficulties or costs of meeting standards.

Various versions of the precautionary principle started to appear in international environmental declarations and agreements in the 1980s. Championed by environmentalists and European governments eager to garnish their green stripes, the principle started appearing in one environmental forum after another. It seems to have made its international debut in 1982 in the United Nations World Charter for Nature, which stated that when "potential adverse effects of activities likely to pose significant risks to nature are not fully understood, the activities should not proceed". In 1987 it appeared in the Second International Conference on the Protection of the North Sea. By the late 1980s and early 1990s, discussions of the precautionary principle or precautionary approach were a staple of international environmental discussions. In 1990 alone, "precautionary" language was, for instance, included in the Final Declaration of the Third International Conference on the Protection of the North Sea, the Bergen Ministerial Declaration at a Conference on Sustainable Development of the U.N. Economic Conference for Europe, and the Second World Climate Conference.

Variations on the Precautionary Theme

By the time the United Nations Conference on Environment and Development (UNCED) met in Rio de Janeiro in 1992, the precautionary principle was well-nigh ubiquitous. At UNCED, somewhat different versions of it were incorporated in the Rio Declaration on Environment and Development, the United Nations Framework Convention on Climate Change, and the Convention on Biological Diversity.

Principle 15 of the Rio Declaration proclaimed, for instance:

"In order to protect the environment, the precautionary approach shall be widely applied by States according to their capabilities. Where there are threats of serious or irreversible damage, lack of full scientific certainty shall not be used as a reason for postponing cost-effective measures to prevent environmental degradation."

The first sentence of this version, by implying that states may constrain their application "according to their capabilities" and that it "shall be widely applied" seems to envision some flexibility and a less-than-absolutist application of the principle. The next sentence is even more tentative. It may be argued that its first part is essentially vacuous, since we can almost never have "full scientific certainty", and therefore—like it or not— actions are almost invariably taken "in the absence of full scientific certainty." This sentence does not require any specific type of action on the part of a state, and it implies that actions that are not "cost-effective" need not be taken. Thus, Principle 15 of the Rio Declaration is not as sweeping as either the Wingspread Declaration or the absolutist interpretation of the precautionary principle requiring limits on any technology unless proven absolutely safe.

Principle 15 is echoed in the Preamble of the Convention on Biological Diversity (CBD) with the exception that it eschews any reference to cost-effectiveness. The precautionary principle included in that Convention states as follows:

"That it is vital to anticipate, prevent and attack the causes of significant reduction or loss of biological diversity at source.... Where there is a threat of significant reduction or loss of biological diversity, lack of full scientific certainty should not be used as a reason for postponing measures to avoid or minimise such a threat."

However, although lack of full scientific certainty may not be used to postpone measures, this version does not preclude rejecting actions that might be economically inefficient (or not cost-effective).

The version of the precautionary principle articulated in Article 3.3 of the U.N. Framework Convention on Climate Change (UNFCCC), although more forthright, is much less absolutist. Like to the CBD version, it commences by explicitly stating that "the Parties should take precautionary measures to anticipate, prevent or minimise the causes of climate change and mitigate its adverse effects". Notably it uses "should" rather than "must." It is similar in a number of other ways to Principle 15, with one critical difference. Specifically, Article 3.3 goes on to state as follows:

"Where there are threats of serious or irreversible damage, lack of full scientific certainty should not be used as a reason for postponing such measures, taking into account that policies and measures to deal with climate change should be cost-effective so as to ensure global benefits at the lowest possible cost. To achieve this, such policies and measures should take into account different socio/economic contexts, be comprehensive, cover all relevant sources, sinks and reservoirs of greenhouse gases and adaptation, and comprise all economic sectors."

This version, unlike the other versions encountered previously, specifies that climate change policies and measures "should" essentially be based on global cost-benefit analyses. However, many commentators have—conveniently or otherwise—overlooked this requirement in the UNFCCC for the design of climate change related policies. But if, and only if, one overlooks this inconvenient detail does the UNFCCC seem consistent with the absolutist interpretation of the Wingspread Declaration.

Out of these various versions of the precautionary principle eventually emerged, in January 2000, the Cartagena Protocol on Biosafety to the Convention on Biological Diversity, which repeatedly uses the precautionary principle as a basis for decision-making and risk assessment with respect to the transboundary transfer (and associated handling and use) of genetically modified organisms that may have adverse effects on the conservation and sustainable use of biological diversity. Rightly or wrongly, the Cartagena Protocol is seen as a major victory for the more absolutist version of the precautionary principle and for its advocates.

A Double-Edged Sword

In keeping with its origins in technological scepticism, the precautionary principle has also been increasingly invoked as justification, among other things, for international controls, if not outright bans, on various technologies, which—despite providing substantial benefits to humanity and, in some cases, to certain aspects of the environment—could also worsen other aspects of the environment or public health. In addition to DDT, other technologies against which the principle has been invoked are fossil fuel combustion, on which much of the world's current prosperity and human well-being are based but which could help cause catastrophic global warming, and genetically modified crops, which promise to reduce global hunger and malnutrition while making agriculture more environmentally sustainable but which have also raised the spectre of "frankenfoods" and "superweeds".

The justifications for these policies have something more than the precautionary principle in common: They also share a common flaw. Each of these justifications takes credit for the public health and environmental risks that might be reduced by implementing the policy, but they overlook those public health and environmental risks that the policy itself might generate or prolong. As a result, these policy prescriptions could be worse for humanity and the environment than the underlying diseases they seek to redress. In essence, we might escape the goblins but be crushed by the jaws of wolves.

A one-sided application of the precautionary principle leads to such a predicament because the principle itself provides no guidance on its application in situations where an action (such as a ban on GM crops) could simultaneously lead to uncertain benefits and uncertain harms. In this regard, the principle is reminiscent of Yogi Berra's admonition, "When you come to a fork in the road, take it."

So what do we do to ensure that in avoiding goblins we do not fall prey to wolves?

It is important to ensure that precautionary policies are not counterproductive for public health and the environment. In the following section, I will develop a framework for applying the precautionary principle in situations in which outcomes might be ambiguous because their benefits might be partly or wholly offset by their harms. The framework is developed with reference to the version of the precautionary principle articulated by the Wingspread Declaration.

A Framework for Applying the Precautionary Principle under Competing Uncertainties

Few actions are either unmitigated disasters or unadulterated benefits, and certainty in science is the exception rather than the rule. How, then, do we formulate precautionary policies in situations where an action could lead simultaneously to uncertain benefits and uncertain harms (or costs) to public health and the environment?

The only way to implement the precautionary principle intelligently under such conditions is to formulate hierarchical criteria and rank various threats based upon their characteristics and the degree of certainty attached to them. Consequently, I offer a set of criteria to construct a precautionary framework.

The first of these is the human mortality criterion—that is, the threat of death to any human being, no matter how lowly that human being may be, outweighs similar threats to members of other species, no matter how magnificent those species. Moreover, in general, other non-mortal threats to human health should take precedence over threats to the environment, although there might be exceptions' based on the nature, severity, and extent of the threat. I will call this the human morbidity criterion. These two criteria can be combined into the public health criterion.

However, when an action under consideration results in both potential benefits and potential harms to public health, additional criteria have to be brought into play. These additional criteria are also valid for cases in which the action under consideration results in positive as well as negative environmental impacts unrelated to public health. I propose five such criteria:

- *The immediacy criterion:* All else being equal, more immediate threats should be given priority over threats that could occur later. Support for this criterion can be found in the fact that people tend to partially discount the value of human lives that might be lost in the more distant future. Although some may question whether such discounting is ethical, it may be justified on the grounds that if death does not come immediately, with greater knowledge and new technology methods may be found in the future to deal with conditions that would otherwise be fatal, and that, in turn, may postpone death even longer. For instance, between 1995 and 1999, estimated AIDS-related deaths in the United States dropped by over two-thirds (from 50,610 to 16,273) even though estimated cases increased by almost half (from 216,796 to 320,282). Thus, if an HIV-positive person in the United States did not succumb to AIDS by 1998 because of advances in medicine, there was a greater likelihood in 1998 that he would live out his normal life span. Accordingly, it would be reasonable to give greater weight to premature deaths that occur sooner. This is related to, but distinct from, the adaptation criterion noted below.

- *The uncertainty criterion:* Threats of harm that are more certain (have higher probabilities of occurrence) should take precedence over those that are less certain if their consequences otherwise would be equivalent.

- *The expectation-value criterion:* For threats that are equally certain, precedence should be given to those that have a higher expectation value. An action resulting in fewer expected deaths is preferred over one that would result in a larger number of expected deaths (assuming that the "quality of lives saved" is equivalent). Similarly, if an action poses a greater risk to biodiversity than inaction, the latter ought to be favoured.

- *The adaptation criterion:* If technologies are available to cope with, or adapt to, the adverse consequences of an impact, then that impact can be discounted to the extent that the threat can be nullified.

- *The irreversibility criterion:* Greater priority should be given to outcomes that are irreversible, or likely to be more persistent.

Ideally, each criterion would be applied, one at a time, to the various sets of public health and environmental consequences of the action under review (minus, for each category, the consequences of persisting with the status quo, or whatever the other options might be). Such an approach could work relatively easily if the factors critical to each criterion were kept constant except the ones related to the criterion under evaluation. But, because the various factors are rarely equal, the net effects (on each of the sets of consequences) usually have to be evaluated by applying several of the criteria simultaneously.

If the results are equivocal with respect to the different sets of consequences, one should apply the human mortality and morbidity criteria. For example, if the action might directly or indirectly increase net human mortality but improve the environment by increasing the recreational potential of a water body, then the action ought to be rejected.

There will obviously be instances in which no cut-and-dried answer is readily apparent. For example, an action might reduce cases of a nonlethal human disease while at the same time potentially killing a large number of animals. In such cases, in addition to considering factors such as the nature, severity, and curability of the disease, the cost of the disease and/or treatment, and the numbers of human and other species affected (factors subsumed in the previously specified criteria, namely, the adaptation, irreversibility, and expectation value criteria), decision-making should also consider factors such as the abundance of the species and whether the species is threatened or endangered.

The New Threats:
Climate Change and Biotechnology

18

A Climate of Uncertainty in
The Greenhouse Century

ROBERT C. BALLING, JR.

The climate of the earth has always been in a state of change, ranging from long periods when the planet was nearly covered by ice to warm periods with no ice caps whatsoever. Millions of years ago, the dinosaurs roamed on earth nearly devoid of ice and snow, while only a few thousands of years ago, woolly mammoths ranged across a North American continent, half of which was covered by a mile of ice. These natural ebbs and flows in the global climate system are closely tied to variations in the earth's orbit around the sun, changes in the output of the sun, periods of unusually high or low volcanic activity, and/or substantial changes in oceanic circulation.

We know about these natural climate variations from research on ice cores, tree rings, ancient pollen spores, and fossil records, all of which tell us the same story: Climate is highly variable, climate can change rapidly, and we should not expect the climate of our day to persist over long periods of time. They also tell us that climate change *has* occurred many times on a grand scale during the history of the earth without any interference from human activities. Despite this rich understanding of the climate history over the past five billion years of earth's existence, we have recently witnessed the emergence of global climate change as one of the pre-eminent environmental issues of the day.

Invention of the Global Warming Scare

The global warming issue had been smoldering in climatology for nearly a century, but in a matter of a few months in 1988, it went from the stacks in the science library to front-page news throughout the world. During the late spring and early summer of 1988, much of the United States was suffering through an exceptional drought and stifling heat. On June 23rd of that year, Dr. James Hansen, then director of NASA's Goddard Institute for Space Studies, told a United States Senate hearing on climate change that the world was warmer than at anytime in the instrumental period (the past 150 or so years) and that at least some part of the warming was related to the build-up of greenhouse gases. In a much repeated and often misunderstood phrase, Hansen told the committee that he was "99 per cent certain" that observed increase in global temperature was related to a global greenhouse effect enhanced by a variety of human activities, most notably the burning of fossil fuels.

Global warming was a leading news story for months to come in 1988, only to be fuelled further by September's wildfires in Yellowstone Park (and throughout the western states of the USA) and Hurricane Gilbert's devastation from the Yucatan to Texas. Throw in a few other calamities from other parts of the world, including a record-breaking windstorm in London, and by the end of 1988 the global warming scare had shifted into high gear.

Acting in response to this scare, the United Nations quickly formed an "Intergovernmental Panel on Climate Change" (IPCC). Soon-to-be US Vice-President Al Gore took a leadership role in the global warming crusade (including publishing *Earth in the Balance* in 1992). Practically every environmental group on the planet climbed on board the global warming bandwagon. The media took any unusual weather event as convincing evidence of global warming, and on and on. Momentum for the global warming scare became so great over such a short period of time that the ongoing debate regarding many key scientific questions and uncertainties was nearly squashed by the calls to "do something" about the problem. This momentum received a boost every January as we would learn that the previous year had been one of the warmest, if not *the* warmest, year on record. That momentum, quite unlike anything seen for any other environmental issue, continues to drive the global warming scare into the 21st century. And while the tragic events of September 11, 2001 may have deflated the importance of the perceived global warming crisis, countries throughout the world continue to push global warming as a centerpiece environmental issue.

The global warming debate enters the new millennium with many scientists and policymakers worldwide believing that without a substantial slow-down or reduction in the consumption of fossil fuels, we will soon witness a significant increase in planetary temperature, a rise in sea level, melting of icecaps and alpine glaciers, and an increase in droughts, floods, and severe storms.

But while some folks see a global warming apocalypse over the horizon, other scientists do not believe that warming is inevitable or that policy actions would significantly impact variations and trends in the global climate system. Furthermore, literally thousands of experiments have been conducted throughout the world showing that elevated levels of atmospheric carbon dioxide (CO_2) cause virtually all plants to increase their photosynthetic rates, water-use efficiency, and resistance to drought and other stresses. The spectrum of opinions runs from CO_2 as a curse threatening the climate system at one end to an inadvertent blessing from the Industrial Revolution at the other end of the spectrum. The "heated debate" is likely to go on for many years to come and uncovering fact verses fiction in such a complicated scientific, political, economic, and social issue is a major challenge unto itself.

How Reliable are Predictions of Climate Change?

The concentration of atmospheric CO_2 has increased from near 280 ppmv (parts per million by volume) at the beginning of the Industrial Revolution to approximately 370 ppmv in 2001, and the consumption of fossil fuels appears to be the largest contributor to this upward trend. At present, approximately 20 per cent of the CO_2 emission comes from the United States leading to a popular observation that a relatively small number of people (approximately 5 per cent of the global population) contribute disproportionately to the build-up of CO_2. But in the first few decades of this new century, CO_2 emissions from developing countries are expected to rise substantially, thereby lowering the proportion of global emissions emanating from the US (the US accounted for nearly half of global fossil-fuel CO_2 emissions in the late 1940s). Humans are currently adding over six billion tonnes of carbon to the atmosphere each year, thereby overwhelming the climate's carbon budgeting system and allowing the atmospheric CO_2 concentration to increase on a global scale.

Other human activities release assorted gases into the atmosphere which also have the ability to trap heat energy that would otherwise escape into space. These other greenhouse gases include methane, nitrous oxide, and various chlorofluorocarbons, and each of these gases has its own unique geography of emissions. For example, rice paddy agriculture in south-eastern Asia is a leading contributor to the global increase in methane. Humans are engaged in activities throughout the world that slightly alter the composition of the global atmosphere; however, even these relatively small changes to the gaseous mixture we call air can produce substantial changes to the climate.

For more than 100 years, climate scientists have calculated that a doubling of the concentration of these many greenhouse gases (expressed as CO_2 equivalents) would raise the planetary temperature by up to 6°C (10.8°F). The naturally-occurring greenhouse gases (water vapour is by far the most important greenhouse gas) act as a thermal blanket and maintain a mean atmospheric temperature 30°C (86°F) to 35°C (95°F) above the planetary temperature we would experience in the absence of these critical gases. The CO_2 and other greenhouse gases increase in atmospheric concentration and absorb heat energy (infrared radiation) given off by the surface of the earth. The addition of these greenhouse gases should warm the earth by themselves, but a water-vapour feedback mechanism nearly triples the overall warming effect. Basically, as the planetary temperature rises in response to higher concentrations of greenhouse gases, more water is evaporated into the atmosphere, and the increased concentration of water vapour drives the global temperature further upward. Sophisticated global climate models, which are giant computer programmes designed to simulate the laws of physics that control the atmosphere, continue to show that a doubling of greenhouse gas concentrations would force the global temperature to rise from between 1°C (1.8°F) to as much as 6°C (10.8°F). Furthermore, these models predict that a warmer world will have an invigorated hydrological cycle leading to an increase in precipitation at most locations across the globe.

There is no doubt that the prediction for higher global temperatures and precipitation levels, given higher concentrations of greenhouse gases, is solidly grounded in the underlying physics that govern the climate system. These are fairly basic principles in the atmospheric sciences that have been known for over a century. Moreover, the prediction for warming comes from numerical climate models developed in major research laboratories in many nations, employing hundreds of the world's leading atmospheric scientists.

But the prediction is not without significant uncertainties. Climate models poorly represent cloud processes that could be critical in determining the overall energy balance of the earth's atmosphere system. Additional high cloud cover in response to global warming would create a positive feedback effect, creating even higher temperatures. However, any increase in low cloud cover would act to cool the earth putting a natural brake on the greenhouse effect. Today's models are not able to resolve critical questions regarding how the global cloud patterns will respond to higher temperatures, and as result of these cloud-related uncertainties, the confidence place on any prediction for future temperature increases remains relatively low.

When the clouds are not well modelled, neither are precipitation patterns, which in turn produce problems with the simulated surface hydrology. And without an adequate representation of soil moisture patterns, the models struggle with the surface energy balance, including the calculation of the near-surface air temperatures. A shortcoming of numerical climate models is that they fail to accurately account for the relationship between the ocean and the atmosphere, which is especially critical for a planet on which land cover is less than 30 per cent of the total global surface area. The numerical models also poorly simulate snow and ice processes, have poor, if any, biological routines, and many require significant flux adjustments (fudge factors) to keep from drifting into unrealistic climate states. One of the greatest uncertainties in model construction is the representation of water vapour levels in the middle atmosphere in the subtropical latitudes which are known to be critical in maintaining the radiation balance of the planet.

Despite these many uncertainties, the models are fabulous achievements in the atmospheric and computing sciences, and their prediction for global warming given a build-up of greenhouse gases should not be taken lightly. The exact amount of warming that will occur as a result of mankind's emissions of greenhouse gases remains unknown but current models, along with those simpler models used for over a century, universally show some amount of global warming.

Figure 1

*Thermometer-based Annual Global Temperature
Anomalies, 1900-2001*

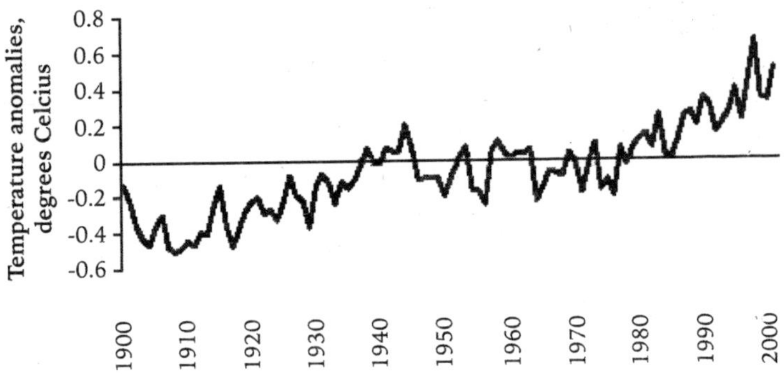

Observed Climate Changes

Given the long-standing prediction for global warming with increasing concentrations of greenhouse gases, it is logical for empirical scientists to examine the temperature record of the earth to determine if warming has in fact occurred during the period of greenhouse gas build-up.

Thermometer-based temperature records from land and sea have been assembled for as much of the earth as possible[1], and indeed, the record shows a linear warming of 0.68°C (1.22°F) over the period 1900-2001, and the warming has accelerated in recent decades (Figure 1).

Scientists have noted that warming has occurred in most areas of the world, with the greatest warming occurring in high-latitude land areas of the Northern Hemisphere, in winter, and at night[2]. While interpretations vary on how to assemble these records and calculate the global temperature, the inescapable fact emerges that the thermometer-based near-surface air temperature record shows statistically significant warming over the past century, with the warmest years occurring in the most recent decades. At first glance, the warming anticipated from model simulations is similarly reflected by the historical temperature record from thermometers around the world. Analyses of global precipitation patterns are far less certain, although some evidence certainly exists suggesting a slight increase in global precipitation over the past century.

Attributing the observed surface warming to changes in atmospheric chemistry is compounded by many factors. First, as noted in the introduction, the climate system fluctuates naturally even without any external forcing, and the observed warming is not outside the bounds of natural fluctuations seen repeatedly in the long-term reconstructions of the global temperature. Also, much of the earth experienced a "Little Ice Age" from approximately 1450 AD to 1850 AD, and the recent warming may be nothing more than a natural recovery from this unusually cool period.

Second, the irradiance from the sun has varied through the past century with a distinctive upward trend (Figure 2) from 1900 to 1960. Many scientists believe that much of the observed warming prior to the 1970s can be ascribed to increasing output from the sun and not to the build-up of greenhouse gases[3]. A more careful inspection of the near-surface global air temperatures (Figure 1) reveals substantial warming from 1910 to 1940, and that warming coincides nicely with the increasing output of the sun.

Figure 2

Solar Irradiance (in watts per square meter), 1900-2001

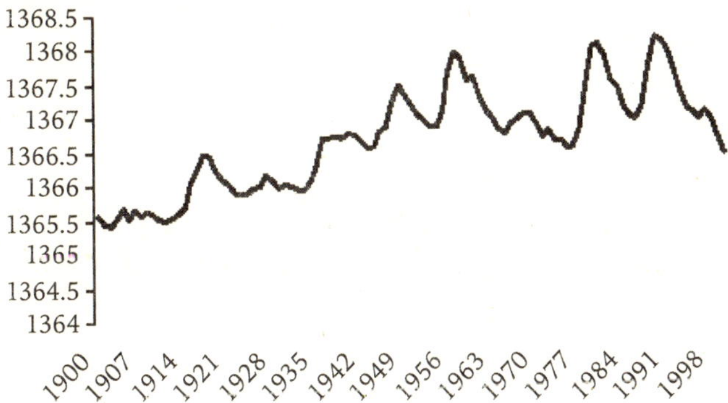

Third, volcanic eruptions and the El Niño/Southern Oscillation (ENSO) patterns exert significant influence on the climate system producing a non-greenhouse forcing on planetary temperature. The periodic disruption to global temperature caused by volcanoes and ENSO complicates any search for the greenhouse "fingerprint" in the records.

Fourth, the thermometer measurements themselves are compromised by local influences (e.g., urban effects), the uneven distribution of the thermometer network, and changes through time in instrumentation and recording practices. We have few if any actual temperature records from huge parts of the planet including much of the South Pacific. Nonetheless, and despite an endless number of criticisms of the thermometer-based near-surface air temperature record, there is compelling evidence that the global temperature has increased on a planetary scale over the past century, and the warming near the surface has accelerated in recent decades. With the models predicting warming, consequent on a build-up of greenhouse gases, and thermometer-based records of the world showing warming, it would seem that any debate on the global warming subject would be one-sided. However, global temperature measurements made from space call into question the recent planetary warming seen in the thermometer records[4].

Figure 3

*Satellite Based Monthly Temperature Anomalies,
January 1979-November 2001*

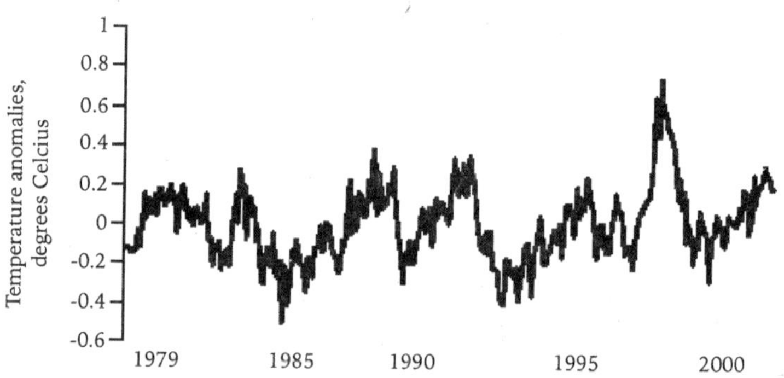

Sounding units aboard polar-orbiting satellites measure microwave emissions from the low atmosphere that are directly related to the temperature of the atmosphere from approximately 1,500 m (4,947 ft) to 8,500 m (28,024 ft). The polar orbits provide true global coverage as the earth rotates underneath the satellites that carry the measuring devices. Unlike the near-surface air temperature record that shows warming over the past few decades (Figure 1), the satellite based observations show no warming whatsoever from 1979 to 2001 (Figure 3) in spite of the very warm El Niño year of 1998. The thermometer-based record and the satellite-based record are well correlated, showing that their annual measurements are similar, but over time the surface record has a significant upward trend while the satellite record has no trend whatsoever.

Obviously, scientists and policymakers want to know which of these two trends is correct, and they search for other means for estimating global temperature. Twice each day at hundreds of locations around the world, balloons are launched through the atmosphere to measure vertical profiles of temperature, wind, and moisture. When the balloon-based measurements are averaged globally for the same 1,500 m to 8,500 m layer measured by the satellites, the two records are nearly identical in terms of variations and overall linear trend (both show no warming from 1979- near present). This finding leads to an argument that the earth is simply not warming, despite the trend seen in the near-surface thermometer record. However, measurements from these same balloons made close to the ground are similar to the thermometer-based temperatures showing approximately the same upward trend. This dilemma has been the focus of many scientific articles and reviews, and most scientists now agree that the surface has been warming over the past few decades while the low atmosphere is not warming at all[5].

These differently-derived temperature trends represent anything but a "draw" in the greenhouse debate. The numerical models of climate used to simulate the response to higher concentrations of greenhouse gases suggest that the low atmosphere should be warming faster than the surface, and that is clearly not the case. The observed temperature trends, warming at the surface and no warming in the low atmosphere, are simply not consistent with model expectations given the build-up of greenhouse gases! Volcanic activity, El Niño events, and even instrumentation changes have all been suggested as causes for the differential warming between the surface and the lower atmosphere, but the answer to this critical matter remains elusive and a major focus of scientific inquiry.

While climatologists have struggled in their attempts directly to link any trends in global temperatures with the build-up of greenhouse gases, other patterns in the climate system have been even more difficult to reconcile.

Precipitation levels are generally increasing in most parts of the world, and the pattern is broadly consistent with the predictions from the numerical models[6]. In the United States and Australia, there is some evidence that heavy precipitation has increased slightly, but the pattern has not been verified in other areas studied to date. Droughts continue to plague different parts of the earth each year, but most scientists are unwilling to attribute drought patterns to changes in greenhouse gas concentration. Analyses of tropical cyclones, mid-latitude cyclones, and/or tornado activity have generally led to the conclusion that no discernible trends in these events are apparent in the historical records. The most recent United Nations Intergovernmental Panel on Climate Change report concludes that "No systematic changes in the frequency of tornadoes, thunder days, or hail events are evident in the limited areas analysed"[7] and that "Changes globally in tropical and extratropical storm intensity and frequency are dominated by inter-decadal and multi-decadal variations, with no significant trends evident over the 20th century."[8]

These conclusions from the scientific community fly in the face of the popularised visions of the greenhouse world that inevitably include images of more severe storms and increased climate variability in the coming decades.

Observed changes in climate over the past few decades likely include (a) warming at the surface particularly at night, in mid-to-high latitudes over land, and during the winter, (b) no warming in the lower troposphere on a global scale, (c) a general increase in precipitation, and (d) no change in extreme events or climate variability. Whatever changes have occurred fall well within the natural variability of the climate system and are difficult to ascribe to the increase in greenhouse gas concentrations.

Some of the changes are broadly consistent with numerical climate model predictions for a build-up of greenhouse gases, others are not. The lack of warming in the lower-troposphere, as measured by satellites and balloons, remains as a major contradiction to expectations from the model simulations, and will likely be the focus of intense research over the next decade.

Complicating Effects

The inability to clearly isolate global or regional climate signals related to the build-up of greenhouse gases is further compounded by other anthropogenic forcing of the climate system.

Burning fossil fuel undoubtedly produces CO_2 that collects in the global atmosphere, but it produces sulphur dioxide (SO_2) as well. The SO_2 enters the atmosphere and quickly transforms to sulphate aerosols that have a widespread cooling effect by (a) reflecting incoming sunlight back to space, (b) brightening clouds, and (c) making clouds last longer. Unlike CO_2, which mixes fairly evenly throughout the entire atmosphere, the sulphur aerosols are short-lived in the atmosphere, thereby producing a regional pattern of high concentrations near major industrial emission sources and low concentrations throughout the rest of the world. This regional structure complicates the ability to model correctly the thermal effects of the sulphur load in the atmosphere. But to date all models show a cooling for elevated sulphur concentrations, though they differ in terms of the magnitude of the cooling.

Humans have degraded drylands throughout the arid and semi-arid portions of all major landmasses, and this degradation has resulted in an increase in mineral aerosols in many parts of the world. The increased load of mineral aerosols appears to have a cooling effect that may be important at the global scale, although significant uncertainties remain as the magnitude of the cooling.

Also, various chlorofluorocarbons deplete ozone in the high atmosphere which results in a significant cooling at the surface, but uncertainties abound in terms of the magnitude of the cooling.

Fossil fuel soot, biomass burning, changes in surface reflectivity caused by land-use patterns, and contrails from high-flying aircraft are all considerations in "predicting" future climate, and yet to date the magnitude of these effects on the climate system remain unknown and highly variable from model to model.

Noted climate modeller James Hansen (the same fellow who was "99 per cent certain" in 1988) and colleagues concluded that "The forcings that drive long-term climate change are not known with an accuracy sufficient to define future climate change."[9] Hansen suggests that even with perfect models and perfect data, climatologists would continue to struggle in an attempt to "predict" climate changes for the next century given the uncertainties regarding the forcings of climate that will be operating 50 to 100 years from present. Overall, the models definitely predict warming in the years to come, but one must appreciate the many uncertainties associated with the prediction.

Hopeless Policy

Uncertainties regarding the earth's climate future will remain for many decades to come. Given these uncertainties, many scientists and policymakers believe that we should cut back on greenhouse gas emissions and avoid conducting an unknown and largely irreversible experiment on the global atmosphere.

The argument that we have only one atmosphere and should therefore avoid making substantial changes to its composition that may significantly alter the "normal" climate system is certainly a compelling one. To that end, the governments of many nations have expressed their grave concern over the threat of global warming and have signed (but not necessarily ratified) the Kyoto Protocol. That Protocol in essence requires that the world return to a level of anthropogenic emission of greenhouse gases equivalent to such emissions in 1990, with the bulk of the reduction coming from the developed nations that have been large emitters for many decades.

Under Kyoto, the United States is required to reduce emissions to seven per cent below:

1990 levels. However, US emissions of greenhouse gases are currently 13 per cent above 1990 levels; emissions of CO_2 are rising quickly (due in no small part to the electrical demand of the Internet), and US President Bush has largely withdrawn from the Kyoto process. The concerns of the Bush administration include uncertainties regarding global warming science and the near-certain negative impact the Kyoto Protocol would have on the United States economy. While greenhouse advocates cry out that we cannot chance our only atmosphere to some global experiment on the effect of elevated greenhouse gas concentrations, opponents in the United States are unwilling to risk their only economy on the outcome of implementing the Kyoto Protocol.

One of the startling facts regarding the Kyoto Protocol is that its implementation with full participation would have a trivial impact on the climate system. The goal of the Protocol is to stabilise *emissions* of CO_2, not the atmospheric *concentration* of CO_2 (and of course the other greenhouse gases). Even if emissions could be stabilised at 1990 levels, six billion tonnes of carbon would be added to the atmosphere annually by human activities. That carbon would build-up in the atmosphere and a doubling of CO_2 would still occur near the middle of this century. If the Protocol went into effect today, greenhouse gas emissions suddenly returned to the 1990 level and remained at that level every year from now until 2050, the entire exercise would "spare" the earth only a few hundredths of a degree of warming. Similar observations have been made by no less a figure than global warming protagonist Tom Wigley. Wigley, a major figure in greenhouse policy circles, is a Senior Scientist at the National Center for Atmospheric Research (NCAR) in Boulder, Colorado, Director of NCAR's Consortium for the Application of Climate Impact Assessments, and was for a long time head of the Climate Research Unit in Norwich, England, which is responsible for assembling and analysing thermometer-based temperature records from throughout the world. Wigley carefully analysed the climate impact of the Kyoto Protocol and concluded that "The influence of the Protocol would, therefore, be undetectable for many decades".[10]

The fact that the Kyoto Protocol would have such a small climate impact is not altogether bad news in many circles. Leaders of the IPCC remind us that the Kyoto Protocol is a framework convention that will lead the way to future international agreements aimed at stabilising concentrations of greenhouse gases below "dangerous" levels. Defining and defending a "dangerous" concentration in this context may be the biggest fight in the greenhouse in the years to come!

Others, particularly those involved in the fossil fuel industry (especially those in the coal industries), may argue that the Kyoto

Protocol will not impact climate, and that there is therefore no need to rush into any implementation that would hurt their enterprise or the economies that depend on their products. These same leaders might be quick to point out that James Hansen (again, of "99 per cent certain" fame) recently wrote in the *Proceedings of the National Academy of Sciences* "But we argue that rapid warming in recent decades has been driven mainly by non- CO_2 greenhouse gases (GHGs), such as chlorofluorocarbons, CH4, and N2O, not by the products of fossil fuel burning".[11] Not only will the Kyoto Protocol have little climatic effect, but it may be firing at the wrong target!

Environmentalists are naturally drawn into the seductiveness of a potential global disaster brought about by the build-up of greenhouse gases emanating largely – at present – from the world's most developed nations. However, they also realise that policies aimed at substantially reducing greenhouse gas emissions are expensive and these same policies will likely have a trivial impact on climate. We have only limited funds for environmental issues, and relatively simple cost/benefit analyses reveal that spending money on the global warming issue produces few if any benefits. Money targeted at global warming is not available for other environmental issues (often local issues) where real return on investment is far more certain.

The global warming "crisis" dropped quickly from the radar screen of public concern during a few short minutes on the morning of September 11, 2001. But as time goes onward, the global warming issue will again find a place in the public consciousness; the greenhouse momentum built up during the 1990s is simply too great for this environmental issue to suddenly vanish for good. Hopefully, cooler heads will prevail in the future, policymakers will base their decisions on climatic and economic facts, not hype and hysteria, and the complexities of the issue will replace the simplistic presentations suggesting that the planet is warming, greenhouse gases are increasing in concentration, and the two must be linked. As we have seen, there is much more to the story, and what might appear to be promising policy options may in fact be useless in impacting the global climate system. Stay tuned...the greenhouse debate has taken some time off, but it will certainly be with us all in the years to come.

Notes

1. Jones, P.D., 1994: "Hemispheric Surface Air Temperature Variations: A Reanalysis and an Update To 1993". *Journal of Climate*, 7, 1794-1802.

2. Balling, R.C., Jr., P.J. Michaels, and P.C. Knappenberger, 1998: "Analysis Of Winter And Summer Warming Rates in Gridded Temperature Time Series". *Climate Research*, 9, 175-181.

3. Lean, J., J. Beer, and R. Bradley, 1995: "Reconstruction Of Solar Irradiance Since 1610: Implications for Climate Change". *Geophysical Research Letters*, 22, 3195-3198.

4. Spencer, R.W., and J.R. Christy, 1990: "Precise Monitoring Of Global Temperature Trends from Satellites". *Science*, 247, 1558-1562.

5. Wallace, J.M., J.R. Christy, D.J. Gaffen, N.C. Grody, J.E. Hansen, D.E. Parker, T.C. Peterson, B.D. Santer, R.W. Spencer, K.E. Trenberth, and F.J. Wentz, 2000: *Reconciling Observations of Global Temperature Change*. Washington, D.C: National Academy Press.

6. Diaz, H.F., R.S. Bradley, and J.K. Eischeid, 1989: "Precipitation Fluctuations Over Global Land Areas Since The Late 1800s". *Journal of Geophysical Research*, 94, 1195-1210.

7. IPCC, 2001: Climate Change 2001: *The Scientific Basis*. Cambridge: Cambridge University Press, p. 5.

8. ibid., p. 5.

9. Hansen, J.E., M. Sato, A. Lacis, R. Ruedy, I. Tegen, and E. Matthews, 1998: "Climate Forcings in The Industrial Era". *Proceedings of the National Academy of Sciences*, 95, 12,753- 12,758, p. 12,753.

10. Wigley, T.M.L., 1998: "The Kyoto Protocol: CO2, CH4 and Climate Implications". *Geophysical Research Letters*, 25, 2285-2288, p. 2288.

11. Hansen, J., M. Sato, R. Ruedy, A. Lacis, and V. Oinas, 2000: "Global Warming in The Twenty First Century: An Alternative Scenario". *Proceedings of the National Academy of Sciences*, 97, 9875-9880, p. 9875.

19

The Attack on Plant Biotechnology

GREGORY CONKO
C.S. PRAKASH

Most people have never even heard of bioengineered crops, and most who have hold a neutral or positive opinion about them. But beneath this otherwise calm surface, there is a growing campaign led by ideological environmentalists against plant biotechnology. "Public interest research" groups argue that bioengineered foods "pose unacceptable risks to human health," risk "spawning new superweeds," or pose hazards to beneficial insects and soil organisms. The activist group Friends of the Earth warns that biotech crops could "seriously threaten biodiversity in agricultural areas" and that they "may also be toxic to humans." And when the United States Agency for International Development sent a shipment of corn and soy-meal that happened to contain some bioengineered varieties in the mix to aid the victims of a cyclone in the Indian state of Orissa, Vandana Shiva, director of the New Delhi-based Research Foundation for Science, Technology and Ecology, argued that, "The U.S. has been using the Orissa victims as guinea pigs for [bioengineered] products."

Other critics are even more shrill. Jeremy Rifkin, a notorious and longtime opponent of all forms of genetic research, calls the introduction of bioengineered plants "the most radical, uncontrolled experiment we've ever seen." Mae-Wan Ho, a biologist at London's Open University argues that biotech crop plants are "worse than nuclear weapons or radioactive wastes." What is it about agricultural biotechnology that inspires such attacks?

Ever since the 1962 publication of Rachel Carson's Silent Spring, ideological environmentalists have warned that mankind's use of modern farming technologies would lead to widespread ecological and human health catastrophes. Then, the villain was synthetic chemicals - particularly the use of insecticides, herbicides, and fungicides on farms to protect growing crop plants. Thirty years later scientific evidence clearly shows that those concerns were wildly exaggerated. Nevertheless, the use of agricultural chemicals can have some negative environmental effects. Ultimately, humanity must choose between using chemicals that can cause some minor harm on the one hand, or sacrificing tremendous gains in food productivity on the other.

For many, the choice is simple. At its heart, all of agriculture requires a never-ending struggle against the destructive forces of nature: pests, diseases, weather, and many others. Despite the steadily growing use of insecticides, herbicides, and fungicides on farms around the world, as much as 40 per cent of crop productivity in Africa and Asia, and about 20 per cent in the industrialised countries of North America and Europe, is lost to insect pests, weeds, and plant diseases. Without any means for controlling those pests, crop losses would climb to as much as 70 per cent. Thus, something clearly must be done to prevent crop losses, or agricultural production would fall dramatically, possibly even subjecting humanity to the widespread famines predicted by Thomas Malthus more than two hundred years ago.

Today, a new crop protection revolution is underway that will help farmers combat pests and pathogens more effectively while also reducing humanity's dependence upon agricultural chemicals. Agricultural biotechnology (alternatively known as bioengineering, genetic engineering, and genetic modification (GM)) uses 21st century advances in genetics and cell biology, to move useful traits from one organism to another, allowing plants to better protect themselves from insects, weeds, diseases, and even from such environmental stresses as poor soils and drought. Biotechnology can also improve the nutritional quality of staple foods like corn and rice by adding healthful vitamins and minerals. The technique is so beneficial that it has been endorsed by dozens of scientific and health associations, including the U.S. National Academy of Sciences, the United Kingdom's Royal Society, the United Nations Development Programme, and many others.

By the year 2000, just five years after their introduction on the market, farmers around the world planted more than 109 million acres (44.2 million hectares) with biotech crops. It's easy to see why. In the

United States alone, bioengineered varieties of corn that are resistant to some insect pests were about five per cent more productive on average than conventional varieties during the period from 1996 to 1999. Biotech cotton varieties generated more than 10 per cent higher yields and simultaneously reduced chemical insecticide use by an average of about 14 per cent during that time. Not surprisingly, farmers have a very favorable view of the development of biotech seeds. By 2001, 26 per cent of all corn, 68 per cent of all soybeans, and 69 per cent of all upland cotton grown in the United States were bioengineered varieties.

Although improved agricultural productivity might seem like a luxury that industrialised countries can do without, it is an absolute necessity for less developed nations. In a report published in July 2000, the UK's Royal Society, the National Academies of Science from Brazil, China, India, Mexico and the US, and the Third World Academy of Science, embraced agricultural biotechnology, arguing that it can be used to advance food security while promoting sustainable agriculture. "It is critical," declared the science academies, "that the potential benefits of [genetic] technology become available to developing countries."

Importantly, the increased productivity made possible by these advances will allow farmers to grow substantially more food and fiber on less land. Such productivity gains will be essential if we are to outpace the projected increase in global population over the coming decades while sparing more land for nature. During the second half of the 20th century, in which the population increased from 3 billion to 6 billion, advances in conventional plant and animal breeding, and improved use of synthetic fertilisers, pesticides, and herbicides allowed food production to grow much faster than population growth. But the average annual per acre increase in cereal yields has been slowing, from 2.2 per cent per year in the late 1960s and 1970s, but only 1.5 per cent per year in the 1980s and early 1990s, to as low as just 1.0 per cent in the second half of the 1990s. More importantly, there has been little or no increase in the theoretical maximum possible yields of rice and corn in a decade.

Worldwide, farmers already use approximately one-third of the Earth's land surface area (excluding Antarctica) for agriculture, of which about one-third, or 5.8 million square miles, is dedicated to growing crops. If the average annual increase in productivity per acre for the cereal grains that make up the bulk of food and animal feed remains at its current rate of around one per cent, the world will have to bring more than 700 million acres of new land into agricultural use by the year 2050 to meet projected demand. Nobel Peace Prize winning plant scientist Norman Borlaug argues that, "Extremists in the environmental movement, largely from rich nations and/or the privileged strata of society in poor nations, seem to be doing everything they can to stop scientific progress in its tracks."

The rate of increase in grain yields is slightly higher on average in less developed countries than industrialised ones, but population growth is higher there as well. And even this average obscures the fact that Africa was almost totally excluded from the productivity gains generated during the Green Revolution. Crop productivity there has much room for growth, but for a variety of reasons, Africa has not been able to take advantage of such production increasing inputs as fertilisers, irrigation, and pesticides. Yields of sorghum and millet in sub-Saharan Africa have not increased since the 1960s. Thus, the productivity gains expected to be generated by biotechnology-enhanced crop plants can not only help to reduce the use of agricultural chemicals, they could save millions of acres of sensitive wildlife habitat from being converted into farmland. Explaining his strong support for biotechnology to a Reuters interviewer, Borlaug said, "You have two choices. You need [biotechnology] to further improve yields so that you can continue to produce the food that's needed on the soil that's well-adapted to agricultural production. Or, you'll be pushed into cutting down more of our forests."

One might expect environmental activists to be pleased with the development of a technology that can make man's footprint on the environment lighter. But ideological environmentalists have launched a global campaign to suppress this vital technology on the specious grounds that it is unsafe for humans and the environment. Bioengineered products are denounced as "Frankenfoods," and claims that the new technology could result in "Andromeda strain"-like plagues abound. Lord Peter Melchett, head of Greenpeace's United Kingdom chapter declared that his organisation's opposition to biotechnology is "a permanent and definite and complete opposition based on a view that there will always be major uncertainties."

Never mind that the weight of scientific evidence does not support such outlandish claims, or the belief of most crop scientists that biotechnology will have substantial benefits for environmental stewardship, as well as for farmers and consumers in poorer regions of the world. Kenyan crop scientist Florence Wambugu believes that biotechnology "can help us increase the production of food and other commodities, lowering their prices to consumers while raising the incomes of poor farmers." That may not be enough to satisfy most ideological environmentalists, though.

What is Plant Biotechnology?

Ever since the dawn of agriculture, which began thousands of years ago with domestication of wild plants and animals from their natural habitats, humans have continuously transformed the crops and animals that we have come to depend upon for food and animal feed. Over many millennia, the crop varieties that were chosen for domestication have been gradually modified by selecting individual plants that grew the best and produced the best grains, vegetables, and fruits. Over time, this process of artificial selection resulted in profound changes in the stature, productivity, and taste of crop varieties. Modern corn is derived from a wild Central American grass plant called teosinte. Through successive generations of selection, breeders developed an entirely new species of plant - corn - that shares very few of its characteristics with the wild teosinte.

Entirely new plant varieties were also developed by crossbreeding plants from different, but related species with one another. The progeny of such hybridisations expressed new traits resulting from the random mixing of literally tens of thousands of genes from the two parent plants. With these "natural" breeding techniques, entirely new proteins and other plant chemicals were routinely introduced into food crops, often from wild species never before part of the food supply. Bread wheat, for example, resulted several hundreds of years ago from the crossing of at least three different species of wild grasses from two different genera. And in the 20th century, wheat and rye, plants from two different genera, were crossed to produce a new variety called triticale, which is used as food and animal feed. Hundreds of useful crop plants were developed with selection and hybridisation techniques. But the flexibility of these techniques is limited by the need for the parent plants to be from species that can breed sexually.

The discovery of genes, chromosomes, and other mechanisms of plant genetics during the 20th century opened up new avenues for modifying plants. Scientists developed many novel tools that expanded the range of modifications that could be used to improve crop varieties. For example, in the late 1940s, agronomists began using x-rays, gamma rays, and caustic chemicals on seeds and young plants to induce random genetic mutations. Such mutations generally kill the plants (or seeds) or cause detrimental changes in the DNA. But on rare occasions, the result is a desirable mutation - for example, one producing a useful trait, such as altered height, more seeds, or larger fruit. In these cases, breeders have no real knowledge of the exact nature of the genetic mutation(s) that produced the useful trait, or of what other mutations might have occurred in the plant. But more than 2,250 mutation-bred varieties of corn, wheat, rice, and dozens of other varieties have been commercialised over the last half century, and they are grown in more than 50 countries around the world.

More sophisticated breeding techniques also permit agronomists to overcome natural barriers to ordinary sexual reproduction. They include methods such as protoplast fusion and embryo rescue, which join cells from sexually incompatible plants in a laboratory and over-come their natural inability to produce offspring. These techniques for genetic modification permit the artificial hybridisation of plants of the same species, different species, and even different genera. "Wide crosses" of plants from different species or genera allow scientists to add into an existing crop species traits for disease and pest resistance, increased yield, or different nutritional qualities. They can even be used to create entirely new plant species. Examples of such artificial wide crosses include a wheat-barley hybrid, a tomato-potato hybrid, and a radish-rapeseed hybrid. Yet, none of these techniques are considered to be bioengineering, so they escape the wrath of ideological environmentalists.

These techniques underpinned the last century's spectacular increases in food productivity in all major crops around the world, including the Green Revolution in developing countries. This dramatic increase in food production has been critical in ensuring an affordable supply of food. For example, U.S. corn growers averaged 134 bushels per acre in 1998 compared to only 26 bushels of corn per acre in 1928. It will be possible to achieve additional productivity improvements through conventional breeding. But these techniques are crude and slow, and the traits that descendant plants eventually carry are not easily predictable. Typically, one or more unwanted traits are transferred to the offspring plants with any of these more conventional breeding techniques, so the breeder's job is not yet done. After the initial modification, agronomists must cross-breed the offspring again and again with the original plant for several generations to eliminate any undesirable traits. And many agronomists believe that we are already nearing the maximum possible gains in yield that can be achieved with conventional breeding. Fortunately, with the advent of modern biotechnology an alternative for boosting crop productivity is now available.

In the 1980s, scientists in the United States and Europe independently developed new and more precise methods for moving single genes directly into plants. This overcame the limits imposed by sexual incompatibility among species and opened up immense possibilities for developing novel crop varieties with improved traits. A naturally-occurring soil bacterium, Agrobacterium tumefaciens, which transfers its own DNA into plants, was modified to deliver desirable genes into plant cells instead of its own infective genes. Subsequently, a few other methods of gene transfer to plants were developed, including a "Gene Gun" that literally shoots gene fragments into the plant chromosomes. Since then, scientists have identified thousands of genes of potential value for agriculture from a wide variety of organisms, and have developed methods to reliably insert genes into every major crop plant. Genes are recipes for producing proteins and those proteins can improve a crop's nutritional value or protect it against pests. These are the various techniques that are now known as genetic engineering, bioengineering, genetic modification, or biotechnology.

In modern biotechnology, the genes coding for specific traits are inserted into plant cells, which are then cultured for development into full plants. The bioengineered plants will then express the new trait - such as resistance to an insect pest. Added genes are taken up into the plant's DNA in random positions, opening biotechnology to questions about unintended and unexpected effects. But such "pleiotropic" effects, brought about by the re-arrangement of DNA, occur even in the conventional breeding of plants from the same species. Compared with the mass genetic alterations that result from using wide-cross hybridisation or mutagenic irradiation, the direct introduction of one or a few genes into crop plants results in much more subtle and far less disruptive changes that are relatively specific and predictable.

The process differs from more conventional breeding methods of hybridisation, induced mutation, and others, in that only one or two specifically identified additional genes are typically introduced into an existing background of tens of thousands of genes. But, because DNA is identical from organism to organism, bioengineering techniques can transfer genes, not just between plants, but from any living organism to any other - such as between plants and animals, or bacteria and plants. This new flexibility aside, scientists see biotech gene transfer techniques as a logical extension of the continuum of methods used to improve crop plants. A report published by the U.S. National Academy of Sciences in 1989 concluded that:

"[Bioengineering] methodology makes it possible to introduce pieces of DNA, consisting of either single or multiple genes, that can be defined in function and even in nucleotide sequence. With classical techniques of gene transfer, a variable number of genes can be transferred, the number depending on the mechanism of transfer; but predicting the precise number or the traits that have been transferred is difficult, and we cannot always predict the [characteristics] that will result. With organisms modified by molecular methods, we are in a better, if not perfect, position to predict the [characteristics]."

Thus, with biotechnology, plant breeders are actually *less* likely to produce unanticipated effects in crops.

To date, more than 70 biotech plant varieties have been commercialised in the United States expressing a range of improved traits, such as heightened resistance to certain insects and diseases, tolerance to herbicides, and longer shelf life. Globally, bioengineered varieties are grown commercially on approximately 109 million acres, in countries ranging from the United States, Argentina, Australia, Brazil, Canada, Chile, China, Mexico, and South Africa. Some critics have suggested that biotech crops are primarily an industrialised country interest. But the proportion of bioengineered crops grown in less developed nations has grown consistently since their introduction, from 14 per cent in 1997, to 24 per cent in 2000.

Some of the most successful crop varieties have been modified by adding a bacterial gene that produces a protein toxic to predatory insects, but not to people or other mammals. By reducing the need for spraying chemical pesticides on crops, such crops are environmentally friendly. Another popular trait is tolerance to a particular herbicide. Herbicide tolerance can be developed in some crop varieties through selection and breeding methods, but biotechnology can achieve the same goal much more quickly and effectively. Today, varieties of canola, corn, cotton, rice, soybean, and sugar beet, have all been bioengineered to tolerate one or another broad spectrum herbicide. Herbicide tolerant varieties allow farmers to control weeds by spraying fields without damaging growing crops. This, in turn, eliminates the need to plow under weeds, which loosens topsoil and contributes to erosion. And because the spraying of herbicides is more efficient, herbicide tolerant crops have even led to a modest reduction in herbicide use.

The purpose of the current generation of bioengineered crops is primarily to improve pest resistance and weed control. In turn, this should reduce the use of crop protection products and/or increase yields.

Herbicide tolerance: The insertion of a herbicide tolerant gene into a plant enables farmers to spray wide spectrum herbicides on their fields killing all plants but the crop.

Insect resistance: By inserting genetic material from the Bacillus thuringiensis (Bt) into seeds, scientists have modified crops, allowing them to produce their own insecticides. For example, Bt cotton combats bollworms and budworms, and Bt corn protects against the European corn borer.

Virus resistance: To date, a virus resistant gene has been introduced into squash, tobacco, potatoes, and papaya. The insertion of a potato leaf roll virus resistance gene protects the potatoes from the corresponding virus, which is usually transmitted through aphids. For that reason, it is expected that there will be a significant decrease in the amount of insecticide used. The introduction of virus resistance genes into other plants may offer similar benefits. Virus resistant papaya varieties have single-handedly revived the Hawaiian papaya industry, nearly totally destroyed by the rampant papaya ring-spot virus.

The Regulation of Biotech Crops

Soon after the creation of the first bioengineered organisms, scientists and policymakers began to ask themselves what type of regulatory oversight would be appropriate. During the last 30 years, dozens of scientific bodies, including the U.S. National Academy of Sciences (NAS), the American Medical Association, the Institute of Food Technologists, and the United Nations' Food and Agriculture Organisation and World Health Organisation have studied the scientific literature and made recommendations about the oversight that is appropriate for bioengineered organisms, arriving at remarkably similar conclusions. The level of risk an individual plant might pose to human health or the ecology has nothing to do with how it was developed; it has solely to do with the characteristics of the plant that is being modified, the specific gene or genes that are added, and the local environment into which it is being introduced.

When introduced into new ecosystems, all types of plants, whether they are wild types or are developed with biotechnology or more conventional breeding methods, pose a danger of becoming invasive weeds and harming local biodiversity. Similarly, both conventional and modern plant breeding involve introducing new genes into established crop plants. Thus, they both pose a risk of introducing potentially harmful proteins and other substances into the food supply, some of which could be allergens or toxins. However, the mere fact that new genes are being added to plants, even from wholly unrelated organisms, does not make them less safe either to the environment or to people.

An analysis published by the Institute of Food Technologists, a professional society of food scientists, concluded that the evaluation of biotech food "does not require a fundamental change in established principles of food safety; nor does it require a different standard of safety" than those that apply to conventional foods. Under U.S. federal law, developers and marketers of all new foods have a responsibility to ensure that the products they sell are safe and in compliance with all legal requirements. Yet, that's where the similarity in regulation of conventional and bioengineered foods ends. Biotech plants are regulated much more stringently, even though scientists agree that the same practices used to regulate new crop varieties produced by means of conventional techniques are sufficient to ensure the safety of plants developed with biotechnology.

For plants developed with more conventional techniques, regulators rely on plant breeders to conduct appropriate safety testing and to eliminate plants that exhibit unexpected adverse traits before they are commercialised. No specific testing is required, nor is pre-market approval necessary, even though new varieties produced with these more conventional methods often contain hundreds of unique proteins and other chemicals that may never have been in the food supply before. Most of those newly introduced substances will be totally unidentified (and unidentifiable) by the plant breeders. But this rarely poses any real danger. Decades of accumulated scientific evidence confirm that even the use of relatively crude and unpredictable genetic techniques for the improvement of crops plants poses minimal risk to human health and the environment.

But bioengineered plants, in which breeders actually know which new genes and proteins are being introduced into the plant, are subjected to heightened scrutiny in every country in the world where they are grown. In the United States, they are regulated by the U.S. Department of Agriculture (USDA), the Environmental Protection Agency (EPA), and the Food and Drug Administration (FDA).

Dozens of new plant varieties produced through less precise techniques like selection, hybridisation, induced mutation, embryo rescue, and other, non-biotech methods enter the market every year without any special pre-market testing requirements. But every single bioengineered plant on the market has been tested and re-tested, going through several hundred - and in some cases, several thousand - different tests to ensure environmental and human health protection. Contrary to the assertions of ideological environmentalists, the regulation of biotechnology is actually far more stringent than necessary to ensure that bioengineered crop varieties are at least as safe as conventional ones.

Are Biotech Crops Safe?

Opponents of biotechnology have long claimed that bioengineered plants are unnatural and dangerous. Complaints range from general charges of random, unintended effects that could make the plants unsafe, to more specific criticisms alleging the possible introduction of new toxins or allergens in the food supply. Ideological environmentalists also claim that bioengineered plants are more likely to have negative environmental impacts, including the destruction of wild biodiversity. But, as mentioned above, all bioengineered crop varieties are subjected to much greater regulatory scrutiny than conventional crops, and the regulatory mechanism has been designed specifically to prevent such potentially harmful side effects.

Because different plant varieties will have different characteristics, and thus, different risks, the regulatory approach for biotech plants focuses on identifying the source of potential hazards to the environment and human health that specific plants might pose. Regulators draw upon the existing risk assessment process for chemicals and novel foods, and factor in additional analyses specific to biotechnology. For example, all methods of crop breeding run the risk of unintended and unexpected disruptions in the normal functioning of specific genes - called pleiotropic effects. So, crop breeders always conduct a number of evaluations to eliminate potentially harmful side effects before commercialisation.

But for biotech plants, regulators require tests to compare the biological, chemical, and agronomic equivalence of the modified varieties with their closest related conventional varieties. This is done to ensure that no pleiotropic effects have changed the new bioengineered plant in a way that would make it unsafe - such as changing the normally existing levels of plant nutrients or other phytochemicals. Modest changes in the level of phytochemicals can occur with any type of breeding, but no bioengineered plants that have shown a significant change in important nutrients or toxins have ever been put on the market. Although several new plant varieties with intentionally altered phytochemicals are now being developed, such as tomatoes, peppers, and rice with added or higher levels of beta carotene, and soybeans with higher levels of vitamin E.

Regulatory evaluations also pay special attention to the genes that are added to bioengineered plants, the source of those genes, the traits that the genes produce, and whether or not they have a history of safe use in the food supply. Scientists generally know a great deal about the safety of genes that come from other plants or microorganisms that are already part of the food supply. For those that are not, additional tests to ensure the safety of the genes and their traits are required. The action of most genes is to help create proteins, which could be toxins or allergens. So, several additional studies are then required to ensure that the proteins are not toxic and to measure the similarity of the proteins with known allergens to ensure that no new allergenic substances are introduced into the food supply. And numerous feed evaluations have shown no adverse effects on livestock, or their meat or milk.

The potential for added genes to make bioengineered plants allergenic is among the most widely cited concerns about biotechnology. Although all forms of plant breeding pose some risk of introducing new allergens into the food supply, biotechnology has been singled out by activists for special attention. Professional scaremonger Jeremy Rifkin argues that, "In the coming years, agrichemical and biotech companies plan on introducing hundreds, even thousands of genes into conventional food crops, raising the very real possibility of triggering new kinds of allergenic responses about which little is known and for which there exist no known treatments." But Professor Steve Taylor, a noted allergen researcher at the University of Nebraska, thinks the risk is very small because, "there are good ways of predicting the potential allergenicity of a genetically modified food." In fact, one of the most important potential advantages of biotechnology is actually to eliminate existing allergens from foods like peanuts, wheat, and milk, by "silencing," or turning off," the genes that generate allergenic proteins. Taylor says, "In the long term, we will have foods that are less hazardous because of biotechnology will have eliminated or diminished their allergenicity."

Just as with human safety, the ecological impact of any new crop depends on the type of introduced trait and the nature of the altered crop. Specific traits are focused on for assessing potential toxicity to beneficial insects, wild birds, and other animals. And the impacts of the whole plants are studied by assessing their similarity to traditional counterparts. New biotech plants are also assessed for their potential to cross-pollinate with wild or weedy plants, which could move the bioengineered traits into wild species with potentially negative consequences. Ecological aspects, such as potential to become problematic weeds and a range of other potential environmental effects, are studied prior to commercialisation in small field trials. These effects are also monitored carefully after commercialisation. Although some complaints have been lodged by farmers regarding the agronomic performance of certain bioengineered crop plants, no genuine environmental problems have yet been identified.

There is a risk that genes from biotech varieties could be transferred to wild plants through cross-pollination, but only in regions where there are closely enough related wild species for ordinary sexual reproduction. Moreover, this "out-crossing" is really only problematic when the genes in question could enhance the reproductive fitness of the recipient weeds: that is, enable weeds to produce and scatter seeds that survive better in the wild. Gene flow between crops and wild plants has been going on for a long time and is by no means unique to biotechnology. It has not been a problem though, because most genes that are introduced into crop plants, conventional or biotech, have little value in the wild. In fact, while some traits added with either bioengineering or conventional breeding methods could provide an ecological advantage, most crop traits tend to make plants less likely to survive the rigors of the wild.

For example, herbicide-tolerant rapeseed plants have been produced with conventional breeding for 20 years, and no unmanageable weed problems have been reported as a result of their use. So, while the transfer of a gene for herbicide tolerance into a wild relative could create a nuisance for farmers, it is unlikely to have any impact on wild biodiversity because the herbicide tolerance trait wouldn't give the wild plant any selective advantage relative to other weeds. Even in the extremely unlikely event that herbicide tolerance genes were transferred to a weed species, it wouldn't run amok in farmers' fields. Farmers could still control it by using other herbicides to which it was not tolerant.

Still ideological environmentalists insist that any out-crossing of genes from bioengineered plants into conventional or wild plants will be negative. In one recent case, ecologists from the University of California at Berkeley reported evidence that genes from bioengineered corn varieties had been transferred into local varieties of corn in Oaxaca state in southern Mexico where no bioengineered varieties have yet been approved for commercial cultivation. Although this report was later shown to be false, concerns arose among some ideological environmentalists that the presence of certain genes could only be explained by cross pollination from bioengineered varieties and that their presence posed a threat to the genetic diversity of the many landrace or heirloom varieties in what is considered to be the birthplace of corn. One Greenpeace activist from Mexico argued that, "It's a worse attack on our culture than if [biotech companies] had torn down the Cathedral of Oaxaca and built a McDonald's over it."

However, Mexican farmers reproduce their varieties by carefully selecting the seed they save from year to year. Thus, if a gene producing an undesirable trait is transferred into certain plants, seed from those crops will not be planted the following year and will be eliminated from the gene pool. This practice has worked very well for millennia and explains why Mexican farmers can plant many different varieties next to one another, without worrying about cross-pollination. Luis Herrera-Estrella, a plant scientist and director of the Center for Research and Advanced Studies in Irapuato, Mexico has noted that "gene flow between commercial and native varieties is a natural process that has been occurring for many decades," so "there is no scientific basis for believing that out-crossing from biotech crops could endanger [corn] biodiversity." Indeed, the presence of certain genes from biotech varieties could actually enhance genetic diversity by improving the ability of landrace varieties to resist pests, making them more productive.

Given concerns about the spread of bioengineered genes, you might think biotech opponents would welcome innovations designed to keep them confined. But when scientists at the U.S. Department of Agriculture and the Delta Pine Land Company did just that, environmentalists were infuriated. The process, called the Technology Protection System (TPS), was designed to make plant seeds sterile by interfering with the development of plant embryos. Hope Shand, research director for the Rural Advancement Foundation International, dubbed it "Terminator Technology." Jeremy Rifkin calls it "pathological," and has spread fears that escape of the TPS genes into weed populations through cross-pollination could destroy great swaths of plant life. But in the remote possibility of cross-pollination with weedy relatives, genes for traits such as herbicide tolerance or pest resistance wouldn't create "superweeds," because the TPS trait would prevent the wild plants from reproducing. Biotechnology companies like TPS because preventing farmers from replanting saved seeds from the prior year's harvest would protect the breeder's considerable investment in the development of new varieties. But critics see TPS as one more facet of global corporate hegemony. Mark Ritchie, president of the Institute for Agriculture and Trade Policy, argues that "It is a threat globally to food security, which is a basic human right."

Like many other concerns about biotechnology, this issue too has a non-biotech analogue. High-yielding hybrid varieties of plants like corn don't breed true, so most crop growers in the U.S. and Western Europe have been buying seed annually for decades. Thus, Technology Protected seeds wouldn't represent a big change in the way many American and European farmers farm. Many farmers in less developed countries have resisted hybrid technology because they prefer to have the option to plant saved seed. Similarly, if farmers didn't want the advantages offered in the enhanced crops protected by TPS, they would be free to buy seeds without the technology protection, just as farmers are free to buy non-hybrid seeds. Nevertheless, some of the biggest biotechnology companies have succumbed to pressures from environmental activists and aid organisations, and have promised not to commercialise the TPS technology. In any case, gene flow from bioengineered crops creating "superweeds" is not very likely.

Also consider that the biotech plants themselves are not likely to "escape" from farm fields and become weeds themselves, because crop plants of all varieties are generally not suited for existence in the wild-they need to be pampered. One noteworthy result of the extensive transformation of wild plants into crop varieties was the loss of many traits required for wild existence and the creation of a true dependency of modern crop plants upon human care for their survival. A ten-year study by British scientists found that neither biotech nor conventional crop plants survive well in the wild, and biotech varieties are no more likely than their conventional counterparts to invade wild ecosystems. Researchers have identified at least 12 genetic traits that are necessary for plants to be successful weeds. And crop plants typically have only six of them. For example, one of the most important traits shared by all weeds is their ability to disperse seeds beyond the immediate area. But crop varieties are bred specifically for their ability to hold seeds, and thus have lost their dispersal ability.

The fact is that modern cultivated plants, such as corn or soybeans, are incapable of invading and taking over forests and meadows.

Naturally, farmers and scientists are nevertheless vigilant against the unlikely chance that plants could out-cross with weeds or that the crop plants themselves could become weedy as a result of adding new traits. But this is the case whether or not a particular plant was modified with conventional or biotech methods. The risk of gene transfer to weeds is similar with both conventional and biotech varieties, and has no relation to the methods used in altering the plants. And because farmers are the first people affected by new weeds, they have a direct and strong incentive to prevent their development. The testing and monitoring of biotech crops, combined with hundreds of years of experience with conventional varieties, provides more than sufficient safeguard that such risks will be minimal and manageable.

The effect of biotechnology on crop biodiversity is another often-cited concern. The popularity of high-yielding varieties has narrowed the genetic variation found in major crops, because more and more farmers are planting the same or similar varieties. But biotechnology, if employed strategically, can reverse this trend by permitting the recovery of older varieties that were discarded for lack of certain features (such as resistance to new disease strains). With modern bioengineering techniques, older heirloom and landrace varieties can be modified to add such traits without destroying genetic diversity. Biotechnology researchers are also developing better methods for the preservation of germplasm in laboratories, such as cryopreservation, where plant cells with valuable genes are being stored and thus saved from extinction.

What is all too often overlooked by anti-biotech activists, however, is the fact that bioengineered crop varieties have substantial positive impacts on the environment. In addition to the significant reduction in chemical insecticide applications mentioned above, the introduction of biotech crops has made agriculture more efficient, promoting the conservation of important resources. Scientists from Louisiana State University and Auburn University found that when farmers plant bioengineered pest resistant crop varieties, fewer natural resources are consumed to manufacture and transport pesticides. Their study, which examined only pest resistant cotton, estimated that in 2000, 3.4

million pounds of raw materials and 1.4 million pounds of fuel oil were saved in the manufacture and distribution of synthetic insecticides. Additionally, 2.16 million pounds of industrial waste were eliminated. On the user end, farmers used 2.4 million gallons less fuel, 93 million gallons less water, and were spared some 41,000 10-hour days needed for applying pesticide sprays.

Perhaps most important is the fact that the increased productivity generated by bioengineered crop varieties will make it easier to conserve valuable wildlife habitat around the world. The loss and fragmentation of native habitats caused by agricultural development in the poorer regions of the world experiencing the greatest rates of population growth is widely recognised as among the most serious threats to the conservation of biodiversity. Thus, increasing agricultural productivity is an essential environmental goal, and one that would be much easier in a world where agricultural biotechnology is in widespread use.

Consider just one example. Rice is the major staple food for about 2.5 billion people, almost all of whom live in the less developed regions of the world where the bulk of 21st century population growth is expected to take place. The International Rice Research Institute estimates that reducing yield losses of rice by just 5 per cent worldwide could feed an additional 140 million people. Highly promising field tests in 1999 and 2000 showed a bioengineered rice variety to produce 28.9 per cent higher yields than conventional hybrid rice varieties. The environmental benefit of just this one biotech variety could be tremendous, if only wrongheaded international regulations inspired by ideological environmentalism do not doom its future.

International Rules

While U.S. regulation of biotechnology is overly strict, it pales in comparison with that in many other countries - particularly those countries that comprise the European Union (EU). Environmental activists in the EU, and in the United Kingdom in particular, have been aided and abetted by a sympathetic media willing to report uncritically activists' scaremongering as a way to sell more newspapers and magazines. The general public in most EU nations has become far more skeptical of biotechnology than the public in the United States.

Theories abound regarding why this suspicion arose. But one thing is certain: The greater public sensitivity to the issue of biotechnology has had a direct and deleterious impact on the development of European regulatory policy.

Beginning in 1990, the European Commission implemented a set of biotechnology regulations for all EU member countries. The rules are far more onerous than those in the United States, and the regulatory process is complex and difficult to navigate. For example, 18 varieties of biotech crop plants - including varieties of corn, canola, cotton, potato, tomato, and soybean - have been approved for commercial cultivation. But only two varieties - one corn and one soybean - have been approved for use in food. None of this matters much, however, because EU rules also require bioengineered foods to be labeled. And, due to the strong negative opinion of biotech foods held by a sizeable portion of the public, few grocery stores will stock products labeled as being bioengineered. Further problems stem from the fact that new bioengineered plant varieties must be approved by all 15 member nations in the European Union before they can be grown by farmers or sold as food.

Although dangerously wrongheaded, the European hysteria over biotech foods initially was seen as a regional problem. Increasingly, however, poor countries in East Asia are taking a far more cautious approach to biotechnology regulation. Japan, which has been a longtime leader in biotechnology research, has recently tightened restrictions on biotech food imports. And the European Union is pushing its overly-strict rules into international treaties affecting countries around the world. The EU was the primary advocate of the Cartagena Protocol on Biosafety, for example, which regulates the planting of bioengineered crops and the international trade in harvested biotech grains, vegetables, and fruits.

Finalised in January 2000, the Biosafety Protocol is intended to ensure that the introduction of bioengineered organisms into the environment is "undertaken in a manner that prevents or reduces the risks to biological diversity." But it also encourages countries to create unnecessarily severe biotechnology regulations based upon the Precautionary Principle that overemphasise biotechnology's very modest risks and ignore its vast potential benefits. Thus, laws enacted under the auspices of the Biosafety Protocol are likely to slow the research and development of new biotech products needlessly. Moreover, by making it easier for countries to create scientifically unjustifiable restrictions, the Protocol will undoubtedly be abused by politicians seeking trade protection for their domestic agriculture and food processing industries.

Importantly, countries whose exporters are adversely affected by biotechnology rules based on the precautionary principle might be able to challenge them through the World Trade Organisation's (WTO) dispute settlement processes. The WTO trade rules generally prohibit countries from restricting trade with environmental or public health laws that are not based upon a scientifically demonstrated risk. For a variety of reasons, however, it is not altogether clear that WTO rules would take precedence over the Biosafety Protocol, nor even that the WTO would be inclined to rule against biotechnology restrictions enacted to meet the Protocol's requirements.

Another important feature of the Biosafety Protocol is its requirement that bulk shipments of harvested agricultural products be labeled if they contain any biotech grains, fruits, or vegetables. To comply, farmers, shippers, and other food handlers would have to create hugely expensive segregation and record-keeping mechanisms, and test the foods at each step of the production process, to isolate conventional varieties from bioengineered ones. The EU's Directorate General for Agriculture estimates that the "identity preservation" costs alone for such a labeling requirement would range from 6 per cent to 17 per cent for commodity grains. The newly proposed European biotechnology law is set to go even further, by requiring not just mandatory labeling, but also "traceability" of biotech foods - an array of technical, labeling, and record-keeping mechanisms that require food processors to keep track of grains, fruits, vegetables, and other ingredients from the plant breeder, to the farm, to the grain handler, and beyond - from dirt to dinner plate.

Ultimately, labeling requirements like those enforced in the European Union represent serious obstacles that could all but destroy the affordability of biotechnology products and impede their adoption in the poorer regions of the world that need it most. The 2001 Human Development Report issued by the United Nations' Development Programme laments that "The opposition to yield-enhancing [bioengineered] crops in industrial countries with food surpluses could block the development and transfer of those crops to food-deficit countries."

What About Labeling?

Regulatory agencies around the world could learn a thing or two from the U.S. Food and Drug Administration's treatment of calls for biotech food labeling. Just as in Europe, some activists in the U.S. have called upon the government to mandate the labeling of all bioengineered foods. They assert that consumers have a "right to know" how their foods have been altered, and that a mandatory label would best allow consumers to choose between bioengineered and conventional foods. Biotechnology proponents and free speech advocates, on the other hand, have argued against mandatory labeling because such a requirement would unnecessarily raise food costs, mislead consumers into believing that the labeled products pose a heightened safety risk, and violate constitutional free speech rights.

Despite harsh attacks and considerable political pressure from environmentalist and consumerist organisations, the FDA has held firm in its respect for the judgment of the scientific opinion about the value of such labeling. In its 1992 statement of policy, the FDA concluded that there was no reason to believe "that bioengineered foods differ from other foods in any meaningful or uniform way."

The American Medical Association, the Institute of Food Technologists, and others have consistently argued that there is no scientific justification for special labeling of biotechnology-derived foods *per se*. Thus, the FDA only requires labeling of biotech foods if the genetic modifications change the food in a way that has a real impact on consumer health. Examples would include alterations in the plants that could increase the level of naturally-occurring but potentially-harmful chemicals; introduce new substances, such as potential allergens, into foods that did not previously have them; or change the nutritional composition or a food's storage or preparation requirements. To date, no bioengineered food products put on the market in the United States have required such labeling. Though, the very first bioengineered fruit, the Calgene corporation's FlavrSavr slow-ripening tomato, carried a voluntary notice that it had been engineered, and it was initially well received by consumers who were willing to pay a premium for the improved flavor promised on the labels.

Similarly, the FDA believes that requiring food labels to indicate the presence of bioengineered ingredients could mislead consumers into believing that the foods differ in safety or nutrition, when they do not. Labels are a valuable source of information for consumers, so U.S. federal law prohibits label statements that are likely to be misunderstood by consumers, even if not technically false. For example, labeling the vegetable broccoli as being "cholesterol-free" could run afoul of the FDA's rules because no broccoli contains cholesterol, and such a statement could suggest to consumers that while the particular broccoli is "cholesterol-free," other broccoli is not. Thus, rather than serving an educational or "right to know" purpose, mandatory labels on biotech foods could be misunderstood by consumers as a warning about some important difference.

Ultimately, though, consumers do not need to rely on mandatory labeling of biotechnology-enhanced foods to truly have a choice. Real world examples show that market forces are fully capable of supplying information about the methods in which foods and other products are produced if consumers truly demand it. Kosher and organic production certification are prime examples. Neither kosher nor organic labels convey relevant information about the safety or nutritional value of those products, but both meet a demand by consumers for information about the way the foods were produced.

The same can be said about biotechnology. Some producers of non-bioengineered products are already making label statements to convey that information to consumers. And the FDA recently published proposed guidelines to assist producers in voluntarily labeling both biotech and non-biotech foods in a way that is not misleading. In addition, under U.S. Department of Agriculture requirements, food products labeled as "organic" can not contain bioengineered ingredients. Consequently, consumers wishing to purchase non-biotech foods need only look for certified organic products.

The Road Ahead

Since the introduction of the very first bioengineered crop plant on the market in 1994, farmers, consumers, and food processors have experienced considerable benefits - from lower production costs to reduced pesticide use. But these benefits are dwarfed by the vast potential of agricultural biotechnology to aid in combating the even more serious problem of global food security.

During the next 50 years, global population may rise by 50 per cent to nine billion people, with nearly all of that growth coming in the poorest regions of the world. Fortunately, mankind will face the extraordinary challenge of hunger and poverty with the very powerful tool of crop biotechnology. As many have noted, the problem of hunger and malnutrition is not now primarily caused by a global shortage of food. At current levels, world food production could provide more than 2,600 calories every day for all six billion people on earth. The primary causes of hunger during this century have been political unrest and corrupt governments, poor transportation and infrastructure, and, of course, poverty. All of these problems and more will need to be addressed if we are to truly conquer worldwide hunger. But ensuring true food security in a world of eight or nine billion will require greater productivity.

As population increases, farmers must be able to grow more and more nutritious food on less land. Biotechnology can provide one very powerful way to do just that. Without such gains in productivity and nutrition, the growing need for food will require plowing under millions of hectares of wilderness - an environmental tragedy surely worse than any imagined by biotechnology's opponents. Furthermore, 650 million of the world's poorest people live in rural areas where agriculture is the primary economic activity. They are highly dependent upon the income that comes from growing and selling crops, so boosting the productivity of their crops would make a tremendous contribution to the battle against hunger and poverty.

Fortunately, the next generation of bioengineered products, now in research labs around the world, is poised to bring improved nutrition, longer shelf life, and greater productivity in the poor soils and harsh climates that tend to be characteristic of impoverished regions. And many of these products are being developed primarily or exclusively for poor subsistence farmers and consumers in less developed countries. Some improved plants include the same or similar traits for resistance to insects and plant diseases that are now used in industrialised countries, but in crops that are grown more typically in less developed nations, including rice, corn, cassava, sweet potato, and tropical fruits, such as bananas and papayas. Other bioengineered traits include faster maturation, drought tolerance, the ability to be irrigated with salty water or to grow in soil contaminated with excess salt, tolerance to extremes of heat and cold, and tolerance to soils with high acidity that are common in the tropics. These traits for greater tolerance to environmental conditions would be tremendously advantageous to poor farmers in less developed countries, and no one more so than in Africa.

Farmers in sub-Saharan Africa never saw the same productivity gains that countries in Asia and South America enjoyed from the Green Revolution. The primary focus of Green Revolution plant breeders was on improving such crops as rice, wheat, and corn, which are not widely grown in Africa. Plus, much of the African dry lands have little rainfall and no potential for irrigation, which play an essential role in productivity success stories of crops such as Asian rice. And the remoteness of many African villages and poor transportation infrastructure in landlocked African countries make it difficult for African farmers to obtain agricultural chemical inputs such as fertilisers, insecticides, and herbicides, even if they had the money to purchase them. Thus, by packaging technological inputs within seeds, biotechnology can provide the same, or better, productivity advantage as chemical or mechanical inputs, but in much more user-friendly manner. Farmers could be able to control insect pests, viral or bacterial pathogens, extremes of heat or drought, and poor soil quality, just by planting their crops.

Still, anti-biotech activists like Vandana Shiva and Miguel Altieri argue that poor farmers in less developed nations will never benefit from biotechnology, because it is controlled by multinational corporations. Altieri says that "Most innovations in agricultural biotechnology have been profit-driven rather than need-driven. The real thrust of the genetic engineering industry is not to make third world agriculture more productive, but rather to generate profits." But that sentiment is not shared by the thousands of academic and public sector researchers actually working on biotech applications in those countries. Cyrus Ndiritu, former director of the Kenyan Agricultural Research Institute, argues that, "It is not the multinationals that have a stranglehold on Africa. It is hunger, poverty and deprivation. And if Africa is going to get out of that, it has got to embrace [biotechnology]."

Researchers are also improving the nutritional quality of plants, by boosting their ability to produce important vitamins, minerals, and proteins. The diet of more than three billion people worldwide includes inadequate levels of many important micronutrients such as iron and vitamin A. Deficiency in just these two micronutrients can result in severe anemia, impaired intellectual development, blindness, and even death. Fortunately, a substantial amount of research into improving the nutritional value of staple crops is well underway. Perhaps the most promising recent advance in this area is the development of a rice variety that has been genetically enhanced to add beta carotene, which is converted in the human body to vitamin A. By boosting the availability of vitamin A in developing world diets, this Golden Rice could help prevent as many as a million deaths per year and eliminate numerous other health problems.

But for critics of biotechnology like India's Vandana Shiva, and New York food journalist Michael Pollan, Golden Rice is just a "Great Yellow Hype" - another ploy by multinational biotechnology corporations to get the world hooked on bioengineering. Never mind that the research, which added genes taken from daffodils and a bacterium to rice, was funded primarily by the New York-based Rockefeller Foundation, which has promised to make the rice available to developing-world farmers at little or no cost. Ismail Serageldin, director of the UN-sponsored Consultative Group on International Agricultural Research, asks opponents, "Do you want 2 to 3 million children a year to go blind and 1 million to die of vitamin A deficiency, just because you object to the way golden rice was created?" Apparently, the critics find it important to oppose biotechnology in any form.

But the benefits of agricultural biotechnology will by no means go exclusively to less developed countries. In industrialised nations such as the United States, consumers and farmers will continue to share in the benefits of improved productivity and reduced agricultural chemicals use. Agricultural biotechnology can also be used to develop healthier cooking oils that are low in saturated fats, vegetables with higher levels of cancer-fighting antioxidants, and foods with better taste and longer shelf life. It is also possible to use bioengineered plants to create biodegradable plastics, better medicines, and to help clean up hazardous wastes.

Although the complexity of biological systems means that some of these promised benefits of biotechnology are many years away, the biggest threats that consumers awaiting the bioengineering revolution currently face are restrictive policies stemming from unwarranted fears that the technology poses unique and dangerous threats to human health or the environment. No one thinks that biotech innovators should not be cautious, as all new technologies have both risks and benefits. But appropriate regulatory approaches involve weighing the risks and benefits of moving into the future against the risks and benefits of forgoing the new technology - not pointing to hypothetical risks and saying no. The bottom line is that scaremongering and over-regulation are slowing progress in agricultural biotechnology and inflating the costs of research and development. Ultimately, this hurts both poor farmers struggling to feed their families and the natural environment upon which we all depend.

Energy and Waste Management

20

Can The Supply of Natural Resources - Especially Energy - Really Be Infinite?

JULIAN SIMON

A professor giving a lecture on energy declares that the world will perish in seven billion years' time because the sun will then burn out. One of the audience becomes very agitated, asks the professor to repeat what he said, and then, completely reassured, heaves a sign of relief, "Phew! I thought he said seven million years!"[1]

Natural resources, properly defined, cannot be measured. Here I draw the logical conclusion: Natural resources are not finite. Yes, you read correctly. This chapter shows that the supply of natural resources is not finite in any economic sense, one reason why their cost can continue to fall indefinitely.

On the face of it, even to inquire whether natural resources are finite seems like nonsense. Everyone "knows" that resources are finite. And this belief has led many persons to draw unfounded, far-reaching conclusions about the future of our world economy and civilisation. A prominent example is the Limits to Growth group, who open the preface to their 1974 book as follows:

"Most people acknowledge that the earth is finite.... Policymakers generally assume that growth will provide them tomorrow with the resources required to deal with today's problems. Recently, however, concern about the consequences of population growth, increased environmental pollution, and the depletion of fossil fuels has cast doubt upon the belief that continuous growth is either possible or a panacea."[2] (Note the rhetorical device embedded in the term "acknowledge" in the first sentence of the quotation. It suggests that the statement is a fact, and that anyone who does not "acknowledge" it is simply refusing to accept or admit it). For many writers on the subject, the inevitable depletion of natural resources is simply not open to question.

The idea that resources are finite in supply is so pervasive and influential that the President's 1972 Commission on Population Growth and the American Future based its policy recommendations squarely upon this assumption. Right at its beginning the report asked, "What does this nation stand for and where is it going? At some point in the future, the finite earth will not satisfactorily accommodate more human beings.... It is both proper and in our best interest to participate fully in the worldwide search for the good life, which must include the eventual stabilisation of our numbers."[3]

The assumption of finiteness indubitably misleads many scientific forecasters because their conclusions follow inexorably from that assumption. From the Limits to Growth team again, this time on food: "The world model is based on the fundamental assumption that there is an upper limit to the total amount of food that can be produced annually by the world's agricultural system."

The idea of finite supplies of natural resources led even a mind as powerful as Bertrand Russell's into error. Here we're not just analysing casual opinions; all of us necessarily hold many casual opinions that are ludicrously wrong simply because life is far too short for us to think through even a small fraction of the topics that we come across. But Russell, in a book ironically titled The Impact of Science on Society, wrote much of a book chapter on the subject. He worried that depletion would cause social instability:

"Raw materials, in the long run, present just as grave a problem as agriculture. Cornwall produced tin from Phoenician times until very lately; now the tin of Cornwall is exhausted... Sooner or later all easily accessible tin will have been used up, and the same is true of most raw materials. The most pressing, at the moment, is oil... The world has been living on capital, and so long as it remains industrial it must continue to do so. This is one inescapable though perhaps rather distant source of instability in a scientific society."[4]

Nor is it only non-economists who fall into this error (though economists are in less danger here because they are accustomed to expect economic adjustment to shortages). John Maynard Keynes's contemporaries thought that he was the cleverest person of the century. But on the subject of natural resources - and about population growth, as we shall see later - he was both ignorant of the facts and stupid (an adjective I never use except for the famous) in his dogmatic logic. In his world-renowned The Economic Consequences of the Peace, published just after World War I, Keynes wrote that Europe could not supply itself and soon would have nowhere to turn:

"By 1914 the domestic requirements of the United States for wheat were approaching their production, and the date was evidently near when there would be an exportable surplus only in years of exceptionally favourable harvest... Europe's claim on the resources of the New World was becoming precarious; the law of diminishing returns was at last reasserting itself, and was making it necessary year by year for Europe to offer a greater quantity of other commodities to obtain the same amount of bread... If France and Italy are to make good their own deficiencies in coal from the output of Germany, then Northern Europe, Switzerland, and Austria...must be starved of their supplies."[5]

All these assertions of impending scarcity turned out to be wildly in error. Millions of American farmers had a far better grasp of the agricultural reality in the 1920s than did Keynes. This demonstrates that one needs to know history as well as technical facts, and not just be a clever reasoner.

Just as in Keynes's day, the question of finiteness is irrelevant to any contemporary considerations, as the joke at the head of the chapter suggests. Nevertheless, we must discuss the topic because of its centrality in so much contemporary doomsday thinking. The argument in this chapter is very counter-intuitive: indeed, science is most useful when it is counter-intuitive. But when scientific ideas are sufficiently far from "common sense," people will be uncomfortable with science, and they will prefer other explanations, as in this parable:

Imagine for the moment that you are a chieftain of a primitive tribe, and that I am explaining to you why water gradually disappears from an open container. I offer the explanation that the water is comprised of a lot of invisible, tiny bits of matter moving at enormous speeds. Because of their speed, the tiny bits escape from the surface and fly off into the air. They go undetected because they are so small that they cannot be seen. Because this happens continuously, eventually all of the tiny, invisible bits fly into the air and the water disappears. Now I ask you: "Is that a rational scientific explanation?" Undoubtedly, you will say yes. However, for a primitive chief, it is not believable. The believable explanation is that the spirits drank it.

But because the ideas in this chapter are counter-intuitive does not mean that there is not a firm theoretical basis for holding them.

The Theory of Decreasing Natural-Resource Scarcity

People's response to the long trend of falling raw-material prices often resembles this parody: We look at a tub of water and mark the water level. We assert that the quantity of water in the tub is "finite." Then we observe people dipping water out of the tub into buckets and taking them away. Yet when we re-examine the tub, lo and behold the water level is higher (analogous to the price being lower) than before. We believe that no one has reason to put water into the tub (as no one will put oil into an oil well), so we figure that some peculiar accident has occurred, one that is not likely to be repeated. But each time we return, the water level in the tub is higher than before - and water is selling at an ever-cheaper price (as oil is). Yet we simply repeat over and over that the quantity of water must be finite and cannot continue to increase, and that's all there is to it.

Would not a prudent person, after a long train of rises in the water level, conclude that perhaps the process may continue - and that it therefore makes sense to seek a reasonable explanation? Would not a sensible person check whether there are inlet pipes to the tub? Or whether someone has developed a process for producing water? Whether people are using less water than before? Whether people are restocking the tub with recycled water? It makes sense to look for the cause of this apparent miracle, rather than clinging to a simpleminded fixed-resources theory and asserting that it cannot continue.

Let's begin with a simple example to see what contrasting possibilities there are. (Such simplifying abstraction is a favourite trick of economists and mathematicians.)

If there is only Alpha Crusoe and a single copper mine on an island, it will be harder to get raw copper next year if Alpha makes a lot of copper pots and bronze tools this year, because copper will be harder to find and dig. And if he continues to use his mine, his son Beta Crusoe will have a tougher time getting copper than did his daddy, because he will have to dig deeper.

Recycling could change the outcome. If Alpha decides in the second year to replace the old tools he made in the first year, he can easily reuse the old copper and do little new mining. And if Alpha adds fewer new pots and tools from year to year, the proportion of cheap, recycled copper can rise year by year. This alone could mean a progressive decrease in the cost of copper, even while the total stock of copper in pots and tools increases.

But let us for the moment leave the possibility of recycling aside. Another scenario:

If suddenly there are not one but two people on the island, Alpha Crusoe and Gamma Defoe, copper will be more scarce for each of them this year than if Alpha lived there alone, unless by cooperative efforts they can devise a more complex but more efficient mining operation - say, one man getting the surface work and one getting the shaft. Or, if there are two fellows this year instead of one, and if copper is therefore harder to get and more scarce, both Alpha and Gamma may spend considerable time looking for new lodes of copper.

Alpha and Gamma may follow still other courses of action. Perhaps they will invent better ways of obtaining copper from a given lode, say a better digging tool, or they may develop new materials to substitute for copper, perhaps iron.

The cause of these new discoveries, or the cause of applying ideas that were discovered earlier, is the "shortage" of copper that is, the increased cost of getting copper. So increased scarcity causes the development of its own remedy. This has been the key process in the supply of natural resources throughout history.

Interestingly, the pressure of low prices can also cause innovation. Improvement in the efficiency of using copper not only reduces resource use in the present, but effectively increases the entire stock of unused resources. For example, an advance in knowledge that leads to a one per cent decrease in the amount of copper that we need to make electrical outlets is much the same as an increase in the total stock of copper that has not yet been mined. And if we were to make such a one per cent increase in efficiency for all uses every year, a one per cent increase in demand for copper in every future year could be accommodated without any increase in the price of copper, even without any other helpful developments.

Discovery of an improved mining method or of a substitute such as iron differs, in a manner that affects future generations, from the discovery of a new lode. Even after the discovery of a new lode, on the average it will still be more costly to obtain copper, that is, more costly than if copper had never been used enough to lead to a "shortage."

But discoveries of improved mining methods and of substitute products can lead to lower costs of the services people seek from copper. Please notice how a discovery of a substitute process or product by Alpha or Gamma benefits innumerable future generations. Alpha and Gamma cannot themselves extract nearly the full benefit from their discovery of iron. (You and I still benefit from the discoveries of the uses of iron and methods of processing it made by our ancestors thousands of years ago). This benefit to later generations is an example of what economists call an "externality" due to Alpha and Gamma's activities, that is, a result of their discovery that does not affect them directly.

If the cost of copper to Alpha and Gamma does not increase, they may not be impelled to develop improved methods and substitutes. If the cost of getting copper does rise for them, however, they may then bestir themselves to make a new discovery. The discovery may not immediately lower the cost of copper dramatically, and Alpha and Gamma may still not be as well off as if the cost had never risen. But subsequent generations may be better off because their ancestors Alpha and Gamma suffered from increasing cost and "scarcity."

This sequence of events explains how it can be that people have been using cooking pots for thousands of years, as well as using copper for many other purposes, and yet the cost of a pot today is vastly cheaper by any measure than it was 100 or 1,000 or 10,000 years ago.

Now I'll restate this line of thought into a theory: More people, and increased income, cause resources to become scarcer in the short run. Heightened scarcity causes prices to rise. The higher prices present opportunity, and prompt inventors and entrepreneurs to search for solutions. Many fail in the search, at cost to themselves. But in a free society, solutions are eventually found. And in the long run the new developments leave us better off than if the problems had not arisen. That is, prices eventually become lower than before the increased scarcity occurred.

It is all-important to recognise that discoveries of improved methods and of substitute products are not just luck. They happen in response to an increase in scarcity - a rise in cost. Even after a discovery is made, there is a good chance that it will not be put into operation until there is need for it due to rising cost. This point is important: Scarcity and technological advance are not two unrelated competitors in a Malthusian race; rather, each influences the other.

Because we now have decades of data to check its predictions, we can learn much from the 1952 U.S. governmental inquiry into raw materials - the President's Materials Policy Commission (the Paley Commission), organised in response to fears of raw-material shortages during and just after World War II. Its report is distinguished by having some of the right logic, but exactly the wrong forecasts:

"There is no completely satisfactory way to measure the real costs of materials over the long sweep of our history. But clearly the man hours required per unit of output declined heavily from 1900 to 1940, thanks especially to improvements in production technology and the heavier use of energy and capital equipment per worker. This long-term decline in real costs is reflected in the downward drift of prices of various groups of materials in relation to the general level of prices in the economy... But since 1940 the trend has been soaring demands, shrinking resources, the consequences pressure toward rising real costs, the risk of wartime shortages, the strong possibility of an arrest or decline in the standard of living we cherish and hope to share."[6]

The commission went on to predict that prices would continue to rise for the next quarter-century. However, prices declined rather than rose. There are two reasons why the Paley Commission's predictions were topsy-turvy: First, the commission reasoned from the notion of finiteness and used a static technical analysis.

A hundred years ago resources seemed limitless and the struggle upward from meagre conditions of life was the struggle to create the means and methods of getting these materials into use. In this struggle we have by now succeeded all too well. The nature of the problem can perhaps be successfully over-simplified by saying that the consumption of almost all materials is expanding at compound rates and is thus pressing harder and harder against resources which whatever else they may be doing are not similarly expanding.

The second reason the Paley Commission went wrong is that it looked at the wrong facts. Its report gave too much emphasis to the trends of costs over the short period from 1940 to 1950, which included World War II and therefore was almost inevitably a period of rising costs, instead of examining the longer period from 1900 to 1940, during which the commission knew that "the man-hours required per unit of output declined heavily".

Let us not repeat the Paley Commission's mistakes. We should look at trends for the longest possible period, rather than focusing on a historical blip; the OPEC-led price rise in all resources after 1973 and then the oil price increase in 1979 are for us as the temporary 1940-50 wartime reversal was for the Paley Commission. We should ignore them, and attend instead to the long-run trends which make it very clear that the costs of materials, and their scarcity, continuously decline with the growth of income and technology.

Resources as Services

As economists or as consumers we are interested, not in the resources themselves, but in the particular services that resources yield. Examples of such services are a capacity to conduct electricity, an ability to support weight, energy to fuel autos or electrical generators, and food calories.

The supply of a service will depend upon (a) which raw materials can supply that service with the existing technology, (b) the availabilities of these materials at various qualities, (c) the costs of extracting and processing them, (d) the amounts needed at the present level of technology to supply the services that we want, (e) the extent to which the previously extracted materials can be recycled, (f) the cost of recycling, (g) the cost of transporting the raw materials and services, and (h) the social and institutional arrangements in force. What is relevant to us is not whether we can find any lead in existing lead mines but whether we can have the services of lead batteries at a reasonable price; it does not matter to us whether this is accomplished by recycling lead, by making batteries last forever, or by replacing lead batteries with another contraption. Similarly, we want intercontinental telephone and television communication, and, as long as we get it, we do not care whether this requires 100,000 tonnes of copper for cables, or a pile of sand for optical fibres, or just a single quarter-tonne communications satellite in space that uses almost no material at all. And we want the plumbing in our homes to carry water; if PVC plastic has replaced the copper that formerly was used to do the job - well, that's just fine.

This concept of services improves our understanding of natural resources and the economy. To return to Crusoe's cooking pot, we are interested in a utensil that one can put over the fire and cook with. After iron and aluminium were discovered, quite satisfactory cooking pots - and perhaps more durable than pots of copper - could be made of these materials. The cost that interests us is the cost of providing the cooking service rather than the cost of copper. If we suppose that copper is used only for pots and that (say) stainless steel is quite satisfactory for most purposes, as long as we have cheap iron it does not matter if the cost of copper rises sky high. (But as we have seen, even the prices of the minerals themselves, as well as the prices of the services they perform, have fallen over the years.)

Here is an example of how we develop new sources of the resources we seek. Ivory used for billiard balls threatened to run out late in the 19th century. As a result of a prize offered for a replacement material, celluloid was developed, and that discovery led directly to the astonishing variety of plastics that now gives us a cornucopia of products (including billiard balls) at prices so low as to boggle the 19th century mind.

Are Natural Resources Finite?

Incredible as it may seem at first, the term "finite" is not only inappropriate but is downright misleading when applied to natural resources, from both the practical and philosophical points of view. As with many important arguments, the finiteness issue is "just semantic." Yet the semantics of resource scarcity muddle public discussion and bring about wrongheaded policy decisions.

The ordinary synonyms of "finite", the dictionary tells us, are "countable" or "limited" or "bounded". This is the appropriate place to start our thinking on the subject, keeping in mind that the appropriateness of the term "finite" in a particular context depends on what interests us. Also please keep in mind that we are interested in material benefits and not abstract mathematical entities *per se.*

Notes

1. Sauvy, Alfred. (1976) *Zero Growth.* Trans. by A. Maguire. New York: Praeger.

2. Meadows, Dennis L.; William W. Behrens, III; Donella H. Meadows; Roger F. Naill; Jorgen Randers; and Erich K. 0. Zahn. (1974) *Dynamics of Growth in A Finite World.* Cambridge, Mass.: Wright-Allen.

3. U.S. The White House. (1972) *Population and the American Future.* The Report of the Commission on Population Growth and the American Future. New York: Signet.

4. Russell, Bertrand. (1967) *The Impact of Science on Society.* London: Allen & Unwin

5. Keynes, John Maynard. (1920) *The Economic Consequences of the Peace.* New York: Harcourt, Brace

6. U.S. The White House. (1952) "Resources for Freedom". 4 vols. *The President's Materials Policy Commission* (The Paley Commission). Washington, D.C.: GPO. June.

21

Law, Markets and Waste

JULIAN MORRIS

Introduction

People often claim that 'the market' is wasteful and that government intervention is necessary in order to reduce waste and increase efficiency. I believe that such claims are based on a false conception of how a true market system functions. I shall argue that the best way to minimise waste is to allow the conventional institutions of the market system - private contracts and civil liability - to define the boundaries of human action. Furthermore, I shall argue that the current socialised system of residuals management is excessively wasteful. If we wish to move towards a more sustainable, less wasteful society, we must reconsider the objectives of policies that are currently directed towards dealing with residuals.

Law

In a pure market system, the production and consumption of goods occur within a legal framework which protects the rights of individuals to own, use and exchange property, enables individuals to enforce contracts, and limits the amount of harm that can be inflicted on persons and property. This legal framework structures what can and what cannot occur in the market system.

The ability to own property enables people to make investments in the future secure in the knowledge that they will reap the benefits of those investments. A farmer who owns his land has a greater incentive to make improvements to that land, than a farmer whose land belongs to the state.

The ability to transfer property contractually enables people to engage in mutually beneficial exchanges. If contracts of sale were not enforceable, people would be reluctant to make such exchanges because the counter-party to any exchange could simply claim back 'their' property.

The ability to protect property from outside interference through civil liability further enhances the incentives to invest in the improvement of that property. If anyone can use agricultural land for whatever purpose they choose regardless of who owns that land and without paying any compensation, then the owner will have little incentive to make any improvements to that land for fear that those improvements might encourage people to destroy his improvements.

The ability to sell the rights to be free from such interference ensures that some activities that are beneficial to society but impose costs on individuals may proceed without in fact harming anyone. So, for example, the right to allow people to come onto one's land in return for a fee ensures that people have access to areas of specific interest without harming the owner or creating a disincentive for him to invest in improvements.

So important is such a legal framework for the existence of markets that it is difficult to imagine what production and consumption would be like without them.

Markets

In pre-market societies, production is limited to those goods that can be manufactured locally. The simplest form of such production is that of the family farm, which is self-sufficient in food. Such methods of production are precarious because, amongst other things, variations in weather patterns can lead to poor harvests. Several strategies are available to avoid such tragedies, of which I shall briefly describe three. First, families can split their land into various sub-plots, each in different parts of the village and each affected differently by the weather. Second, farmers can pool their resources and share out the proceeds at the end. Third, farmers can plant more than is necessary for survival. Of these strategies, the first two are typically adopted in the early stages of production, when markets are little developed and the failure of a crop could mean death. However, over time markets for surplus produce develop and the risk of adopting the third strategy declines. As a result, people specialise, adapting their productive systems to the environment and developing technologies that increase efficiency, such as ploughs, tractors, fertilisers, pesticides, and new varieties of crops.

Waste

The production of goods in this kind of extensive market system is clearly beneficial for the participants: it is often less risky than share-cropping or strip-farming, and it also typically results in greater wealth creation. However, it also causes a change in the way that the residuals of production and consumption are managed. In primitive societies, residuals, such as manure from horses and oxen, and chaff from wheat, are used as fertiliser or fuel within the village itself, whilst unwanted residuals are dumped nearby; this type of residuals-management might be called 'closed-loop'. In the extended market order, residuals that are produced in one place are often transformed or disposed of in another; this type of residuals-management might be called open-loop. Some people seem to think that closed-loop residuals management is preferable to open-loop residuals-management. I suspect it is because they ask themselves the wrong questions. The question should not be, 'how can I minimise the transportation of residuals?' or even, 'how can I ensure that the maximum amount of a particular type of residual is recycled?' rather, it should be, 'how can I ensure that the residuals-management system results in the least waste of resources?'

Minimising Waste

The fundamental objective of any business enterprise is to create added value - to sell products at a price greater than the costs of manufacture. So the entrepreneur is always vigilant for ways of improving product performance and reducing costs. Cost reductions can be made in numerous ways, including by reducing the use of raw materials and from using by-products more efficiently.

If it is possible to save money by utilising the by-products of manufacture rather than paying to have them disposed, then entrepreneurs will generally discover those uses and, over time, adjust their manufacturing processes to enable such utilisation. So important is it that these efficiency gains be realised that in the 1970s management consultants such as Arthur D Little developed procedures for carrying out internal 'life cycle analyses' (LCAs), in which all the most important inputs to and outputs from any production process are assessed in order to discover potential efficiency gains. Of course, carrying out an LCA and transforming manufacturing processes in light of the findings are costly – they consume resources – and so some apparent improvements will not take place (at least in the short term). But that in itself is not a criticism of the current system; it is merely a consequence of a world of imperfect knowledge.

The notion that this system can be improved upon by government intervention – as many argue – is implausible. The government's (or regulator's) knowledge of what use of resources is most efficient is likely in most cases to be less complete than that of the individual manufacturers, who must day after day assess the costs of inputs and prices of outputs.

With regard to material use per unit of manufacture, the situation is equally clear. Consider the example of packaging, which serves both to improve the quality of products and to reduce costs. Packaging enhances the shelf life of food products and means that less food will be wasted on the journey from the producer to the consumer. This means that the products can be sold at a lower price, satisfying more consumers and increasing the profits of the manufacturer and retailer. In addition, consumers benefit from being able to store their products at home for longer – saving them trips to the supermarket, which might have involved the use of some fossil fuel-based transport system.

Nevertheless, packaging itself uses resources, so over time entrepreneurs have developed packaging systems that use less material. Compare the heavy glass bottles that were the predominant means of packaging milk and other soft drinks twenty years ago with the lightweight plastic bottles and laminated cartons used today. These modern alternatives are not only cheaper to produce, but their lighter weight and more rectangular shape also reduce transport costs. Moreover, in the case of fruit juices, the use of aseptic laminated containers (the brick-like packs made of layers of plastic, paper and aluminium) has dramatically reduced the quantity of resources consumed during storage, since it is no longer necessary to refrigerate them. Similar advances have been made in other areas. Cables carrying information long distances are now typically made of glass-fibre rather than copper: a cable made from 60 pounds of silica can carry 1000 times as much information as a cable made from a tonne of copper.[1] Computers offer perhaps the most startling example of this 'dematerialisation'. In the 1950s computers were the size of a two-bedroom flat and could process only about 1000 instructions per second. Today, computers the size and weight of a book can process 200,000,000 instructions per second. These advances in computer technology have also led to more efficient use of resources in other areas. For example, all the world's telephone numbers can now be stored on a few easily searchable compact discs, rather than in hundreds of cumbersome and poorly cross-referenced books. Letters and manuscripts can now be sent electronically from England to New Zealand in a few seconds, whereas before they went by fossil fuel-guzzling aeroplane or boat and took days or weeks.

It is clear, then, that entrepreneurs have strong incentives to reduce their consumption of resources over time. However, these incentives are often distorted by interventions in the market. For example, where municipalities operate a monopolistic solid waste management system, companies and individuals are unable to decide which type of residuals-management system would be most appropriate. This situation is made worse if the municipality charges a flat fee, since this erodes even the marginal incentives to limit the generation of solid waste that is created by unit pricing and distorts the companies residuals-management system towards the over-production of solid waste. Similar distortions are created by the existence of statutory licenses to emit substances into the atmosphere or watercourse.

These licenses typically over-ride civil liability, so that companies need no longer pay affected parties for the costs that they impose on them. As a result, the residuals-management system might be distorted in favour of excessive use of emissions to air or water. On the other hand, the cost of licenses and fines for exceeding emissions limits may be greater than the price that private individuals would charge for the right to pollute the air, in which case there would be a distortion in favour of excessive recycling and the over-production of solid waste.

The problem with such a socialised system of residuals management is that we do not know how individuals value their environment and so we cannot know whether the implicit prices charged for use of that environment are correct. This problem applies equally, of course, to pollution taxes and to tradable emissions permits, although such instruments may have certain efficiency advantages over a simple command and control system.

Privatising Pollution

A true market solution to these distortions would entail moving back to a system of civil liability for protection of private property and to private contracting for waste services. I shall briefly adumbrate how such a system would function. Consider first the problem of pollution. If A, intentionally or unintentionally, emits a substance that damages the property of B, then A should pay compensation to B. For example, in the case of *St. Helen's Smelting Co vs. Tipping*[2], the owners of a smelter were forced to pay compensation for causing physical damage to Mr Tipping's shrubs and trees. This seems a fine rule where the damages may reasonably be estimated by a third party.[3] However, where the infringement is one affecting the reasonable enjoyment of land, compensation may be more difficult for a third party to calculate, so it may be desirable, in addition to awarding compensation, to enjoin activities which cause such infringements. For example, in *Aldred's case*, from 1611,[4] Aldred owned a property abutting a pig farm, which caused an unbearable stench. Aldred sued the owner of the pig farm and was granted an injunction.[5] Note that if this were the rule, then the person harmed could decide to sell his right not to be polluted if he so desires.

General application of such private rights to be free from pollution, including removal of the defence of statutory authority, would, I believe, enable individual's subjective valuations of the environment to be better expressed. Of course, such a system is unlikely to be perfect. In particular, where there are many parties affected by pollution, the costs of bargaining with the polluter would be high, in which case there might be an inefficiently low level of pollution. Moreover, where there are many parties causing the pollution, it may be difficult to identify the specific impacts of any particular polluter, so the level of compensation may be too low and the level of pollution inefficiently high. In this latter case, private landowners might make agreement amongst themselves setting general rules governing permissible levels of pollution. However, these rules are unlikely to satisfy everyone, so the system remains imperfect. Nevertheless, the question remains whether public regulation is a solution to these problems, or whether it would be better to allow the level of harm that results from private ordering.

With regard to management of solid waste, the solution I would advocate is to devolve management entirely to private contractual arrangements. If all individuals and companies were responsible for disposing of their own solid waste, within the context of the above-mentioned system of civil liability for damage to property, then they would discover the most cost effective – that is to say the least wasteful – ways of disposing of their residuals.

How do we get from the current system to a private system of residuals management? I would suggest that governments do the following: First, remove all instances of statutory protection from civil liability for damage to property. Second, remove all mandatory duties on municipalities to provide waste collection and disposal services and remove all mandatory restrictions on private contracting for waste management services. Third, remove all mandatory controls on end-of-life management of specific products.

Notes

1. Scarlett, L., "Packaging, Solid Waste and Environmental Trade-Offs", Illahee, vol. 10 (1), 1994, pp. 15-33.

2. (1865) 11 All ER 1483

3. However, a rule of compensation for physical damage might encourage inappropriate land-uses; for example the possibility of claiming damages might encourage farmers to grow more valuable crops in order to claim greater damages. Coase, R. (1960) "The Problem of Social Cost", Journal of Law and Economics, 4, 1- 40.

4. Aldred's Case (1611) 9 Co. Rep. 57

5. Individuals might, in exceptional circumstances, be able to prevent an activity even before it is begun if they can show that the activity would cause a nuisance. So, for example, in Ireland a judge was asked to rule on whether a company should be prevented from constructing a landfill, on the grounds that the landfill might cause a nuisance to local residents. The judge in that case decided that there was insufficient evidence that the landfill would indeed cause a nuisance, however he noted that if the landfill ever did cause a nuisance then it should be enjoined (McGrane vs. Louth County Council, Unreported, 9 December 1983).

22

Coercive Recycling, Forced Conservation, and Free-market Alternatives

JULIAN SIMON

Some people get pleasure from recycling. That's fine, even if the impulse springs from the wrongheaded belief that recycling is a public service. Coercing people to recycle is something very different, however.

Bette Bao Lord's description of forced-labour recycling in China during the "Cultural Revolution" is instructive:

"One of the most frequent duties was to sort the city garbage dump. The job was a big one and we were usually excused from school for the day. We had to sort the trash into piles of paper or leather or scrap steel, etc., which were in turn reprocessed for further manufacture. Another frequent detail was cleaning the streets of used paper and other litter. Each student was given a stick with a nail at one end for spearing and a paper bag for collection. After a day of this type of labour detail, I suffered from a backache. When the school collected all our full bags, it would turn them over to the factory for making new paper to ease the paper shortage."[1]

In poor societies, it may be worthwhile to recycle. But in China people not only recycle because they are poor, they are also poor because they recycle. The recycling mindset in China in the late 1950s and early 1960s was inimical to economic development. The notion of making progress by forcing people to recycle, when they could be doing more productive things, makes societies poor. This mistake is sometimes compounded by the glorification of primitive production techniques (as Mao tried to do with backyard steel making and Gandhi with household textile manufacturing), rather than using the same amount of effort to learn how to do the work more efficiently and to gradually build a more modern system. The recycling mindset can also be counterproductive in rich countries. People praise saving trees by recycling newspapers, and they condemn those who cut down trees. But they do not cheer those who grow the trees in the first place—trees deliberately planted and grown to make paper. This is like praising people for "saving" a field of wheat by not eating bread, while ignoring the farmers who grow our food. It acts to suppress the creative impulse (There is a further perversity in thinking here. People are focusing only on harvesting and death, not on planting and growth. It is much the same with respect to population growth - wringing of hands over someone who may die prematurely or unnecessarily, but no cheering of those who create and nurture new life).

This is not to say that all recycling is misguided. Recycling is one thing when it is economically worthwhile for the person who is doing the recycling, but another thing when it is done purely for symbolic purposes, as now mostly in the United States. Please reflect on the fact that in our kitchens we voluntarily recycle ceramic plates but not paper plates. Would it make sense to require households to recycle paper plates and cups by washing and reusing them? It would make no economic sense whatever (and it's doubtful that the paper could survive washing), though for some it might make symbolic sense - and those people are entitled, of course, to their private rituals.

The logic is the same with community recycling. People voluntarily recycle valuable resources and throw away less valuable items that take more effort to recycle than they are worth. Coercive recycling is actually more wasteful than throwing things away. It wastes valuable labour and materials that could be put to better use - creating new life, new resources, a cleaner environment.

Why Do People Worry So Much About Wastes?

Why do so many people worry so greatly about what seems to be such an easy problem to handle, given time and resources? My best guess is simply that the worriers 1) lack understanding of how an economic system responds to an increased shortage of some resource, 2) lack technical and economic imagination, together, and 3) harbour a moral impulse that says recycling is inherently good, and we should be willing to suffer a bit for it. Let's touch briefly on all three.

Lack of Understanding of How a Market Economy Works

Persons who are not economists - and perhaps especially those persons such as biologists who work with entities other than human beings - may underestimate the likelihood that people and organisations will make characteristically human adjustments to daily life and society. Such persons may then have too little faith in the adjustment capacities of human beings. Concepts that are appropriate for non-human organisms — niche, carrying capacity, etc. — are inappropriate for the creative aspect of human beings, which is the central element in long-run economic activity. The 1960's and 1970's generalisation of the work of Calhoun on Norwegian rats to policy recommendations for human society has been a classic example of this muddle.

For example, in regard to the possibility of fusion energy, Paul Ehrlich recycled his quote about energy in general by saying that cheap, inexhaustible power from fusion is "like giving a machine gun to an idiot child." Presumably the pronouncer of such pronouncements arrogates to him/herself greater wisdom than the people being pronounced about. Garrett Hardin says, "People often ask me, well don't you have faith in anything? And I always have the same answer... I have an unshakable [belief] in the unreliability of man. I know that no matter what we do, some damn fool will make a mess of it." This sort of thinking blinds one to the possibility that spontaneously coordinated market responses will deal with the wastes our society generates.

Lack of Technical Knowledge or Imagination

Many people who do not already know a technical answer to a given problem do not try to discover possible solutions, and they do not even imagine that others can dream up solutions. For example, when trains came along, some worried about the problem of providing toilets; where would the wastes go? Then when long-distance buses came along, people again worried, because they knew that the ship and train solution - dump over the side or from the bottom - was not possible on the roads. Then some worried about the same problem in planes. Then when the problem was solved for airplanes, the problem seemed insuperable in spacecraft. Yet in all these cases the problem simply required some diligent thought and engineering skill. And so it is with just about every other waste-disposal problem.

"But you are depending on a technical fix," some say. Yes, and why not? A "technical fix" is the entire story of civilisation. Yet our solutions are not limited to the physical-technical; new economic-political arrangements can solve problems. How can a community reduce the amount of garbage that people put out? Charge households and businesses by the pound or the litre of waste collected. So it is for other waste problems, too. Imagination and economic-organisational skill are necessary. But there is every reason to believe that once we direct those skills to a waste problem, solutions will be forthcoming.

The combination of technical imagination plus a free-enterprise system constitutes the crucial mechanism for dealing with waste - turning something of negative value into something of positive value. Slag from steel plants was long thought of as a nuisance. Then there arose entrepreneurs who saw the slag as "man-made igneous rock, harvested from blast furnaces," which could be used to provide excellent traction in road surfaces. And old railroad ties are not just clumsy chunks of wood to be burned, but now decorate retaining walls for lawns.

The Moral Feeling About Recycling

In a hearing on taxing disposable diapers, California State Senator Boatright pointed to the committee members and said that they "all raised our kids with cloth diapers. And you know, it didn't kill 'em at all...Got your hands a little dirty, maybe, but it didn't kill 'em. It's a hell of a lot more convenient to use disposable diapers, but you know what - you save a lot of money when you use cloth diapers."

Like a fraternity initiation - I suffered, why shouldn't you? When I tell my neighbour that I don't want to wash bottles and cans in order to recycle them, he tells me that "it takes practically no time at all". But when I ask him, in light of his assessment, if he would regularly do the job for me, he demurs.

Pros and Cons of Waste Disposal Policies

There are three ways society can organise waste disposal: a) commanding, b) guiding by tax and subsidy, and c) leaving it to the individual and the market. The appropriate method depends on the characteristics of the situation, and depends to a considerable extent on whether there are difficult "externalities" - effects that go beyond the individuals involved and that cannot be "internalised" (a rare situation).

Almost all solid household waste is a perfect case for untrammelled free enterprise.

Garbage is one of the easiest of waste disposal problems for society to deal with because there are no externalities that are difficult to deal with.

In Urbana, Illinois, we had the best and cheapest garbage collection I have ever heard of, each homeowner contracting with any one of the private haulers who worked in the area. They gauged their charges according to how much garbage you put out, the standard rate being augmented by special charges for special hauling. If the service was sloppy or unpunctual, you changed haulers at the end of the month. The hauler in turn made a deal with a private landfill operator. The community could be neutral and inactive, and there were no significant externalities. There is no sound reason why such a system cannot operate almost anywhere.

Problems arise when the community gets involved, either through contracting with a single hauler, or the municipality paying the hauler, or the municipality providing the service directly. There is then no incentive for homeowners to reduce the amount of waste they create, which leads to high waste output. And the community becomes subject to voters who have ideological views about waste and recycling, which distorts the best economic choice. And as a result of individuals' political values, together with their ignorance of the facts stated above, communities turn to recycling.

There are two drawbacks to recycling. The first is that it costs the taxpayers more money, as noted below. The second drawback is that it usually injects an element of coercion which is antagonistic both to basic values as well as to the efficient working of a free-market system. Both will now be discussed.

The Resource Costs of Recycling

Unnecessary recycling and conservation cannot simply be dismissed as a harmless symbolic amusement. There is a major cost in other activities that are foregone.

People could be building instead of recycling — building roads, public buildings, parks, even improving their own private shelters and spaces.

People's value for building may have been affected by the role of government. In the eight years that I have lived in a suburb of Washington, D.C., we have had scores of people come to the door soliciting funds. Never once has anyone solicited funds for a building project — never for trees to be planted, a cathedral to be built, or a concert to be given. Almost every solicitation has been for funds to lobby government to do something. And in most cases a large proportion of the funds collected goes toward soliciting more funds to do the same thing. In decades past it was not so.

People solicited funds to build private hospitals, and religious and cultural institutions, and other great works for the benefit of the public. But now people expect the government to perform those activities. Hence, funds are solicited mainly to bend the government's will to support people's private interests.

If government did a better job in building and administering, perhaps there would be no grounds for complaint. But we know from much solid evidence that government performs these activities very poorly — at high cost, and with poor service. These are part of the costs of a conservation mentality — of saving whales and Chesapeake Bay.

Another unreckoned cost is the coerced time of private persons. When a town calculates the cost of a recycling programme, it ignores the minutes spent by householders in separating types of trash, putting them in separate places, and stocking separate containers, and learning where and when the various kinds of trash must be placed for pick-up. These are non-trivial costs.

The Cost in Coercion

As of 1988, six states in the U S had mandated household separation of types of waste, and as of 1991, 28 states have recycling or waste reduction "quotas." As of 1990, the District of Columbia - like a large proportion of the 2700 communities that have recycling programmes - requires recycling of newspapers, glass, and metals in all apartment buildings. Not only does this law require individuals and firms to do what they would not do voluntarily, with fines for offenders, but it invites some of the practices of a totalitarian society wherein people are invited to meddle in their neighbours' lives.

And there is provision "to report to the city's 'trash police', a team of 10 inspectors hired to search through trash and identify culprits". Not only are such practices odious (at least to me), but policing expenditures are a high cost that surely are not included when an account is made of the total burden of recycling.

On a larger scale, the state of Rhode Island prevented a hauler from taking waste out of the state to dumps in Maine and Massachusetts. The state-owned Central Landfill is now the only licensed facility in the state, and its monopoly prices are higher than elsewhere. The state prevented the hauler from going elsewhere so that it would not lose the revenue. If a private firm were to somehow manoeuvre itself into a similar monopoly position, it could not use the law to keep customers from going elsewhere. Furthermore, the Federal Trade Commission would break up its monopoly. But government can get away with abuses that private enterprises cannot.

The problem of junked abandoned cars is not well handled by the market, however.

One knows in principle what should be done: Create proper rules that will compel producers of the waste to pay appropriate compensation to those who suffer from the pollution, either as individuals or as groups, so as to "internalise the negative externalities", as we say in economistese. But crafting and legislating such rules is not easy; Friedrich Hayek asserts that this is the most difficult of intellectual tasks, and the most important. And there often are strong private interests that militate against remedial actions. The outcome of this sort of pollution, then, depends largely on the social will and on political power, as well as upon our wisdom.

Price and Value

The most complex and confusing conservation issues are those that are dollars-and-cents questions to some people, but to other people are matters of aesthetics and basic values. Consider saving old newspapers in this connection. In some cases it makes sense to save and recycle paper because the sales revenue makes the effort profitable. In World War II the price of waste paper rose high enough to make the effort worthwhile for many householders. And where shredded newspapers are used as insulation after treatment with fire-retardant chemicals, paper collection can be a fund-raising device for community groups such as Boy Scout troops. But nowadays, in most communities the cost of recycling less the sales price of paper seems greater than the cost of landfill disposal or incineration. (Indeed, the price is now negative; by 1989 the amount of recycled paper had grown so large that communities had to pay $5-25 per tonne to have paper brokers take the stuff.

Here we must consider, what is the economic meaning of the market price of waste paper? That price will roughly equal the price of new paper less the recycling cost of making the old paper like new. In turn, the price of new paper is the sum of what it costs to grow a tree, cut the tree, and then transport and convert the wood to paper.

If the cost of growing new trees rises, so will the prices of new and used paper. But if the cost of growing trees goes down, or if good substitutes for trees are developed, the price of used and of new paper, and of wood, will fall (An increase in recycling also drives the price down). And that is what has been happening. The total quantity of growing trees has been increasing and their price falling, and the newspapers report the successful development of kenaf as a substitute for paper. So why bother to recycle newspapers?

Conservationists, however, feel that there is more to be said on this matter. They feel that trees should be saved for reasons other than the dollars-and-cents value of pulp or lumber. They argue that it is inherently right to try to avoid cutting down a tree "unnecessarily." This argument is based on aesthetic or even religious values.

The conservationists believe that stands of trees are unique national or international treasures just as Westminster Abbey is for Englishmen and as the Mosque of the Golden Dome is for Muslims. Perhaps we can express this by saying that those who do not directly use the treasure are willing to pay, in money or in effort, so that other people - now and in the future - can enjoy the good without paying the full price of creating it (Additionally, there is the argument that, even if future generations will be willing to pay the price, they will be unable to do so if we don't preserve it for them). Some people even impute feelings to nature, to trees or to animals, and they aim to prevent pain to those feelings.

There is no economic argument against a person holding these points of view. But neither is one person's aesthetic taste or religious belief an economic warrant for levying taxes on other people to pay for those preferences.

If one is willing to urge that the community pay in order to recycle, it is useful to know the price that has to be paid. The state of Florida has promised to subsidise firms to produce products out of recycled material and then sell them to the state, which will guarantee to buy the goods. Florida is just one among thirty states that are willing to buy recycled products (mainly paper) that cost 5 per cent to 10 per cent more than comparable non-recycled products. For perspective, perhaps three to five years of economic growth are necessary to make up that much of a reversal in progress by way of cost increase and productivity reduction. The governor of Florida calls this a "no-brainer for companies" because it is so easy for them to decide that a guaranteed market makes sense. Sometimes an argument is made for conservation on the grounds of national security and international bargaining. It may well make sense for a country to stockpile enough oil and other strategically sensitive resources for months or even years of consumption. But these political matters are beyond the scope of this chapter and of the usual discussions of conservation.

Conservation and the Government System

Do our treasured resources fare better in a private-enterprise system or a socialist centrally planned system?

There is no evidence that public officials are better stewards of any class of property - including wilderness and parks - than are private owners (Witness the state of upkeep of public and private housing). Consider the private (not-for-profit) case of Ravenna Park in Seattle, Washington, which was a public treasure from 1887 to 1925, selling admission to see the giant Douglas firs at very modest prices. Then the city took it over by legal action. The park fell victim to corrupt tree-cutting for the sale of firewood.

Government ownership permits over-use of resources not only because of common ownership but also because of incompetent management. Low prices encourage heavy use of a government-controlled resource. The private market would not "fail" to charge and collect appropriately-high prices for access to grazing lands, timber, and so on.

I cannot prove that a combination of academics and politicians will never design a social-political economic system that will better conserve nature and produce an aesthetic environment than will an undesigned system that relies upon people's spontaneous impulses and a market system governed only by very general rules. I cannot even assert that each and every ecological feature would be better served in an undesigned system. It may be that if governments were not to preserve the giant pandas in zoos, the animals would die out. Indeed, no private firm now exhibits pandas for profit (though P. T. Barnum's great museum did this sort of thing). It might well be, however, that if government were not to do so, interested individuals might take steps to preserve the pandas through voluntary organisations.

In the 19th century, when hawks were being shot as predators and were in danger of extinction, private individuals created a preserve that is now Hawk Mountain Sanctuary in Pennsylvania, which eventually became not only a preserve but also a centre for research on raptors. Nature lovers should not underestimate the likelihood that, in the absence of government action, others like them will act spontaneously to protect the values they hold dear (Indeed, that was the original nature of such organisations as the Audubon Society before they turned much of their effort to attempting to increase government intervention). But government activity often drives out private activity, as government intervention in the provision of hospital services has reduced the role of religious organisations that built and maintained hospitals in earlier years in the U S.

Now follows the general vision which does not promise that every particular aspect of ecology will fare better than a planned system, but which does promise that - taking everything together - humans will prefer the overall outcome to the overall outcome of a system where officials plan and dictate how resources will be treated.

Those who seek to plan our ecology argue that while markets may successfully bring us autos and tennis matches, markets "fail" with respect to wilderness, plant and animal species, and aesthetics of the landscape. In support of their vision of a designed system they point to the slaughter of the buffalo and the passenger pigeon, the befouling of our rivers and lakes for many decades, and ugliness of some inexpensive housing developments. And they argue than humanity changes the planet further and further away from the pure pristine world that they imagine existed before humans became many.

In rebuttal, the free-market environmentalists analyse such regrettable events and argue that the source of the destruction usually is a system lacking property rights that would prevent such events. And they point to countervailing examples - beef cattle preserved by individual ownership, facilitated by the branding system and the barbed wire invented to resolve the problem of common grazing; the growth of elephant herds in Zimbabwe; the habitat preservation on farms where owners sell the rights to hunt; the beauty of fishing streams in Great Britain which are privately owned or which clubs own fishing rights to; and the well-stocked fish ponds that sell fishing rights in old "borrow pits" by the sides of highways in Illinois, dug by the highway builders to get building materials.

Free-market environmentalists also cite the insults to human dignity and constraints upon individual freedom that planned conservation brings about - "trash police," "water cops," farmers being fined and going to jail for making their farms more productive at the expense of migratory ducks; neighbours and people's own children bringing pressure on homeowners to separate types of trash.

These examples (and the theory behind them, to be discussed below) are compelling, in my view. But as the designed system advocates offer counter-examples, the issue must also be discussed at the level of the overall vision, and the test is the analogous comparative experience of planned versus free-enterprise economies.

"Common sense" argues that specific planning for particular kinds of outcomes results in a better general outcome than an undesigned system. Yet the evidence is incontrovertible that with respect to ordinary economic matters, an undesigned free-enterprise system does much better than does a planned socialist system.

Investigations of the performance of socialist economies, and the breakup of communism in the 1980s and 1990s entirely vindicate the theoretical visions of David Hume, Adam Smith, Karl Menger, Ludwig von Mises, and Friedrich Hayek. And even worse than the economic effects of a planned system are the effects upon human welfare in non-economic aspects - liberty and health, for example.

Telling evidence is found in the economic comparison of North and South Korea, and of East and West Germany, and also of Taiwan and China. The communist countries use(d) much more energy and produce(d) much greater amounts of pollution both per person and per dollar of GNP. One might say that this is because the communist countries are poorer. But this relative poverty is itself part of the story: centrally controlled economies do less well economically, which is part of the reason they pollute more per unit of GNP, and even per capita.

The same sort of result would undoubtedly hold if the investigation were widened to include all countries, as Scully (1988) has done for the effect of state system on per person income.[2]

Before the 1980s there was only anecdotal evidence for the comparative success of market-based decentralised systems versus centrally-planned systems, just as there is today with respect to conservation and pollution of the environment. And the same kinds of common-sensical arguments for the advantage of a "rationally" planned system were given in favour of communism - economies of scale, avoidance of duplication, control by the cleverest intellectuals instead of less-educated "ordinary" persons, and so on. But the transformation of countries to a planned command-and-control system based upon this theorising was demonstrably the greatest human blunder of all time. This is the same vision that leads to such an "activist" command-and-control approach as this suggestion for the Alaskan Wildlife Refuge: "You drive a stake in the ground and say, 'You don't cross the Canning River'."

It must be said, however, that if one wants a planet the way it was 20,000 years ago - not just "stop the world I want to get off," but "reverse the world to the way it used to be" - a free society will not move in that direction. The only way to obtain that result is to kill off almost all of humanity, and destroy all libraries so that the few remaining humans cannot quickly learn how to rebuild what we now have.

If someone wishes this sort of "conservation", no arguments are relevant.

Some ecologists urge on us a "mixed" system where government only does what is "necessary". It is impossible to disagree with this in principle - the panda being a case in point. But those persons should be aware that in the analogous situation of the "mixed" economy, almost all the activities that were regarded as "necessary"— government-owned steel mills and mines and telephone systems, for example - turned out not to have the desirable properties that in advance they were touted to have. The "mixed system" is mainly a mirage and an intellectual illusion, and sometimes a Trojan horse for wider planning. And the idea that government can manipulate prices to simulate a well-working market has in Eastern Europe been shown conclusively to be a figment of unsound theorising.

The first recourse of policymakers is still to command-and-control systems. Since the 1980s, however, there has been growing recognition, even among non-economists, that direct controls are usually inferior to market-based economic incentives as ways to manage pollution. For example, the law now makes provision for firms such as electrical utilities to buy and sell rights to emit quantities of pollutants. The Chicago Board of Trade even has laid the groundwork for a public market in "smog futures" along the lines of commodity and securities markets, and environmental organisations have approved.

In the 1980s, these economic ideas were refined into a body of applied work. The free market environmentalist writers emphasise the concepts of property rights, and of allowing price rather than command-and-control devices to allocate goods. They discuss "non-market government failures" as a counterpart to the "market failures" that many see at the root of pollution problems.

The free-market environmentalists point to the successes of market systems in the cases of conservation mentioned earlier in the chapter. Proven successes with market systems for controlling pollution are less conspicuous. But such conservation cases as private fishing streams may be seen as the opposite side of the coin from the issue of pollution in streams and rivers, where privatisation of rights has been shown to be effective.

Conclusion

If you want society to force you to waste effort and money because sacrifice is good for you, mandatory recycling fits the bill. But if you want people to have the best chance to live their lives in the ways that they wish, without imposing waste costs on others, then allowing private persons to deal with their private problems privately should be your choice.

The key question for social decisions is how to get the optimal level of recycling and waste disposal, pollution, and environmental preservation with a minimum of constraint upon individuals and economic activity. Economic incentives working in free markets often are a better system than command-and-control. Aside from the economic costs of inefficient controls, there is usually excessive, overzealous coercion - as in the regulations that eventuated in King Edward executing the man who polluted the air with sea coal, or the milder coercion of recycling zealots today.

Notes

1. Lord, Bette Bao, *A Younger Sister's Story of Growing up in China During the Cultural Revolution*, as told to Lord.

2. Scully, Gerald W. (1988) "The Institutional Framework and Economic Development", *Journal of Political Economy*, Vol 96, 3, June. pp 652-662.

Free Trade and Green Trade

23

Thinking Clearly About the Linkage
Between Trade and the Environment*

JAGADISH BHAGWATI

Introduction

The question of linkage between trade and environmental issues, indeed between trade and labour standards and between trade and human rights, has reached centre stage as several NGOs (non-governmental organisations) have demanded that the WTO formally incorporate such a linkage through, for example, a Social Clause on labour standards in WTO and through as-yet-unspecified mechanisms as far as environmental standards are concerned.

Within the environmental arena, the GATT itself, and now the WTO (GATT's successor), have been the focus of much agitation by environmental groups that see this trade institution as an obsolete obstacle to environmental progress. The anti-GATT feeling materialised first when the celebrated Dolphin-Tuna decision was announced, declaring Mexico the winner in the dispute over the U.S. legislation that sought to proscribe access to Mexican tuna caught in purse seine nets. A throwback to that sentiment occurred recently when the Shrimp-Turtle Panel decision also went against the United States over its legislation that mandated unilaterally a denial of access to shrimp harvested without the use of TEDs (the turtle excluding devices).[1]

These cases reflected one of a number of different ways in which the work of the WTO interfaces today with the environmental

---

* Based on a paper presented at the Conference on Environment and Trade at the Kiel Institute for World Economics, Kiel, Germany, Summer 1999, and at the Columbia University Conference on The Next Trade Negotiating Round: Examining the Agenda for Seattle, New York City, July 22–23, 1999.

questions and agendas. The main and unifying essence of both cases was this question: Should suspension of market access be allowed automatically to a nation that objects unilaterally (i.e., without obtaining a multilateral consensus) to other countries exporting products to it when those products are made by using processes that the nation objects to on "values" grounds?

In the Dolphin-Tuna case, the US government objected to the use of purse seine nets in harvesting tuna because these nets kill dolphins (which Americans have voted to protect, presumably because they are "cute" and a great draw at zoos) gratuitously and cruelly. In the Shrimp-Turtle case, the objection was similar: it related to what are called in GATT jargon PPM objections, namely, objections to process and production methods. Similar cases can arise if nations object in much the same way to the importation of, say, chickens produced in batteries, or hogs produced in crowded pens, or fur harvested from animals caught in leghold traps, and many more instances of what some nations, but not all or most, consider to be "values"-wise unacceptable PPMs.

But while these cases are the subject of high-profile, high-octane attacks on the GATT and the WTO, and they raise questions that necessarily involve an interface between the WTO and the environmental groups, other issues are equally the object of demands by environmental groups on the WTO but, in my view, are not necessarily ones that belong to a WTO or trade-treaty agenda. In particular, I believe this to be true of the demands for harmonisation or upgrading of environmental standards in developing countries if they wish to export products, even when the pollution involved is "local" and has no global environmental externalities as with global warming or ozone layer depletion or acid rain.

Unlike in the popular debate, which tends often to blur necessary distinctions among different types of problems, and where the environmentally sensitive lobbies often are unwilling to make the distinctions anyway because they would weaken coalition building for political action, I propose here (since I am asked to address the "trade and the environment" linkage) to make these distinctions very sharply.

In particular, using these distinctions, I set myself the task of providing a road map that is aimed at dividing the current "linkage" demands between trade (whether institutions or negotiations) and environmental questions into those that are "necessary" and those that are not. In the latter case, as when trade access is used as a way of pushing environmental agendas abroad on altruistic grounds, I will also propose *alternative* ways in which such agendas may be pursued outside the trade context and institutions (e.g., in UNEP rather than WTO), thus raising the question of what I like to call the design of "appropriate governance" (i.e., what agenda to pursue where).

A Necessary Taxonomy

I first provide a necessary taxonomy so that the issues concerning linkage of trade and environmental questions can be analysed with clarity and optimal policy solutions designed with the aid of such analysis. This taxonomy can be built essentially around two sets of distinctions:

i.   whether the environmental damage or pollution is "domestic" or "international"; and

ii. whether the country addressing it follows egotistical (i.e., its own advantage) or altruistic (i.e., others' advantage) objectives.

The former distinction was introduced principally in the 1992 GATT Report on Trade and the Environment, though it must have been used simultaneously by many researchers, I am sure. If I pollute a lake in India that only (even then just a few) Indians have heard of, the pollution is of concern to Indians at risk. But if I pollute a river that flows into Bangladesh, or produce acid rain in the United States that goes across and hurts Canadians, the problem is clearly international. When global warming and ozone layer depletion are involved, the problem is actually global. The international/global problem is clearly one where externalities are at stake unless a "market" is already in place to internalise these externalities (e.g., by having tradable permits, suitably devised).

The latter distinction is appreciated by few, including economists who write about globalisation, about fixing the world trading system, and so on without any real understanding of the complexity of the I issues at hand.[2] Thus, we must distinguish between, say, objecting to the import of child labour-produced carpets from India because we object to being put at a competitive disadvantage with other nations because we prohibit, and they allow, cheaper child labour, and objecting to such imports instead with a view to reducing or eliminating the use of child labour abroad because we think that our cessation of such imports will help bring that about. In the former case, we are "egotistic": we are simply interested in maintaining our competing industries. In the latter case, we are being "altruistic" in thinking of children's welfare even though they are abroad in other nations, and we are using consequentialist ethics, hoping to effect change abroad. The latter is there fore a matter of seeking to advance social agendas abroad; the former is a matter of protecting our industries, for our own benefit. In assessing the demands for prohibiting the imports of products made with child labour, our evaluation of the proposal and the design of appropriate policy instruments will clearly have to be different, depending on which of these two motivations we are confronting.[3]

Once these two sets of distinctions are made, we have four sets of problems: domestic environmental problems with egotistic and with altruistic objectives by nations, and international environmental problems with egotistic and with altruistic objectives again. In this chapter, I devote myself to the domestic environmental issues, which are among the trickier ones where a great amount of confusion reigns.

The *international environmental issues* are understood much better, including in their interface with the WTO's functioning, and I shall eschew a discussion of them here. Let me just say, in regard to them, that the interface with WTO comes principally insofar as the MEAs (Multilateral Environmental Agreements such as Basel and the Montreal Protocol) seek to use trade sanctions against defectors and against free riders, and that these two sets of nations, when WTO members, could claim WTO-defined rights against the use of such sanctions. These questions are not easy to settle since we must raise questions such as: Is the MEA efficiently and equitably designed?[4] If the scientific evidence in support of it is disputed, can you treat a nation that does not wish to join as a free rider when in fact it may simply be opting out of getting on to the bus? In this context, let me say that, while Kyoto is a useful step forward (though I share some of the misgivings about its design that the economist Richard Cooper expressed in his article in *Foreign Affairs*,[5]) I have been surprised that none of the models that I have seen seem to do the obvious if you know the domestic environmental scene in the United States:

- For the "stock" problem, that is, the damage done in the past, a clearly defined responsibility must exist for the polluters: this is a principle that has been accepted in the Superfund approach and in the torts claims addressed to past polluters for phenomena such as the Love Canal disaster. Why is it not accepted at the international level as well for past damage to the environment on global warming (principally, of course, by the developed nations)? The question then must be not whether there is responsibility for past environmental damage, but how to assess it and the specific ways in which the levy can be used to reduce the global warming problem—for example, by financing the creation of new environment-friendly technologies and their subsidised diffusion across the world.

- For the "flow" problem, as to what to charge for emissions, the conceptually clear answer has to be to put all such emissions (net of absorption services through, for instance, your forests) into the pot of world demand for such pollution and then to determine, with a suitable utility function defined positively on goods and services and negatively on pollution, the shadow price of a unit pollution. That would then define the cost that the nation must pay for its contribution to the global warming problem. Needless to say, that cost would be vastly higher for the rich than for the poor countries. Instead of doing this, the rich countries are opting for an international variant of the principle that we refer to in the United States as the "PSD" Principle—namely, prevention of significant deterioration. In plain English, this means that those who pollute a lot as part of the initial condition can get away with it: burdens are to be prorated to marginal changes in pollution!

So, what we have therefore in the global warming debate is a cynical and virtual denial by the rich countries of the Superfund principle whose incidence would hurt them, and an adoption of the PSD principle that would help them. Not bad, indeed. As I read the Kyoto arguments and policy papers, it seems to me therefore that the developing countries have an intuitive sense of what I am saying above but no conceptual clarity or technical work to back it. Instead they talk inchoately about how the *flow* burden should be far less on them both because the stock damage was due to the developed countries and also because they are poor and hence should not be asked to bear any burden. But as soon as they do that, arguing their case on these grounds, they are shot down in the U.S. Congress as countries that are doubly wrong because they wish to be free riders and are also guilty of trying to exploit the "guilt" angle!

Again, one needs to consider whether, given the hostility manifest between the more vociferous environmentalists and free traders on many other fronts, it would not be wise to "grandfather" the existing MEAs and to leave the contentious question of the WTO-compatibility of *future* MEAs to further consensus building among the WTO membership: we may prudently decide that this was one major battle from which we could withdraw. This is certainly an issue with which the WTO must come to immediate grips.

Domestic Environmental Issues

So, let me turn to the purely domestic environmental problems, dealing first with the egotistical objective and next with the altruistic one.

Egotistical Objective

Here, I deal with the following distinct aspects of the demands for "greening the WTO":

Contention 1

WTO should allow importing countries to countervail "social" dumping, namely, when a product is produced with differential tax burdens in different countries and the exporting country has a lower tax burden. This is what T. N. Srinivasan and I have called *CCII (cross-country-intra-industry) harmonisation of tax burdens.*[6] Clearly, this is wrong. With different fundamentals, there is no good reason for such harmonisation to occur or to be demanded. We may demand that every nation adopt a Polluter Pay Principle; but the pollution tax, for the same carcinogen in the same industry, will generally be different.

Contention 2

We nonetheless may object to others having lower tax burdens because that will result in a "race to the bottom" that hurts our standards even if we do not care otherwise what standards others have on a CCII basis. Therefore WTO should allow countervailing duties to offset "social dumping."

Unlike the previous argument, this is a theoretically sound one. But it is an argument for a cooperative solution that will nonetheless not be characterised in general by harmonisation. Besides, a "race towards the top" can occur, as John Wilson points out in his chapter on the subject in the Bhagwati-Hudec volume.[7] Moreover, the argument depends on capital taxation being suboptimal. Finally, the empirical evidence for such a race does not seem to be strong since (1) multinationals do not seem to respond to lower environmental burdens (not just because the differences among different locations are small, since these could rise) for a variety of reasons including reputational ones, though a couple of recent papers detect some elasticity of response to differential environmental burdens within the United States across states; and (2) the evidence that poor countries lower environmental regulations to attract MNCs is not plausible when democratic countries are involved: the competition for capital/ MNCs is really through tax breaks, tax holidays, and land grants—all of which amount to a race to the bottom in taxation that hurts the competing countries, most analysts believe. Few democratic countries are going to offer facilities to pollute freely as a way of attracting MNCs.

So, *for contentions 1 and 2, the demand for harmonisation and/or legitimation at the WTO of countervailing duties on so-called social dumping seems to me to be not the way to green the WTO. We should resist such demands.*

Instead, I recommend two other solutions:

1. I have argued that MNCs must be asked to adopt the environmental standards of their home countries when they go abroad. If they tend to do so anyway, as argued above empirically, then this mandate will not hurt and will buy environmentalists' approbation at very little deadweight loss.[8] I think now of this as a *mandatory code,* unlike the voluntary code approach discussed below that is complementary in my view.

2. We can also go ahead with devising *voluntary labelling schemes* like the SA8000, that firms can sign on to and several have indeed recently. This defines conditions of work and includes

independent monitoring. This means that all the signatory firms from every country would have to adopt the common minimum standards, whereas the mandatory Bhagwati-style code above would permit differences among firms from different countries.

"Values-Related" PPMs

Next, there is the Shrimp-Turtle and Dolphin-Tuna type of problem. U.S. consumers simply feel that the United States should be allowed to prohibit imports of products using morally objectionable PPMs.

Evidently, we cannot force such imports down people's throats. Indeed, economists are well aware of the legitimacy of PPMs as something that enters our utility functions: after all, the way we produce something is part of the characteristics of the vector that defines a product. The problem is not that we free traders have not realised that PPMs are legitimate and must be dealt with,[9] but how do we deal with them when there is no consensus on that "value"? Do we allow automatic unilateral shut-off of such products?

Here again, the 1992 GATT Report, correctly in my view, argued that the grant of automatic market access suspension rights in such ethical or moral or "values"-related cases would be a slippery slope: how would we draw the line? Moreover, we do know that protectionist intent will occasionally underlie environmental legislation, often in the specifics of the design of the environmental regulation (as in the Dolphin-Tuna case and the Ontario-U.S. beer can case). Are we simply to ignore that by saying an environmentally aimed prohibition on imports cannot be challenged at all?

Therefore, I would say that the precise way in which the WTO deals with such values-related PPM problems should not be along the lines of automaticity. Nor should we accept an ill-considered proposal such as Rodrik's that an administrative procedure like antidumping be devised to ensure that the moral preference is genuine and widely shared within the country, after which the imports should be unilaterally shut off (as if the enactment of the Turtle and Dolphin legislation itself was not the expression of such a widely shared preference and as if an administrative body could sit in judgment over a legislative outcome).

Rather, we should proceed along the lines of *labelling*. This raises a number of questions that UNCTAD has been considering and that we know about from U.S. experience as well. For example, who determines the label, how "alarmist" or "realistic" should it be, and so on. Also, what are the problems for small producers in developing countries that have few facilities for such labelling? But it is still the way to explore and go, giving consumers information and choice.

Equally, I think it is necessary to ensure that, if the WTO continues to object to automaticity of such suspensions of access, as I believe it should, then the remedy when a country has lost such a case and still wishes to maintain the import suspension and is unwilling to accept a labelling solution should not be to slap on retaliatory measures (as the United States favours, if recent examples in the hormone-fed beef and the bananas cases reveal a trend) but rather to go for a cash compensation that reflects the gains from trade lost. There is no point in disrupting trade yet further: it is time that the economists weighed in on this aspect of the dispute settlement procedures and remedies.

Altruistic Argument for Linkage

But suppose that we seek linkage because of altruistic reasons, treating trade treaties and/or institutions as mere instrumentalities through which we hope to effect change in morally offensive practices abroad.

Those who wish to use trade treaties and institutions to do this, as do the proponents of a Social Clause at the WTO, seem to me to use one instrument (i.e., trade policy) to achieve two targets (i.e., the liberalisation of trade and the promotion of their social agendas). But typically, as we know from the theory of economic policy, we need generally two instruments to best achieve two objectives. A Social Clause illustrates this well. Dividing the supporters of trade liberalisation into those who oppose and those who support the Social Clause undermines trade liberalisation, as happened with the Seattle debacle. Making many, especially in developing countries, suspicious that the true objective of the proponents of a Social Clause is protectionism rather than advancement of social agendas also undermines the legitimacy of the morally motivated groups and their case.

Instead, I propose that the social agendas are best pursued by nontrade means. This can be done at the ILO for child labour eradication, for example. In fact, the ILO has an excellent International Programme for the Eradication of Child Labour that does the heavy lifting necessary to do effective work on the problem, whereas the WTO has literally no competence in the area.

As elsewhere, I consider this issue best solved by assigning the resolution of a problem to an appropriate agency (rather than overloading one institution like the WTO with all problems); hence, I have called it a question of "Appropriate Governance" in my recent writings.

Such a policy would enable us to pursue both free trade (which I consider to be an important social and moral agenda as well, since the prosperity it brings is essential for removing poverty worldwide) and other social and moral agendas. To sacrifice one to the other gratuitously, when both can be pursued together, would be inexcusable. It is time for our political leadership to act forcefully on that insight instead of succumbing to populist demands for solutions that yield less than the best.

Notes

1. In each of these cases, there have been more than one panel findings; in the latter case, the new Appellate Court also ruled after the initial panel finding. The precise grounds on which the United States lost in both the cases have therefore varied.

2. Here, the culprits include my good friend, Dani Rodrik, and his publisher, the Institute for International Economics in Washington, D.C., which has published yet other authors such as the political scientist Mac Destler, whose knowledge of the economics of international trade policy questions seems to be exclusively based on reading what the institute brings out, advocating linkage to facilitate fast-track renewal and the start of the Millennium Round.

3. In this chapter, I cannot discuss these distinctions fully. A systematic and deep analysis is provided in my and others' contributions to Bhagwati and Robert Hudec (eds.), *Fair Trade and Harmonisation: Prerequisites for Free Trade?* Vol. 1, (Cambridge, Mass.: MIT Press, 1996); and also in three chapters in part VI of my recent book, *A Stream of Windows: Unsettling Reflections on Trade, Immigration, and Democracy* (Cambridge, Mass.: MIT Press: 1998).

4. Thus, in relation to the NPT and CTBT, India refused to sign them because it did not accept the division of the world into the *status quo* of those who had nuclear weapons and those that did not, and backed instead a *universal* nuclear disarmament plan. The moral incoherence of the nuclear nations is manifest from Britain's condemnation of India's nuclear tests when Britain, an admirable nation in other ways, holds on for no reason whatsoever to its own nuclear stockpile when it could instead make an important moral and effective gesture by bringing the great unilateral nuclear disarmament advocate Vanessa Redgrave (leader of the unilateral-British-nuclear-disarmament CND movement) out of the mothballs and putting her in charge of a rapid unilateral destruction of Britain's stockpile! One might also note that the United States itself has

not ratified the CTBT yet. The mere fact that a certain powerful group of nations, and its NGOs with their vast resources compared to those situated in the poor nations, support an MEA is no proof that it is equitable, free from the power play that distorts priorities and burdens from an objective point of view, and that those who refuse to sign on to it are therefore "free riders" or "rejectionists." At least we economists need to look at such claims with a cynical eye.

5. See Richard N. Cooper, "Toward a Real Global Warming Treaty," *Foreign Affairs* 77, 2 (Mar./Apr. 1998).

6. See Bhagwati and T. N. Srinivasan, "Trade and the Environment: Does Environmental Diversity Detract from the Case for Free Trade?," Chap. 4 of *Fair Trade and Harmonisation*.

7. See John Douglas Wilson, "Capital Mobility and Environmental Standards: Is There a Theoretical Basis for a Race to the Bottom," Chap. 10 of *Fair Trade and Harmonisation*.

8. Bhagwati and Robert Hudec (eds.), *Fair Trade and Harmonisation: Prerequisites for Free Trade?* Vol. 1, (Cambridge, Mass.: MIT Press, 1996)

9. Rodrik, in his IIE pamphlet on globalisation, argues as if we are so unmindful. But this is to betray ignorance of the extensive debate over the problem, including that in the 1991 GATT Report on Trade and the Environment.

24

Freedom to Trade Protects
the Environment

KENDRA OKONSKI

For years environmental groups have promoted the view that trade is bad for the environment. They argue that businesses want free trade so that they can escape regulations, and that trade leads to a "race to the bottom" for environmental standards. According to this view, corporations are causing poverty and environmental degradation. They want consumers in wealthy countries to feel guilty for consuming goods which were produced by people in developing countries. Such myths must be put to rest.

Trade Improves the Environment

1. Trade helps to eliminate local environmental problems
2. Trade encourages better use of natural resources and less environmental stress
3. Resource-saving technologies change the way that goods are produced,
4. Legal, financial and physical infrastructure encourages better use of natural resources
5. Rising incomes of poor people lead to better environmental awareness. Wealth changes people's priorities from day-to-day survival, and allows them to focus on protecting the environment.

Trade Helps to Eliminate
Local Environmental Problems

The majority of environmental problems in poor countries are local: indoor air pollution, dirty water, and a lack of sanitation are the biggest environmental problems, leading to millions of people suffering from disease, and millions of premature deaths each year:

- Over 1 billion people lack access to clean drinking water

- Over 2 billion people lack access to improved sanitation, and the majority of these people are in Africa and Asia, mostly in rural areas.[1]

- There are 4 billion cases of diarrhoea each year, resulting in 2.2 million deaths, most of them children under the age of five. "This is equivalent to one child dying every 15 seconds, or 20 jumbo jets crashing every day."[2]

- "Indoor air [pollution] increases the risk of acute respiratory infections, [and is] one of the leading causes of infant and child mortality in developing countries. In Asia, such exposure accounts for between half and one million excess deaths every year. In sub-Saharan Africa the estimate is 300,000-500,000 excess deaths."[3]

- Poor countries have often abandoned the needs of their poorest citizens, by adopting strategies and policies that make it more difficult for people to have access to clean water and sanitation, electricity, and technologies which would improve their lives.

- Dirty water and poor sanitation is a primary vehicle for poverty diseases: Water-borne diseases, including intestinal worms, guinea worms, schistosomiasis, trachoma, cholera, typhoid, hepatitis A, and dysentery, affect at least half a billion poor people.

- Another half-billion people are affected by water-related diseases, such as malaria and dengue fever.[4]

- 2 billion people worldwide rely on biomass, including wood, dung, and crop residues, for cooking, heat, and light, mostly burned in poorly-flued indoor stoves.[5] These are a primary source of indoor air pollution.

How does Trade Help to Eliminate Environmental Problems?

There is much sound evidence and no sound opposition, to the idea that "wealthier is healthier" and "wealthier is cleaner". The diseases of poor countries once plagued wealthy countries, but they were eliminated because of economic development. Health and development go hand in hand, and frustrating economic growth will only compromise the ability of countries to improve human health, adopt new technologies, afford better drugs and achieve improvements in healthcare that rich countries enjoy.

By allowing people to exchange without intervention, trade also means that people have more incentives to innovate and create new ways to solve problems.

Poverty means that people's priority is mere survival rather than protecting the environment. When incomes rise and people become wealthier, they can think about the future. It is then that they demand better environmental protection and use better technologies, such as more efficient fuels like gas and electricity, more efficient cars, and better sanitation, which lead to the saving of other resources and fewer local environmental problems.[6]

How Does Trade Lead to Development?

Trade is the process by which people voluntarily exchange goods and services, which are produced with natural and human resources. People in poor countries have been restricted in their ability to trade, both by foreign countries through protectionist trade barriers, agricultural subsidies, and by their own governments, through protectionist trade barriers, corruption, a lack of property rights, and oppressive regulations. Free trade is an important step towards giving poor people access to external markets, and towards forcing corrupt governments to reform their policies.

How do Trade Restrictions Perpetuate Poverty?

Trade and exchange is fundamental to development, which then enables people to adopt higher standards for production. But to require the latter (regulations) before the former (development) will only prolong poverty. The effect of trade restrictions is to prevent voluntary exchange, and to replace it with regulations and restrictions enforced by the state.

Such restrictions are often motivated by economic interest groups who beg for government protection from external competition.

Environmental groups seek to restrict trade on the basis that some goods produced by poor people do not meet the regulatory standards of wealthy countries, based on arbitrary criteria such as the "precautionary principle", or based on the fact that a country has not signed up to international environmental laws.

Five Myths about Trade and the Environment

Myth: Trade is a "race to the bottom" and leads to a reduction in environmental standards.

Reality: Trade is beneficial for everyone. It leads to higher production standards, better use of resources, innovation and new technologies.

While it may be an intuitively appealing argument that companies exploit resources and the environment in poor countries, it is false. Local producers in poor countries often have no choice but to use old production processes and goods which may use resources inefficiently and may cause environmental harms. The process of trade brings investment and better technologies which replace these, leading to more efficient use of resources.

Trade also leads to higher wages, and people may cease environmentally harmful activities because they shift into other professions, and because they can afford new technologies which are more environmentally benign.

Myth: Trade sanctions and restrictions lead to better environmental protection.

Reality: Sanctions do not provide the right incentives for environmental protection, and may perversely lead to worse environmental consequences.

Trade bans (sanctions) in international environmental agreements are not effective at achieving environmental goals[7] and the same applies to environmental provisions in trade rules. Banning trade in certain goods, especially against producers in poor countries, does not mean that environmentally harmful production will cease. Rather, those producers will continue such activities, which largely take place in the informal economy. Those producers will remain poor and less able to replace old technologies with new ones, and less able to invest, thus environmentally harmful activities may continue.

Moreover, the value of some goods (wildlife and its products, for instance) is higher when trade is restricted or banned. This creates a perverse incentive to produce and sell that good (i.e. endangered species), rather than protect it. This is especially true when trade restrictions undermine the ability of poor people to control their own land and resources. Making trade illegal also means that trade in banned goods is not transparent, and it is thus more difficult to control. Regarding wildlife, "There have been remarkably few, if any, species extinctions that can be attributed to exploitation for international trade."[8]

Trade sanctions could also drive up production costs, discouraging companies from investing and producing in poor countries – whether they are small local producers or multinationals. Moreover, trade sanctions distort prices, leading to less efficient uses of natural resources.

Myth: The process of trade leads to an unsustainable consumption of natural resources, dumping of wastes in poor countries, and contributes to global warming.

Reality: Sustainable consumption results from wealth and trade, and poor people should be able to choose whatever lifestyle they desire, rather than having those choices imposed on them.

Environmental groups seek regulations which would force people to consume fewer resources, and consume products which are made locally rather than transported, on the grounds that this is "sustainable" consumption. This makes little sense, though, as it would result in the overuse and exploitation of some resources, and underuse of others. Moreover, this may lead to different environmental problems – such as local air and water pollution.

By denying technology transfer and economic development that comes from open, rules-based trade, businesses and people in poor countries will continue to use technologies that cause harm to the environment and themselves. Without trade, investment and economic development, poor people would not use more efficient stoves or more efficient fuels. They would not use chlorine-based chemicals to rid their water of cholera, hepatitis and other deadly bacteria. If allowed to continue, environmental restrictions on trade could lead to more poverty.

It is simply patronizing to suggest that people – whether poor or wealthy – should not be allowed to buy cars (or refrigerators, hair dryers, lipstick and mobile phones), products that improve their lives, on the basis that this might cause environmental problems. The problems that poor people face currently are related to a lack of consumption of energy and resources, rather than too much of it.

Myth: Corporations and wealth cause poverty and environmental degradation, and government intervention is the only way to prevent this.

Reality: Poverty, and resulting environmental problems, is caused by a lack of trade, investment, and supporting institutions. Most poor country governments have prolonged this situation through restrictions on trade.

The goal of businesses (whether local or foreign) is not to destroy the environment or to deplete resources. Business uses resources to fulfill consumers' needs and desires, through competition and prices. Involving government with trade encourages a system based on power and privilege, rather than on the best use of resources.

In essence, this simply perpetuates an existing situation in poor countries, where entrepreneurship is stifled, market access is restricted (both internally and externally), and the use of natural resources is decided with a political process – lobbying, bribery, corruption – rather than by individuals, consumers and actors in the market.

This system was formerly called mercantilism: "To achieve its objectives, the mercantilist state granted privileges to favored producers and consumers by means of regulations, subsidies, taxes and licenses."[9] While a few people and businesses do benefit from mercantilism (or its modern equivalents), the majority of people are stifled by it and in poor countries it will perpetuate poverty.

Myth: Environmental regulations should take precedence over trade rules, and these should be harmonised and enforced by international agencies.

Reality: If regulations are imposed on poor countries, it will slow their path of development and will prolong poverty.

Environmental groups want to make international environmental rules superior to trade rules, and they want to "link" trade rules with environmental rules. Their goal is to stop trade which they deem to be environmentally harmful. Vested interests (such as businesses) support these measures because they can also be used as trade protectionism. The purpose of trade rules is to eliminate discrimination – either blatant or inherent.

If international regimes are used to impose environmental standards on poor countries, their poverty will be prolonged. Instead of focusing on the desires of environmental groups, poor countries should pursue a strategy to eliminate poverty. This means enabling everyone – not just corrupt politicians and the elite – to create, innovate, and build wealth. Free trade and the removal of onerous regulations which prevent exchange between people are two fundamental steps in this strategy.

If specific environmental goals are deemed important enough to merit international action, then these policies should exist alongside trade agreements, without a need to make one or the other superior. The purpose of trade rules is to facilitate trade between people, and the purpose of environmental rules is to protect the environment. These are mutually supporting policies, because trade leads to wealth creation, which leads to better environmental protection.

Recommendations

Regulations, trade restrictions, and less economic growth are not the best way to protect the environment, because they do not strike at the heart of environmental problems. "Environmental protection" has different meanings in wealthy and poor countries. In wealthy countries, it often means more government regulations and protection of endangered species, less intensive agriculture and 'renewable' energy.

To people in the developing world, such 'standards' are absurd: what matters to their wellbeing is access to clean water and sanitation, reliable energy sources, and access to technologies that will improve their productivity and help them to generate income, thus to escape poverty.

The best way to eliminate poverty *and* protect the environment is to encourage a regime of open trade, along with other supporting policies. Freedom to trade is an important step towards eliminating poverty and towards better environmental protection.

Notes

1. *Global Water Supply and Sanitation Assessment 2000 Report.* World Health Organisation. http://www.who.int/water_sanitation_health/Globassessment/Global1.htm#1

2. *Ibid.*

3. http://www.who.int/inf-fs/en/fact187.html

4. World Health Organisation, Fact Sheet No. 112. http://www.who.int/inf-fs/en/fact112.html

5. http://www.who.int/inf-fs/en/fact187.html

6. Indur Goklany "Richer is Cleaner: Long-term Trends in Global Air Quality" in Ronald Bailey (ed), *The True State of the Planet*, New York: Free Press, 1995.

7. The UN Conference on Trade and Development (UNCTAD) reviewed several multilateral environmental agreements, and it reported that in no case had trade bans secured the environmental purpose of the MEAs.

8. Huxley, C. (2000), p. 4. "CITES: The Vision," in *Endangered Species, Threatened Convention: The Past,Present and Future of CITES*, London: Earthscan. Cited in Jonathan Adler, "Do conservation conventions conserve?" in Sustainable Development: Promoting Progress or Perpetuating Poverty? ed. Julian Morris, Profile Books 2002.

9. Hernando de Soto, *The Other Path* (New York: Harper and Row, 1989), p.202.

Environmentalism:
The New Imperialism

25

The New Cultural Imperialism:
The Greens and Economic Development

DEEPAK LAL

Introduction

The late Julian Simon spent his professional life collecting data and producing analyses that showed that there was no evidence that rapid population growth harmed economic development. The plausibility of the contrary view in the public mind is just due to an arithmetic relationship; as per capita income (which is usually taken as a measure of a country's economic welfare) is defined as the ratio of GDP to population. So that with no change in the numerator, a rise in the denominator will arithmetically reduce per capita income. But the absurdity of this view can be seen from the fact that if a cow has a child per capita income goes up, but if born to a human it goes down. But, as Julian Simon eloquently argued, men are not merely receptacles for output, they are also producers.

As the crude Malthusian fears subsided, with the growing recognition that the burgeoning populations in the Third World were part of a 'demographic transition', similar to what had exorcised the Malthusian sceptre in the industrialised countries, the doomsters changed tack. Beginning with the infamous Club of Rome's The Limits of Growth, the argument became that even though the world's population might stabilise, as economic growth in the Third World made parents – as it had in the West – choose quality over quantity in their desired family size and thence lower fertility, the expected world population – with the high standard of living which would have triggered the demographic transition – was unsustainable, as the natural resources which were required to provide this higher global income would run out.

Simon (maintaining that if the doomsters were right, then we should see a sustained rise in the prices of these natural resources) famously wagered Paul Ehrlich (one of the leading doomsters) that resource prices would be lower at the end of the 1980s than at the beginning despite rapid increases in world population and output. Simon was right and Ehrlich paid up, but Simon never cashed the cheque, framing it as a memento of his victory.

This did not stop the Greens from announcing various other doom-laden scenarios. They were thus playing on an ancient human fear of the Apocalypse. While most such resource scares have been subsequently disproved, the ones which have stuck concern what is now anthropomorphically called 'the environment'. They have even led to public action in the form of various transnational treaties – many of which India has signed – and continuing Green agitation for more. They pose a serious threat to the economic health of developing countries, in particular India and China, and that is the subject I would like to discuss in this paper.

Facts or Values?

Openly Green political parties which have contested elections in the public arena – outside Germany and some of the Scandinavian countries – have not had much public support. They have, therefore, adopted another tactic to push their agenda. Organised into non-governmental organisations (NGOs) who are the self-proclaimed voice of an international civil society, they have sought to push their agenda through various trans-national organisations like UNEP, and increasingly the World Bank and the WHO.

Their aim is to push through international treaties and conventions sponsored by these organisations to regulate various aspects of economies, particularly of the Third World. The following international treaties have either been concluded or are under negotiation: the Biodiversity Convention; the Basel Convention; the Convention to Combat Desertification, the POPs (Persistent Organic Pollutants) Treaty, and of course, the Kyoto Protocol. In all these cases the Green NGOs having failed, by and large, to gain political legitimacy for their viewpoint through the ballot box, are attempting to legislate it through the unelected bureaucracies of transnational institutions. The big prize they seek, and which is still not entirely in their grasp, is the WTO, where they would like to see trade sanctions being used to further their agenda.

So what is their agenda? Even though their shifting scares have been countered by rational and scientific arguments, it has had no effect on the Greens. But the Greens' position is not based on reason: it is a new secular religion. Take just one example.

Presented with evidence that some purported environmental threat is extremely unlikely and uncertain, they resort to a stock argument, the precautionary principle: 'It is better to be safe than sorry.' This has some resonance with the public, as it has echoes of Pascal's famous wager about the existence of God, viz. if God did not exist one would only have eschewed the finite pleasures from forsaking a sinful life, but if he did exist a sinful life would lead to damnation and the infinite pain of Hell. In expected utility terms (as economists would call it), it was better to give up the finite pleasures from a sinful life for even an infinitesimally small probability of burning forever in Hell.

Ehrlich's well-known restatement of this wager: "If I'm right we'll save the world by curbing population growth. If I'm wrong, people will still be better fed, better housed and happier, thanks to our effort. Will any-thing be lost if it turns out later that we can support a much larger population than seems possible today."

But, as Julian Simon points out in his riposte to Ehrlich, note that "Pascal's wager applies entirely to one person. No one else loses if she or he is wrong. But Ehrlich bets what he thinks will be the economic gains that our descendants and we might enjoy against the unborns' very lives. Would he make the same wager if his own life rather than others' lives were at stake?" So it does come down to a question of values after all, not facts or logic.

A New Secular Religion

Some time after my foray into the snake-pit of the environmental debate it became clear that what we are witnessing here is another crusade, reminiscent of those which led to western imperialism in the past.

Recently, the *Sarsangchalak* of the RSS, K.S. Sudarshan attacked the Christian Church and fundamentalist Islam as agents of destabilization of India. This was misplaced for two reasons. Leaving aside the question of Islamic fundamentalism, the attack on the Christian Church seems to have been provoked by Pope John Paul II's address in India last year, in which he declared that, in the third millennium the Church's aim would be to evangelise the whole of Asia as it had Europe in the first and the Americas and Africa in the second. But this is an old objective of the West. In his still relevant book on Asia and Western Dominance, the late K.M. Panikkar had charted this aspect of the encounter between the old Eurasian civilisations and the newly resurgent West, and shown how despite repeated attempts at converting the people of Asia, even with the aid of gunboats and diplomatic pressure, the Christian missionaries failed in their evangelical mission. Whatever converts they made were from the lower social classes and were looked down upon by their compatriots – being contemptuously labeled as 'rice Christians' and 'secondary barbarians' by the Chinese. Even when the Christians tried various forms of syncretism by claiming their religion was compatible with local traditions, the indigenous cultures were too strong and did not accept the cultural superiority of the West.

So, if, even at a time these missionaries could count on State support for their operations, they failed, it is difficult to see that they are likely to succeed today, when the playing field is more level. The Pope's hopes of converting Asia are likely to be as frustrated as those of St Francis Xavier who died on a rocky island off the Kwangtung Coast in 1552, attended only by his Chinese servant, trying vainly to get to Beijing in the hope of repeating the early Church's victory in the Roman Empire through the conversion of Constantine. As Panikkar remarked, even in Goa, where conversion by force was undertaken, "The attempt to Christianise was not a complete success, . . . [as] the majority of the population after 430 years of Portuguese rule is still non-Christian."

Second, with the death of the Christian God in the minds of many in the West, the Christian cosmological beliefs have found expression in many different secular religions which they are now seeking to impose on Asia. It is these, and not the overt strictly religious evangelism, which pose the threat of a new western cultural imperialism.

First, note that western cosmological beliefs, to the extent they are coherent and commonly shared, are still deeply rooted in Christianity, particularly its theological formalisation in St. Augustine's "City of God". There are a number of distinctive features about Christianity, which it shares with its Semitic cousin Islam, but not entirely with its parent Judaism, and which are not to be found in any of the other great Eurasian civilisational religions, past or present. The most important is its universality. Neither the Jews nor the Hindus or the Sinic civilisations had religions claiming to be universal. You could not choose to be a Hindu, Chinese or Jew, you were born as one. This also meant that unlike Christianity and Islam these religions did not proselytise. Third, only the Semitic religions being monotheistic have also been egalitarian. The others have believed in Homo Hierarchicus. An ethic which claims to be universal and egalitarian and proselytises for converts is a continuing Christian legacy even in secular western minds, and is the basis for the moral crusade of 'ethical trading'.

It would take us too far afield to substantiate this argument in any detail but since Augustine's "City of God", the West has been haunted by its cosmology. From the Enlightenment to Marxism to Freudianism to Eco-fundamentalism, Augustine's vision of the Heavenly City has had a tenacious hold on the western mind. The same narrative with a Garden of Eden, a Fall leading to original Sin and a Day of Judgement for the Elect and Hell for the Damned keeps recurring. Thus the philosophers displaced the Garden of Eden by classical Greece and Rome, and God became an abstract cause, the Divine Watchmaker. The Christian centuries were the Fall, and the Christian revelations a fraud as God expressed his purpose through his laws recorded in the Great Book of Nature. The Enlightened were the elect and the Christian paradise was replaced by Posterity (see Becker). By this updating of the Christian narrative the eighteenth-century philosophers of the Enlightenment thought they had been able to salvage a basis for morality and social order in the world of the Divine Watchmaker. But once (as a result of Darwin) he was seen to be blind, as Nietzsche proclaimed from the housetops at the end of the nineteenth century, God was dead, and the moral foundations of the West were thereafter in ruins.

The subsequent attempts to found a morality based on reason are open to Nietzsche's fatal objection in his aphorism about utilitarianism: "moral sensibilities are nowadays at such cross purposes that to one man a morality is proved by its utility, while to another its utility refutes it" (Nietzsche, 1881-1982, p. 220). Nietzsche's greatness lies in clearly seeing the moral abyss that the death of its God had created for the West. Kant's attempt to ground a rational morality on his principle of universalisability – harking back to the Biblical injunction "therefore all things whatsoever ye do would that men should do to you, do even so to them" – founders on Hegel's two objections: it is merely a principle of logical consistency without any specific moral content, and worse, it is as a result powerless to prevent any immoral conduct that takes our fancy. The subsequent ink spilt by moral philosophers has merely clothed their particular prejudices in rational form.

The death of the Christian God did not, however, end variations on the theme of Augustine's "City". It was to go through two further mutations in the form of Marxism and Freudianism, and the most recent and bizarre mutation in the form of Eco-fundamentalism. As both Marxism (in its post-modern form) and Eco-fundamentalism provide the ballast for ecological imperialism it is worth noting their secular transformations of Augustine's Heavenly City.

Marxism like the old faith looks to the past and the future. There is a Garden of Eden, before "property" relations corrupted "natural man". Then the Fall as "commodification" leads to class societies and a continuing but impersonal conflict of material forces, which leads in turn to the Day of Judgment with the Revolution and the millennial Paradise of Communism. This movement towards earthly salvation being mediated, not as the Enlightenment sages had claimed through enlightenment and the preaching of good will, but by the inexorable forces of historical materialism. Another secular "City of God" has been created.

Eco-fundamentalism is the latest of these secular mutations of Augustine's "City of God" (Lal 1995). It carries the Christian notion of *contemptus mundi* to its logical conclusion. Humankind is evil and only by living in harmony with a deified Nature can it be saved.

The environmental movement (at least in its "deep" version) is now a secular religion in many parts of the West. The historian of the ecological movement, Anna Bramwell notes that in the past western man was able to see the earth as man's unique domain precisely because of God's existence. . . . When science took over the role of religion in the nineteenth century, the belief that God made the world with a purpose in which man was paramount declined. But if there was no purpose, how was man to live on the earth? The hedonistic answer, to enjoy it as long as possible, was not acceptable. If Man had become God, then he had become the shepherd of the earth, the guardian responsible for the earth (Bramwell, p. 23).

The spiritual and moral void created by the death of God is, thus, increasingly being filled in the secular western world by the worship of Nature. In a final irony, those haunted natural spirits that the medieval Church sought to exorcise so that the West could conquer its forests (see Southern) are now being glorified and being placed above Man. The surrealist and anti-human nature of this contrast between eco-morality and what mankind has sought through its religions in the past is perfectly captured by Douglas and Wildavsky who write: "The sacred places of the world are crowded with pilgrims and worshippers. Mecca is crowded, Jerusalem is crowded. In most religions, people occupy the foreground of the thinking. The Sierra Nevada are vacant places, loved explicitly because they are vacant. So the environment has come to take first place" (p. 125). The guilt evinced against sinning against God has been replaced by that of sinning against Nature. Saving Spaceship Earth has replaced the saving of souls! But why should the rest of the world subscribe to this continuing Augustinian narrative cloaked in different secular guises?

Towards World Disorder

There are ominous parallels between the last decades of the nineteenth century and the present century.

In both periods it seemed that a world increasingly closely knit through foreign trade and capital flows would bring universal peace and prosperity. This dream came to an end on the fields of Flanders. The First World War (which has been aptly described as a wholly unnecessary war) put an end to the first Liberal International Economic Order (LIEO) created under British leadership. It took nearly a century to resurrect a new LIEO under the United States.

One of the causes of the First War was the imperial competition for colonies. This imperialism was fuelled by the territorial imperative as well as the "white man's burden", to save the heathen souls. In nineteenth-century India, as Stokes demonstrated, there was an unholy alliance of Evangelicals – with their belief in the Gospels, and Utilitarians and Radicals – with their faith in reason, who believed in the superiority of western ways, religious and secular. Their attempts to transform Indian "habits of the heart" led to the nationalist backlash of the 1857 Mutiny. Today we see a similar alliance between some scientists and the eco-fundamentalists with a similar imperialist form though differing content. But history never repeats itself. Whereas the nineteenth-century battles for "hearts and minds" were fought within and between 'nation-states' the arena for today's imperialist project are various transnational organisations.

It is instructive to see how this has happened and its likely consequences.

Stephen Toulmin's (1990) brilliant reconstruction of the origins of the "modernity" project provides the necessary clues. Toulmin argues that there were two strands in modernity. The sceptical humanism of the late Renaissance epitomised by Montaigne, Erasmus and Shakespeare, and the rationalism of the late sixteenth century epitomised by Descartes search for certainty, which underpinned the triumphs of the scientific revolution as well as the methods of mechanistic Newtonian physics as the exemplary form of rationality. Toulmin's most original insight is that the rationalist project was prompted by the Thirty Years War that followed the assassination of Henry IV of France in 1610. Henry's attempt to create a religiously tolerant secular state with equal rights for Catholics and Protestants mirrored the skeptical humanism of Montaigne. Henry's assassination was taken as a sign of the failure of this tolerant Renaissance scepticism. With the carnage that followed the religious wars in support of different dogmas, Descartes set himself the project of overcoming Montaigne's skepticism, which seemed to have led to such disastrous consequences by defining a decontextualised certainty. This rationalist project, which created the scientific revolution, found resonance, argues Toulmin, in the coterminous development of the system of sovereign nation states following the peace of Westphalia.

The ascendancy of these two "systems" continued in tandem till the First World War. But chinks were appearing in the armour of the rationalist Cartesian project with its separation of human from physical nature with the developments in the late nineteenth century associated with Darwin and Freud. Despite the replacement of Newtonian physics by the less "mechanistic" physics of Einstein and his successors, the political disorder of the 1930s led as in the 1630s to a search for certainty and the logical positivist movement was born.

The final dismantling of the scaffolding of the rationalist project begun with the peace of Westphalia, according to Toulmin, occurred in the 1960s – with Kennedy's assassination being as emblematic as Henry IV's. With many hoping that Kennedy was about to launch a period ending the Age of Nations and beginning one of transnational cooperation through trans-national institutions. Thus, since the 1960s, the world has been trying to reinvent the humanism of the Renaissance that was sidelined by the rationalist Cartesian project of the sixteenth century. As he writes, by the 1950s there were already the best of reasons, intellectual and practical for restoring the unities dichotomised in the seventeenth century: humanity vs nature, mental activity vs its material correlates, human rationality vs emotional springs of action and so on.

He then goes on to argue that the post-war generation was the first to respond: "because they had strong personal stakes in the then current political situation." The Vietnam war shocked them into rethinking the claims of the nation, and above all its claim to unqualified sovereignty. Rachel Carson had shown them that nature and humanity are ecologically interdependent, Freud's successors had shown them a better grasp of their emotional lives, and now disquieting images on the television news called the moral wisdom of their rulers in doubt. In this situation, one must be incorrigibly obtuse or morally insensible to fail to see the point. This point did not relate particularly to Vietnam: rather what was apparent was the superannuation of the modern world view that was accepted as the intellectual warrant for "nationhood" in or around 1700. (Toulmin, p. 161).

This is the place to introduce the insights of Douglas and Wildavsky concerning the cultural and political characteristics of the environmental movement. They define a hierarchical centre which has been characteristic of the nation state – much as Toulmin does. Opposing this has been what they call "border" organisations. They comprise "secular and religious protest movements and sects and communes of all kinds" (p.102). They argue that the "border is self-defined by its opposition to encompassing larger social systems. It is composed of small units and it sees no disaster in reduction of the scale of organisation. It warns the centre that its cherished social systems will wither because the centre does not listen to warnings of cataclysm.

The border is worried about God or nature, two arbiters external to the large-scale social systems of the centre. Either God will punish or nature will punish; the jeremiad is the same and the sins are the same: worldly ambition, lust after material things, large organisations." (p. 123) Like Toulmin, they see the Vietnam War, and Watergate undermining support for the centre in the US, and giving greater legitimacy to the border – particularly to the segment which emphasises Nature.

There are various more complex reasons – which we cannot go into on this occasion – why the moral authority of the centre in many western states has been under-mined. This has given rise to sources of moral authority outside the hierarchical structure of the nation state, which echoes a return to pre-modern western medieval forms. As Toulmin notes, "one notable feature of the system of European Powers established by the Peace of Westphalia . . . was the untrammeled sovereignty it conferred on the European Powers. Before the Reformation, the established rulers . . . exercised their political power under the moral supervision of the Church. As Henry II of England found after the murder of Thomas Becket, the Church might even oblige a King to accept a humiliating penance as the price of its continued support" (*Ibid.* p. 196).

With the undermining of the moral authority of western nation states, Toulmin notes that this moral authority is increasingly being taken over by non-governmental organisations (NGOs) like Amnesty International, and in many cases the environmental NGOs. This unravelling of the Westphalian system and a partial reversion to the world of the Middle Ages, poses in my view the real threat of eco-imperialism, modeled less on the model of the 19th century scramble for Africa, than the Crusades.

For while the West may be turning its back on modernity and its associated untrammelled sovereignty of nation states, the Rest have no intention of giving up the latter and are eagerly seeking to adopt the technological fruits of the former, without giving up their souls. Hence even religious fundamentalists in the Rest recognise the need for economic progress, if for no other reason than to acquire the ability to produce or purchase those arms which they feel are essential to prevent any repetition of the humiliation they have suffered at the hands of superior western might in the past. As the then Indian Defence Minister is reported to have said when asked about the lesson he learned from the Gulf War: "Don't fight the United States unless you have nuclear weapons" (cited in Huntington, p. 46). Numerous developing countries for good or ill have, or are rushing to acquire this new countervailing power. The attempts by the eco-moralists to curb their development of the industrial bases of this power, to save Spaceship Earth will be fiercely resisted.

This has ominous consequences for the various trans-national organisations like the United Nations, the World Bank and the World Trade Organisation. As Huntington notes, "global political and security issues are effectively settled by a directorate of the United States, Britain and France, world economic issues by a directorate of the United States, Germany and Japan . . . to the exclusion of lesser and largely non-western countries. Decisions made at the UN Security Council or in the International Monetary Fund that reflect the interests of the West are presented to the world as reflecting the desires of the world community. The very phrase 'the world community' has become the euphemistic collective noun . . . to give global legitimacy to actions reflecting the interests of the United States and other western powers" (p. 39).

It is not surprising therefore that the ecologists should seek to influence the agenda of these international organisations (see above). But as I have argued given the globally divisive nature of their agenda, how long will it be before the frictions it causes will destroy these institutions?

What Should India Do?

The time has surely come to take on these new cultural imperialists. The first point of resistance is to recognise what they are seeking to do. Bluntly, they would like to perpetuate the ancient poverty of the great Eurasian civilisations – India and China – with, as they see it, their burgeoning unwashed masses increasingly emitting noxious pollutants as they seek to make their people prosperous, and achieve parity with the West.

For as economic historians have emphasised, it was not till the Industrial Revolution that mankind found the key to intensive growth – a sustained rise in per capita income – which, as the example of the West and many newly industrialising countries have shown, has the potential of eradicating mass structural poverty – the scourge which in the past was considered to be irremediable (e.g. the Biblical saying that the poor will always be with us). For in the past, most growth was extensive – with output growing in line with (modest) population growth (Reynolds, 1983). As pre-industrial economies relied on organic raw materials for food, clothing, housing and fuel (energy), whose supply in the long run was inevitably constrained by the fixed factor, land, their growth was ultimately bounded by the productivity of land. For even traditional industry and transportation – depending upon animal muscle for mechanical energy, and upon charcoal (a vegetable substance) for smelting and working crude ores and providing heat – would ultimately be constrained by the diminishing returns to land that would inexorably set in once the land frontier was reached. In these organic economies (Wrigley, 1988), with diminishing returns to land conjoined with the Malthusian principle of population, a long run stationary state where the mass of the people languished at a subsistence standard of living seemed inevitable. No wonder the classical economists were so gloomy!

But even in organic economies there could be some respite, through the adoption of market "capitalism" and free trade defended by Adam Smith. This could generate some intensive growth as it would increase the productivity of the economy as compared with mercantilism, and by lowering the cost of the consumption bundle (through cheaper imports) would lead to a rise in per capita income. But if this growth in popular opulence led to excessive breeding the land constraint would inexorably lead back to subsistence wages. Technical progress could hold the stationary state at bay but the land constraint would ultimately prove binding.

The Industrial Revolution led to the substitution of this organic economy by a mineral-based energy economy. It escaped from the land constraint by using mineral raw materials instead of the organic products of land. Coal was the most notable, providing most of the heat energy of industry and with the development of the steam engine virtually unlimited supplies of mechanical energy. Intensive growth now became possible, as the land constraint on the raw materials required for raising aggregate output was removed.

Thus the Industrial Revolution in England was based on two forms of "capitalism", one institutional, namely that defended by Adam Smith – because of its productivity enhancing effects, even in an organic economy – and the other physical: the capital stock of stored energy represented by the fossil fuels which allowed mankind to create in the words of E.A. Wrigley: "a world that no longer follows the rhythm of the sun and the seasons; a world in which the fortunes of men depend largely upon how he himself regulates the economy and not upon the vagaries of weather and harvest; a world in which poverty has become an optional state rather than a reflection of the necessary limitations of human productive powers" (Wrigley, 1988, p. 6).

The Greens are of course, against both forms of "capitalism" – the free trade promoted by Smith, as well as the continued burning of fossil fuels – leaving little hope for the world's poor.

Kyoto Protocol: India along with China is therefore to be commended for standing firmly at Kyoto against any restriction of their CO_2 emissions. With the recent collapse of the negotiations for a Climate Change Treaty, largely because of differences between the US and Europe, this is perhaps an issue which the Greens will no longer be able to exploit. But India has already signed various international ecological treaties inimical to her interests.

Basel Convention: Thus, for example, India is a signatory to the Basel Convention, which by defining various metals as 'hazardous,' controls trade in waste, scrap and recyclable materials. Greenpeace is using the treaty to organise a total embargo on trade with developing countries, excluding them from global scrap metal markets. This is already having deleterious effects. There are recent press reports that a highly profitable industry, shipbreaking, at Alang in Gujarat is likely to fall foul of this convention.

Shipbreaking was till the 1970s performed with cranes and heavy equipment at salvage docks in big shipyards. When labour costs and environmental regulations made this un-competitive, the industry shifted to Korea and Taiwan. But in the 1980s, enterprising businessmen in India, Pakistan and Bangladesh realised that to wreck a ship they did not need expensive docks and tools; they could just drive the ship up onto a beach as they might a fishing boat and tear it apart by hand. The scrap metal obtained can be profitably sold in South Asia with its insatiable demand for low-grade steel, mainly for ribbed reinforcing bars used in constructing concrete walls. These rods are produced locally from the ships' hull plating by small-scale re-rolling mills of which there are close to a 100 near Alang alone. Today nearly 90 per cent of the world's annual of 700 condemned ships are wrecked on South Asian beaches, nearly half of them at Alang. The economic effects are substantial. "Alang and the industries that have sprung from it provide a livelihood however meagre for perhaps as many as a million Indians" (Langewiesche, 2000). It is not ironic, therefore, that some of these NGOs have formed a coalition called BAN (Basle Action Network) to monitor the implementation of the various prohibitions on trade and banish millions of people to perpetual poverty. This industry is sought to be destroyed by Greenpeace under the auspices of the Basle Convention. India should walk away from this convention.

POPs and DDT: Among the two other treaties under negotiation, which India should have nothing to do with, are the POPs (Persistent Organic Pollutants) Treaty and the Biodiversity Convention. These are attempts to ban DDT and GM food. As both are of vital interest to India's future, it may be worth saying something more on these.

The Persistent Organic Pollutants (POPs) Framework Convention is being negotiated under pressure from environmental groups, who want a binding treaty to ban 'persistent organic pollutants': defined as pesticides, industrial chemicals and their by-products. DDT is sought to be banned under the treaty. If India foolishly signs this convention it will seriously damage the nation's health. For DDT is the most cost-effective controller of diseases spread by insects like flies and mosquitoes that has ever been produced.

The US National Academy of Sciences estimated it had saved 500 million lives from malaria by 1970. In India, effective spraying had virtually eliminated the disease by the 1960s, so much so that the mosquito nets which were ubiquitous in my childhood had disappeared from urban houses by the time I was at University in the late 1950s. DDT spraying had reduced the number of malaria cases from 75 million in 1951 to around 50,000 in 1961, and the number of malaria deaths from nearly a million in the 1940s to a few thousand in the 1960s. But then in the 1970s largely as a result of an environmental scare promoted by Rachel Carson's book *Silent Spring*, foreign aid agencies and various UN organisations began to take a jaundiced view of DDT, and the use of DDT declined. Not surprisingly, the mosquitoes hit back and endemic malaria returned to India. By 1997 the UNDP's Human Development Report 2000 estimates there were about 2.6 million malaria cases.

The same story of a decline and rise in disease with the increase and decrease in DDT spraying can be told about kala-azar, which is spread by the sand fly. DDT largely rid India of kala-azar in the 1950s and 1960s. But, with the subsequent decline in DDT use it has come back. The State Minister of Health in Bihar recently informed the Assembly that, 408 people had died, and 12,000 were afflicted with the disease in 30 districts of Northern Bihar.

So why did DDT fall out of favour despite its demonstrated merits? It was Rachel Carson in 1962 who started the DDT scare with her claim that its use had devastating effects on bird life, particularly those higher up the food chain. It was also claimed it caused hepatitis in humans. Numerous scientific studies showed these fears to be baseless. It was shown to be safe to humans, causing death only if eaten like pancakes! In 1971 the distinguished biologist Philip Handler as President of the US National Academy of Science said, "DDT is the greatest chemical that has ever been discovered." Commission after commission, expert after Nobel Prize-winning expert has given DDT a clean bill of health (see E.M. Whelan, 1985: *Toxic Terror*).

Yet in 1972, President Nixon's head of the US Environmental Protection Agency, William Ruckelshaus banned DDT against all the expert scientific advice he had been given. He argued that, the pesticide was "a warning that man maybe exposing himself to a substance that may ultimately have a serious effect on his health." Most developed countries followed the US and banned the chemical for all uses. Many developing countries followed suit by banning the pesticide in agriculture, and some for all uses. USAID which along with the WHO had been at the fore-front of the mosquito eradication programmes based on house spraying with DDT, now turned their backs on DDT. USAID has maintained that, as DDT is not registered by the EPA for use in the US, foreign assistance is not available for programmes that use DDT. Thus despite the WHO's Malaria Expert

Committee's ruling that DDT is safe and effective for malaria control, since 1979 the WHO itself has championed a strategy which ignores the causal link between decreasing numbers of houses sprayed and increasing malaria, by emphasising curative and demphasising preventive measures. Instead of fighting malaria by the only effective method known, the WHO is spending its limited resources instead on the politically correct and highly dubious campaign against smoking (sec Lal 2000). The decline in full house spraying, created DDT-resistant mosquitoes. But even then, when DDT was vigorously used, as in Mexico, malaria rates declined despite the increasing DDT resistance of mosquitoes.

Moreover, DDT is now increasingly needed as the Anopheles mosquito has become resistant to the pesticides (synthetic pyrethroids) currently used. The favoured WHO strategy of distributing pesticide impregnated mosquito nets, and using chloroquine to treat the disease is vitiated by two factors. First, distributing and monitoring the use of mosquito nets is even more complicated than house spraying, and likely to be much less effective. Second, the chloroquine resistance built up by mosquitoes in the 1960s in South East Asia and South America has subsequently spread to most malarial countries around the world. There are some promising new drugs on the horizon, but the hope of a malaria vaccine is at least seven years away. While clearly curative measures must continue to form part of a malaria control programme, preventive measures are just as important, and for this, killing the mosquitoes with DDT remains the most efficient and cost-effective measure.

If both science and economics favours DDT, why has this growing ban on DDT spread? Ruckelshaus' reasons for his un-scientific decision to ban DDT in the early 1970s provides the clue. The environmental movement's supposedly key concept is 'sustainable development'. This was endorsed by the World Commission on Environment and Development's report, Our Common Future whose chair was the then Norwegian Prime Minister, Gro Harlem Bruntland. The notion of sustainability – at least in its strong form – asserts that natural capital, such as forests, wildlife and other natural resources cannot be substituted by manmade capital. As pesticides are assumed to have adverse effects on natural capital they are inconsistent with sustainable development. Hence, instead of using them to control bugs, the alternatives of mosquito nets and drugs to fight the disease should be used.

The argument that there is no scientific evidence that DDT spraying to kill mosquitoes damages natural capital is once again countered by the so-called "precautionary principle". Once again, the environmentalists are willing to ban DDT because they are willing to sacrifice human lives for those of birds.

This underlying misanthropy of the environmentalists is explicitly brought out by the following statement by Ehrlich about India: "I came to understand the population explosion emotionally one stinking hot night in Delhi. . . . The streets seemed alive with people. People eating, people washing, people sleeping, people visiting, arguing, and screaming. People thrusting their hands through the taxi window, begging. People defecating and urinating. People clinging to buses. People herding animals. People, people, people."

Not surprisingly many environmentalists have argued since the 1950s that, in the words of one: "It maybe unkind to keep people dying from malaria so that they could die more slowly of starvation. [So that, malaria may even be] a blessing in disguise, since a large proportion of the malaria belt is not suited to agriculture, and the disease has helped to keep man from destroying it – and from wasting his substance on it." Or more recently: "Some day anti-malarial vaccines will probably be developed, which may even wipe out the various forms of the disease entirely, but then another difficulty will arise: important wild areas that had been protected from the dangers of malaria will be exposed to unwise development" (cited in Tren and Bate: *Malaria and the DDT Story*, IEA, London, 2000).

Biodiversity and GM Foods: The recent scare about GM (genetically modified) food equally needs to be resisted. The Green Revolution having disproved the doomsters predictions that the world would not be able to feed a burgeoning population, they are now attempting to stop the next stage in the agricultural revolution offered by bio-technology. As the father of the Green Revolution Norman Borlaug, Nobel Peace Prize Laureate, has recently noted: though "the Green Revolution is [not] over, as increases in crop management productivity can be made all along the line: in tillage, water use, fertilisation, weed and pest control and harvesting, however, for the genetic improvement of food crops to continue at a pace sufficient to meet the needs of the 8.3 billion people projected to be on this planet at the end of the quarter century both conventional technology and biotechnology are needed" (Borlaug, 2000).

In 1995 there were 4 million acres of biotech crops planted, which had risen to 100 million in 1999. In the US 50 per cent of the soybean crop and more than one-third of the corn crop were transgenic in 1999. These GM crops provide major economic benefits as they have reduced pesticide applications, higher yields and lower consumer prices. (Krattiger 2000) They have been readily adopted where they have been introduced. Yet, particularly in Europe, the Greens – again led by Greenpeace – have created mass hysteria about these crops, labelling them as Frankenstein foods.

But if GM crops are the creation of a Frankenstein, so is virtually everything we eat. Any method that uses life forms to make or modify a product is biotechnology: brewing beer or making leavened bread is a 'traditional' biotechnology application. As Borlaug states: "The fact is that genetic modification started long before humankind started altering crops by artificial selection. 'Mother Nature' did it, often in a big way. For example, the wheat groups we rely on for much of our food supply are the result of unusual (but natural) crosses between different species of grasses. Today's bread wheat is the result of the hybridisation of three different plant genomes, each containing a set of seven chromosomes, and thus could easily be classified as transgenic. Maize is another crop that is the product of transgenic hybridisation. . . Neolithic humans domesticated virtually all our food and livestock species over a relatively short period 10,000 to 15,000 years ago. Several hundred generations of farmer descendants were subsequently responsible for making enormous genetic modifications in all our major crop and animal species. To see how far the evolutionary changes have come, one only needs to look at the 5000-year-old fossilised corn cobs found in the caves of Tehuacan in Mexico, which are one-tenth the size of modern maize varieties. Thanks to the development of science over the past 150 years, we now have the insights into plant genetics and plant breeding to do what 'Mother Nature' did herself in the past by chance. Genetic modification of crops is not some kind of witchcraft; rather it is the progressive harnessing of the forces of nature to the benefit of feeding the human race."

For what biotechnology merely does is to isolate individual genes from organisms and transfer them into others without the usual sexual crosses necessary to combine the genes of two parents.

Nor is there any danger to health or the environment from GM food as has been repeatedly noted: by a 2100 signatory declaration in support of biotechnology by scientists worldwide, by the US National Academy of Science, by the US House of Representatives Committee on Science and by a Nuffield Foundation Study in the UK. Since 1994, more than 300 million North Americans have been eating several dozen GM foods grown on more than 100 million acres, but not one problem with health or the environment has been noted. (Whelan, 2000). Yet the hysteria continues. To see the misanthropy at its heart, there is no better example than that of the miracle 'golden rice'.

Scientists from the Swiss Federal Institute of Technology (Zurich) and the International Rice Research Institute (Philippines) have successfully transferred genes producing beta-carotene, a precursor of Vitamin A, into rice to increase the quantities of vitamin A, iron and other micronutrients. As the GM rice produces beta carotene it has a bronze-orange appearance, hence its name 'golden rice'. It promises to have a profound effect on the lives of millions suffering from Vitamin A and iron deficiencies which lead to blindness and anaemia respectively. It has been estimated that more than 180 million children, mostly in developing countries suffer from Vitamin A deficiency, of whom two million die from it each year.

About a billion people suffer anaemia from iron deficiency. The new golden rice is being distributed free of charge to public rice breeding institutions around the world. Millions will be able to reduce their risks of these disabling costs at little or no cost.

Yet as the inventor of 'golden rice' Professor Ingo Potrykus has noted, though it satisfies all the demands of the Greens they still oppose it. As he notes, the new rice has not been developed by or for industry; benefits the poor and disadvantaged; provides a sustainable, cost free solution, not requiring other resources; is given free of charge and restrictions to subsistence farmers; can be re-sown each year from the saved harvest; does not reduce agricultural biodiversity;

does not affect natural biodiversity; has no negative effect on the environment; has no conceivable risk to consumer health and could not have been developed with traditional methods.

But, notes Prof. Potrykus: "The GMO opposition is doing everything to prevent 'golden rice' reaching the subsistence farmer. We have learned that the GMO opposition has a hidden, political agenda. It is not so much the concern about the environment, or the health of the consumer, or the help for the disadvantaged. It is a radical fight against technology and for political success" (Potrykus, 2000).

Conclusions

There we have it. The Green movement is a modern secular religious movement engaged in a worldwide crusade to impose its 'habits of the heart' on the world. Its primary target is to prevent the economic development which alone offers the world's poor any chance of escaping their age-old poverty. This modern-day secular Christian crusade has exchanged the saving of souls for saving Spaceship Earth. It needs to be fiercely resisted.

First, by standing up to the local converts – the modern day descendants of what the Chinese called 'rice Christians' and 'secondary barbarians' – the Arundhati Roys, Vandana Shivas and Medha Patkars of this world. Their argument that their views are in consonance with Hindu cosmology are reminiscent of those used by the proselytising Christians promoting a syncretised Christianity in the nineteenth century, and are equally derisory.

Second, by refusing to accept the transnational treaties and conventions which the Greens are promoting to legislate their ends. As many of the environmental ministries have become outposts of their local converts, the economic ministries must play a central role in resisting this Green imperialism, by insisting on having the last say on any transnational treaty India signs. As China has shown, through its continuing production and use of DDT and continuing development of GM technology, there is no need to give into this latest manifestation of western cultural imperialism, and in this fight, as the shining example of Julian Simon shows, there are still many in the West itself, who have not been infected with this secular Christian religion, and will join in showing up the Greens and their agenda as paper tigers, much as the Christian missionaries found in the last phase of western imperialism.

References

Becker C.L. (1932), *The Heavenly City of the Eighteenth Century Philosophers*, Yale University Press, New Haven.

Borlaug N.E. (2000), "Ending World Hunger: The Promise of Biotechnology and the Threat of Antiscience Zealotry", *Plant Physiology*, Vol. 124, pp. 487–90.

Bramwell A. (1989), *Ecology in the Twentieth Century: A History*, Yale University Press, New Haven.

Douglas M. and A. Wildavsky (1983), *Risk and Culture*, University of California Press, Berkeley.

Dumont L. (1970), *Homo Hierarchicus*, Weidenfeld and Nicholson, London.

Ehrlich P. (1968), *The Population Bomb*, Ballantine, Baltimore.

E. Gellner (1993): *The Psychoanalytic Movement: The Cunning of Unreason*, Northwestern University Press, Evanston, Illinois.

Huntington S.P. (1993), "The Clash of Civilisations", *Foreign Affairs*, Vol.72, No.3.

A.F. Krattiger (2000), "Food Biotechnology: Promising Havoc or Hope for the Poor?" *Proteus*, 17:38.

Lal D. (1988), *The Hindu Equilibrium*, 2 Vols., Clarendon Press, Oxford.

—————. (1998a), "Social Standards and Social Dumping" in H. Giersch (ed.): *The Merits of Markets*, Springer, New York.

—————. (1990), *The Limits of International Cooperation*, Twentieth Annual Wincott Lecture, Institute of Economic Affairs, London, reprinted in Lal (1994).

—————. (1994), *Against Dirigisme*, ICS Press, San Francisco.

—————. (1995), "Eco-fundamentalism", *International Affairs*, Vol. 71, 1995. reprinted in Lal (1999).

—————. (1997), "Ecological Imperialism: The Prospective Costs of Kyoto for the Third World", in J.H. Adler (ed.): *The Costs of Kyoto*, Competitive Enterprise Institute, Washington D.C.

—————. (1998), *Unintended Consequences*, MIT Press, Cambridge Mass, and Oxford University Press, New Delhi, 1999.

—————. (1999), *Unfinished Business*, Oxford University Press, New Delhi.

—————. (2000), "Smoke Gets in Your Eyes: The Economic Welfare Effect of the World Bank–WHO Global Crusade Against Tobacco" in *War on Tobacco: At What Cost?*, Liberty Institute, New Delhi.

Langewiesche W. (2000), "The Shipbreakers", *The Atlantic Monthly*, August 2000.

F. Nietzsche (1881/1982), *Daybreak: Thoughts on the Prejudices of Morality*, Cambridge University Press, Cambridge.

Nordhaus W.D. (1994), *Managing the Global Commons*, MIT Press, Cambridge Mass.

Panikkar K.M. (1953), *Asia and Western Dominance*, Allen and Unwin, London.

Potrykus I. (2000), "The 'Golden Rice' Tale", *Aghioview*, 23 October, archived at www.agbioview.listbot.com.

Reynolds L.G. (1985), *Economic Growth in the Third World*, Yale University Press, New Haven.

Schneider S. (1989), *Global Warming*, Sierra Club, San Francisco.

Sen A.K. and B. Williams (1982), *Utilitarianism and Beyond*, Cambridge University Press, Cambridge.

Simon J.L. (1996), *The Ultimate Resource 2*, Princeton University Press.

Southern R.W. (1970), *Western Society and the Church in the Middle Ages*, Penguin, London.

Stokes E. (1959), *The English Utilitarians and India*, Oxford University Press, Oxford.

Toulmin S. (1990), *Cosmopolis*, University of Chicago Press, Chicago.

Webster R. (1995), *Why Freud Was Wrong*, Harper Collins, London.

Whelan E. (2000), "The Case for Genetically Modified Food", in *Nutrinews*, archived at www.agbioview.listbot.com.

Wrigley E.A. (1988), *Continuity, Chance and Change*, Cambridge University Press, Cambridge.

Epilogue

26

Environment Management in India[*]

SOMNATH BANDYOPADHYAY

The notion that the environment could be, or should be, managed, is relatively recent in India. Like any other old civilisation, Indians traditionally revered nature even while they gradually learnt to manipulate it to an extent that met its own basic needs of food and shelter. However, the degree of manipulation seldom achieved extraction of resources beyond subsistence levels, either from the farm-lands or from the forests, thanks to their ecological characters, whose variability remain largely unpredictable even today.

Tropical location, Himalayan mountains to the north and open seas to the south create unique conditions of air circulation over the sub-continent that lead to a seasonal distribution of rainfall, referred to as the "monsoons". The variability of the rainfall pattern was not only the major external determinant during the early development of human habitations in the sub-continent, but continues to influence the economic growth of modern India significantly.

The long-term variation of rainfall over space and time led to the development of distinct vegetation patterns over the sub-continent, dominated by various forest types. The original inhabitants of these forests – often referred to as *vanvasis* or *adivasis* – were predominantly hunter-gatherers, who extracted a wide variety of plant and animal resources offered by the forests themselves. Fire, along with other crude implements, was used as a key management technique to extract resources that were relatively abundant, protect resources from

---

[*] This article only intends to provide a perspective on environmental management in India, and therefore refrains from using specific references and examples. The purpose is to provoke a whole new set of inquisition that transcends boundaries constraining our present sphere of thinking. Proper environmental solutions in India will necessarily be "home-grown" for which appropriate mind-space is to be created that will test and nurture new ideas.

other wild competitors and fulfil the consumption needs of a human population which had a very limited life-span and even limited demands.

Gradual increase in life-span, improved knowledge and consequent diversification of demands for natural resources brought with it a variety of fundamental changes. Basic social groups and rudimentary institutional forms emerged in order to share the extraction and use of these resources. Elementary norms were evolved for the protection of certain species, particularly during their breeding seasons or when their numbers needed to recover and justify a certain degree of hunting effort. Norms were gradually established through forms – such as cultural and religious rituals – that defined personal food habits as well as community outlook on local resources such as village tanks and groves. Although constrained by a lack of any written scripts, there is increasing evidence to indicate that individual and community rights had evolved among indigenous communities even on the basis of oral traditions.

Even though written by "outsiders", references to *"vanvas"* in the early texts and folklores provide glimpses of the lives of the original inhabitants of the sub-continent. Resource extraction under such conditions has been as variable as the environmental conditions that determined its availability. This had, in turn, left the dynamics of the human population at the mercy of sudden food shortages (famines), sudden emergence of high population of competing species (like locusts) and sudden development of conditions that sustained pathogens (like cholera and plague). In short, native Indians seemed to have survived more as an integral part of the ecological systems, learning to accept (and, often, revere) the eccentricity of Nature's bounties rather than dominate these systems by actively controlling their production processes in his own favour.

Conflicts and the Phenomenon of Hereditary Occupations

The quest for an active control of the production processes began to succeed first in the agricultural settlements of the Indo-Gangetic plains. Any agricultural system, like the Prairies and Steppes in the temperate regions of the world, is more uniformly productive, that allows for the rapid fulfilment of basic needs, creates surpluses and provides opportunities to diversify the economic base. The Indo-

Gangetic river system, however, distinguished itself for being far less uniform and predictable in its flooding patterns, responding as it were to the unique variability in the monsoon rains of the sub-continent.

Shifts in river courses are known to have eliminated an entire civilisation along the Indus valley, wiping in its course amazingly advanced systems of trade, industry and urban settlements. Although less dramatic, similar shifts in the courses of rivers in the Indo-Gangetic plains have often led to large-scale disputes over the possession of fertile lands – the key productive asset in an economy dominated by agriculture – resulting in a deeply fractured and caste-ridden social system.

Mythology and ancient history of the sub-continent provide enough evidence of strife in the chequered polity of the Indo-Gangetic plains. However, its relations with control over economic resources are not as easily apparent. Communities had to use force to maintain control over transient productive tracts in the floodplains. Communities also had to use elaborate trading skills and systems to stabilise the consequences of extremely variable farm productions. And, above all, success of communities was linked to institutions of knowledge for expansion and improved control of the productive systems. This complex social organisation – or the *varnas* – that developed in ancient India to manage the natural production systems has been studied from an ecological perspective by a few sociologists only.

The quest for expansion of the agricultural systems introduced a new dimension to the range of conflicts over control and management of natural systems. Fire – a management tool of the forest-dwellers – was used in a more devastating form to clear the forests on a more permanent basis. While the traditional resource base of the forest-dwellers was being rapidly eroded, the expansion of agriculture was not as rapid thanks to the limiting nature of water, whose availability, unlike land, could not be enhanced so easily. Vast tracts of forest land, therefore, got converted into intermediate grassland ecosystems, where pastorals gained control.

The necessity of specialised occupational training for the management of complex agrarian systems, coupled with a predominantly oral tradition and an elitist formal system of education, ensured greater dependence on family as the dominant social institution. The complex mosaic of castes and sub-castes are nothing but extended patrilineage families. While such hereditary occupational pattern ensured the perpetuation and enrichment of traditional knowledge and skills, it was also believed to provide incentives for sustainable management of resources by creating monopolies over a partitioned resource-base.

Unfortunately, hereditary occupation also provided incentives to perpetuate conflicts by promoting a culture that accepts fate and resists any active quest for change of livelihood. Forest-based occupations thus came in direct conflict with agriculture-based occupations. The politics involved in major irrigation schemes, particularly in the manner in which forest lands are submerged, original inhabitants are rehabilitated and benefits to farmers in the command area are distributed seem to suggest that little has changed in the nature of these basic conflicts even today.

Hereditary occupations also served to limit the horizons of human endeavours to predefined professions, reducing any active pursuit for excellence. A limit to individual growth inevitably leads towards growth by association. Economic growth is then measured in terms of land area controlled by the cultivator or the number of cattle-heads controlled by the pastoral. It must be noted that the forest-dwellers had no such cultural compulsions, but were increasingly being marginalised in the overall scheme of social structures dominated by the landed gentry.

In sum, the ancient property rights regimes followed three distinct patterns – the caste Hindus who owned and managed riparian agricultural lands, the forest-dwellers whose rights were increasingly ignored in the political context dominated by the caste Hindus and the pastorals who owned cattle-heads that grazed in community-owned lands vacated by the forest-dwellers but not occupied by the caste Hindus.

Colonial Influence and the Emergence of Scarcity

Throughout most of the medieval periods, a variety of kingdoms organised a system of managing resources that maximised the control of wealth by the crown. Aggression and annexation of fertile lands and cattle was a popular method. Land, the most important natural resource, was primarily in the public sector, with the State holding all virgin land, forests and water resources. Arable land, however, was both in the public and the private sectors. While public lands were either leased or doled, income from private lands were taxed. Apart from arable land, sometimes even other immovable property like fields, embankments, water tanks and reservoirs were privately owned and available for transfer through sale. Mining and fishing were also in both sectors. Animals like cows, sheep and goat were in both sectors and individual rights over captured animals (deer, birds, wild animals) were protected.

Much of the public resources were used to maintain armies, administration and ornaments of the monarchy. During times of relative peace and prosperity, public resources would also be spent on art, architecture and other cultural domains. Wealth was, thus, largely ploughed back to the local economy in a variety of manners, depending on the priorities of the crown. This may not have been a particularly efficient economic system but also did not lead to any serious erosion of wealth or natural resources.

Except for a few waves of marauders, primarily from Central Asia, much of the national wealth remained within the sub-continent till the advent of the Europeans. The diverse and abundant natural resources in India were coveted by the western European nations during their early stages of industrialisation. Sea routes were discovered and naval strengths were enhanced to secure and expand trade relations. Military prowess was a key determinant of the terms of trade during the early periods of colonisation. The eventual dominance of the British East India Company tilted the balance decidedly.

Unequal trade relations, which promoted the flight of capital and rapid erosion of wealth in India, were institutionalised through colonial governance. In fact, Britain more than made up for its own limited natural resources by organising an economic system that drew on its colonies for raw materials like plants, animals and minerals for feeding its industrial hubs. What began with trade in spices and condiments swiftly progressed towards large-scale plantations of cash-crops like cotton, indigo, tea etc. on forcibly acquired farmlands and virgin forest lands, freely using the natural resources and exploiting the cheap labour. These provided a competitive edge to the British manufacturing units, whose products began to find a global market, including India.

The colonial superiority was not just evident in military terms, but manifest itself in manufacturing and trade as well, particularly through the use of technology and professional management of businesses. The rapid development of scientific knowledge and imaginative use of technology helped in the diversification and modernisation of the British manufacturing sector. However, it was only in their interest that science and technology was used in the colonies to enhance productivity of natural resources, particularly of those that fed the British manufacturing units, rather than invest in manufacturing *per se*.

Timber, for example, was coveted by the British economy – in particular the Royal Navy – for which large-scale felling (as opposed to earlier destruction by fire) was introduced in the country for the first time. Use of modern felling technology, a planned approach to timber extraction and the introduction of new species were clearly based on much superior knowledge of manipulation and, coupled with the fact that it was being promoted by the ruling class, became legitimate "scientific" management of forests. Felling became acceptable and the debate shifted to the "manner" of felling and ancillary activities like raising nurseries and planting certain species.

The colonial influence on Indian forests had far reaching consequences. Most of the Princely States crudely emulated the processes of felling in order to generate revenue for their States, particularly when threatened with dispossession. Independent India developed a cadre to systematically extract the forest resources. Emulating the British in adopting "scientific" management systems, contrasting its controlled extraction with the perverse clear-felling followed by several Princely States and the genuine need to "develop" a modern Nation (even if accompanied by some "sacrifices") entrenched the notional legitimacy of the State to control and extract forest resources till date.

In the process, vast tracts of forests were also converted into grasslands (where re-afforestation did not succeed), agriculture lands (where encroachments happened as well) or plantations (where afforestation succeeded partially). The systematic depletion of forest area and decimation of forest resources have led the vast indigenous populations dependent on these resources to misery. Such dispossession, however, was hardly ever recognised, except when large-scale submergence took place in reservoir regions (when the "sacrifice for the larger good" argument was put forward).

Investing in Food Security

During Independence, India found herself burdened with a rapidly growing population (with increasing life expectancy) but little means for enhancing food production. Food security of millions in India was not a priority for the colonial government and hence little investments were made in irrigation infrastructure and agriculture development prior to Independence. Chronic food shortages and recurrent famines were, thus, the major considerations of post-Independence India. A slew of policies and investment programmes were initiated by the State to boost agriculture production that would have far reaching implications on land-use patterns and water management.

Land was viewed as a food-producing entity, albeit with many location-specific constraints, chiefly water regimes, that needed to be understood and overcome. Availability of water at definite periods to irrigate the crops was identified as the single most important factor to boost agriculture. For this purpose, the largely unpredictable and seasonal flooding patterns of the Indian rivers had to be "tamed" to provide a more assured supply of water. Western models of engineering solutions were adopted once again for constructing dams, barrages and weirs to control the natural flow of water, channels and embankments to contain the natural spread of water and a network of canal systems to direct the waters into croplands rather than the seas.

The early successes of large dams provoked our first Prime Minister to proclaim these as the "temples of modern India", enabling a rapid expansion of the State Departments for irrigation primarily by employing civil engineers. Development of water resources meant extraction and supply rather than management of an important resource. To make matters worse, inter-State water sharing treaties usually left a clause for review after 25 years, on the basis of demand. Expansion of canal networks served to indicate a growing demand and position the States for a higher stake in river waters while at the same time provide a rationale for continued State support for a burgeoning cadre of technocrats. Water management, thus, became a "technical" issue that could "solve" flood problems, irrigate drought-prone areas and "develop" groundwater resources.

The negative impact of all these efforts towards "rapid development of water resources" is only now being recognised. Rivers have dried up while adjacent croplands are waterlogged. Promises, rather than water, have reached the tail-enders in irrigation command areas across the country. Free, or heavily subsidised, power is provided to farmers in lieu of water, providing strong incentives to tap the groundwater resources. Depletion of groundwater has reached alarming levels in several areas, often accompanied by deteriorating quality. Sinking bore-wells to tap the groundwater is a major gamble for farmers in semi-arid, drought-prone regions of the country, often resulting in piling debt burdens and an increasingly alarming rate of suicides. Populist responses push towards further power subsidies, rather than attempting to pull out intensive farming from non-irrigated lands. Food shortages were replaced now with water shortages and while famines have become history, recurrent droughts have come to haunt modern India.

Investing in Food Security – Version II

The "green revolution" was another basket of technology – improved varieties of seeds, chemical fertilisers and pesticides – that was introduced in order to boost agriculture. Naturally, the benefits chiefly went to those pockets that had the basic irrigation infrastructure already in place, usually in floodplains of large perennial rivers. Even with a limited geographic spread, national food production was enhanced significantly, in the process shedding its (often humiliating) dependence on external aid for food.

A more confident nation was now willing to play a decisive role in procurement and distribution of these foodgrains as well, justified by its earlier investments in irrigation infrastructure, input subsidies and the felt need for equitable distribution. The unintended consequences of these actions were many but a couple of these merit attention in this context. The centralised monopoly policies in procurement and distribution of foodgrains severely restricted wealth creation of farmers, particularly in the more entrepreneurial grain baskets of Punjab and Western UP. The growing public discontent in these areas was addressed politically by re-organising governance units, countering violent secessionist movements and dealing with growing emigration and human trafficking.

The second consequence, however, was more pervasive. An overzealous approach to the notions of self-sufficiency, coupled with a desire to control prices, led to policies that strongly discouraged any private exchange – the traditional lifeline of farmers in India – forcing the rural masses to depend fully on a public distribution system (PDS) for supplying a fixed ration of "essential goods", particularly food-grains. The primary means of production (natural resources such as land, water, vegetation and financial resources such as credit) were also placed under State controls, virtually eliminating any possibility of local enterprise or economic development.

The growing disconnects between local demands and a centralised supply system was hardly acknowledged even as the issue of inefficiency and corruption in PDS gained centre-stage. The poor, meanwhile, had the onerous task of building a rural economy literally from scratch. Slopes were cultivated, wetlands were reclaimed, rocky wastelands were scratched, deserts, saline lands and even dry river-beds were brought under the plough, *albeit* mostly illegally. Shifting dependence towards subsistence-level agriculture, growing political disenchantment of the rural masses and a run-down State delivery mechanism finally forced a re-think on agriculture policies and investments.

Since canal-irrigation was not an option for this vast rural population, the State devised the watershed development programme (WDP) – a more decentralised, micro-irrigation based agriculture system for the masses. At a fraction of the costs for any major irrigation project, the State was able to contain the growing disenchantment of the masses by providing them with a "sense" of ownership of the local natural resources as well as public funds. NGOs too loved the opportunity to decide on a share of public funds while nursing their constituencies of rural communities and show-casing their 'models' of development and poverty reduction.

The WDP was also an answer to the environmentalists' growing concerns over issues such as excessive use of fertilisers and pesticides, soil degradation due to over-irrigation (water-logging) and monoculture of high yielding varieties. It advocated the ecologically sound "ridge to valley" approach, promoted small check-dams and highlighted conservation as a stated objective. In reality, however, the WDP proved to be more of a sop to millions of small and medium farmers who felt they had missed the mega-irrigation and "green revolution" bus. The ridges were hardly touched, ostensibly due to land ownership by the Forest Department, where yet another related programme for Joint Forest Management (JFM) remained largely on paper. The recent thrust on agriculture extension through the Watershed Plus programme reinforces the notion that what we are witnessing is yet another round of technology infusion in irrigation and agriculture, *albeit* at a much wider level.

Emerging Issues in Rural India

The main environmental threat to sustained agricultural productivity in India is the growing alkalinity (and salinity) of waterlogged soils in the major irrigated croplands, particularly in the Indo-Gangetic plains. However, this problem is usually associated with relatively affluent farmers who have used excess water to grow cash crops. It is therefore more of a private concern that is linked to individual profits rather than a societal concern over future food security. Government programmes are generally for technical assistance that leverage private infusion of resources for the purpose. With increasing involvement of the corporate sector in farming activities in these areas, it is likely that substantial resources would indeed be mobilised and the issue would be treated solely as a private problem.

The scope of corporate involvement is likely to remain limited for quite some time in most of the non-irrigated croplands, where government programmes will continue to play a major role. Soils degraded by improper management (excess of water, fertiliser or pesticides) in these croplands is therefore likely to be interpreted more as a programmatic failure rather than as a more rational desire of the small and marginal farmer to derive short-term benefits before shifting to other occupations. Some of these lands would eventually convert into regular irrigated agricultural lands, while some others would be used for non-agricultural purposes.

The bulk of dry lands, however, may not be used for either intensive agriculture (simply for ecological reasons) or converted totally to non-farm uses (for want of policy support, at least in the short term). This is where indigenous environment management skills would indeed be tested and enhanced in order to provide sustainable economic dividends. A broad basket of such skills would include organic cultivation, dry land agricultural practices, and integration of diverse (but locally appropriate) farm-level interventions such as horticulture, floriculture, cultivation of medicinal plants, sericulture, bee keeping, animal husbandry, aquaculture etc. It may be interesting to note that the promotion of innovative, environmentally sound, farm-based livelihood practices are indeed succeeding in areas where conscious efforts are being made to use local conditions and skills and link the produce with niche markets, rather than merely appeal to the logic of sustainable yield.

Managing diversity, particularly of biological resources in indigenous ecosystems, is indeed the single most formidable challenge in India. However, with greater individual stakes and consequent infusion of knowledge and other resources, the farm sector is set to diversify and optimise its productivity vis-à-vis its ecological milieu, thereby achieving the broad landscape level of biological diversity.

The species (and genetic) levels of biological diversity are less well understood but more talked about. Therefore, the precautionary principle is being used to deny rights to local communities, deter corporate involvement and deny infrastructure development in forests, wetlands and other areas demarcated as "protected areas" or as "fragile ecosystems". The government has taken upon itself the task of protecting nearly 6 per cent of biological species found worldwide (which also provide the indigenous gene pool that is supposed to protect our national food production systems). As a first step, it has already committed itself to increase the proportion of land notified as "forest area" from a quarter to a third of the country's territory. Despite the sops of eco-development projects, the cycle of dispossession, alienation and conflict is set to continue for the most marginalised communities of forest dwellers and grazers.

Poverty, Pollution and Disconnected Laws

Pollution is, in general, a broad societal concern. In a developing economy, however, the poor seem to be disproportionately afflicted by it. The living and working spaces of poorer communities are particularly vulnerable to the scourges of environmental degradation and pollution, often having a direct bearing on their health. Quality of drinking water, sanitation and garbage are universal problems of the poor, be it the rural settlements or the peri-urban slums.

Similarly, the work environment of the poor – be it the small-scale industry, a quarry or even a household hearth – remain the most polluted zones. Pollution in the work environment is generally dealt with as an occupational health issue in India, usually as an appendage to an elaborate set of laws that regulate almost every aspect of industry.

The first global conference on environment was held at Stockholm in 1972 and was attended by the heads of only two nations – India and hosts Sweden. The shared zeal and vision unfortunately failed to consider the contextual differences in developing and developed economies. Problems of drinking water quality and sanitation were unknown to the western world and stringent laws for occupational health issues were already in place. The western world-view on environmental problems, therefore, focussed on industrial emissions and discharge that affected people not related to the cause of pollution.

This form of pollution was, at best, only a marginal concern in India, when stringent laws were introduced and an elaborate enforcement system was created in the late '70s and early '80s. The disconnect was not just in the focus of legislative actions but also in their content. The laws, largely adapted from the USEPA for industrial pollution standards or from the European laws for vehicle emission standards, were idealistic rather than realistic, and the State seldom had the resources to implement these laws or absorb the cost of implementing them. The inevitable result was a widening gulf between what was desired and what was achieved.

Pollution control in India was, thus, faced with a strange set of issues because of this gap. Large manufacturing units were mostly under State control, which either used taxpayers' money to install costly pollution control equipment or eluded any individual accountability, which was essential for any prosecution. Smaller manufacturing units, usually in the private domain and the real polluting ones, were too numerous for any effective monitoring of compliance. Prosecution of defaulters is usually under criminal laws, which is both difficult and drawn out in our legal system. Direct actions, like disconnection of power and water supplies, have been allowed more recently, but even then enforcements have proved to be difficult to monitor.

Ineffective laws tend to get trivialised, both by the law enforcer as well as the intended law abider. Institutional redundancy breeds complacency and inefficiency. Other power structures then intervene in important matters, which helps these institutions to further abdicate their responsibilities.

Development, Rights and Civil Actions

In general, the environmental concerns of civil society were, on the other hand, much broader and more apt. When the State was busy importing (and imposing) environmental solutions in the early '70s, the home-grown "Chipko" (hugging a tree) movement was being resurrected by communities in the Himalayan foothills to prevent felling of trees. This resulted in a ban on felling in the forest areas of several States.

Many forest areas, however, had already been degraded, where, rather than felling, the real issues pertained to regeneration of vegetation, along with retention of top-soil, recharge of groundwater and conservation of wild species. Unlike urban and other external economies that tend to primarily value only the timber, local communities are largely dependent on a variety of goods and services offered by a forest ecosystem. Rigid controls on resource extraction, therefore, severely affected traditional livelihoods, particularly of the poorer communities, rather than the large companies who worked out other supply sources in bamboo and plantations.

The first models of community-managed watersheds emerged in the dry parts of Maharashtra during the mid '70s. Ralegaon Siddhi proved that communities could take charge of their lives even under extreme scarcity of resources. Other initiatives like the Gokul project became the precursor of highly successful initiatives of the Indo-German watershed projects. Poor communities in Mendha-Lekha (Gadchiroli) attempted the regeneration of highly degraded forests. The success of such Joint Forest Management (JFM) initiatives was evident when the communities in Sukho Majri actually paid an income tax on their incomes from the forests.

The issue of water rights was seriously raised through the Narmada movement in the '80s, to be later joined by the Tehri movement. The notion of "sacrifice for the greater common good" was seriously questioned, and unprecedented questions were raised on environment and social issues. Land rights, riparian rights and environmental impacts were discussed seriously for the first time, eventually leading to fundamental changes in the policies of government and multi-lateral funding of large irrigation projects.

Pollution issues gained prominence among civil society in the aftermath of the Bhopal gas tragedy in 1984 that killed over 3000 people and maimed many more thousands for life. Notwithstanding the shortcomings of our legal system, the episode spawned considerable interest in environmental laws among legal practitioners. The judiciary, particularly the apex institution, adopted a much more active role thereafter in moving *suo moto* petitions on environment and admitting class suits as public interest litigations (PILs).

While the executive activism of the '70s did lead to some of the most prominent environmental actions in India (like the scrapping of the Silent Valley project, promulgation of the coastal zone regulations and the pollution control acts), judicial activism opened the doors for civil society to participate in the environmental debate. Support from the judiciary led to several successful campaigns such as those launched to protect the Chilika lake of Orissa and other coastal areas from industrial aquaculture. Subsequently, public hearings on environmental issues were introduced, which provided further opportunities for environmental action. Civil society groups working in the industrial belts of south Gujarat, for example, used the option effectively to prevent discharge of untreated effluents into the river Narmada.

Environmental awareness and education played a very significant role in this entire process. Nature clubs, initiated in India in 1976 by the WWF (now the Worldwide Fund for Nature), and the writings of Salim Ali and other naturalists in the Bombay Natural History Society (BNHS) drew the interest of young minds towards the beauty of nature. But the real breakthrough came with the landmark publication of the Citizen's Reports of the Centre for Science and Environment (CSE) in the early '80s. It was the first effort to draw popular attention to the real environmental crisis in India, refreshingly different from the official confusion on the subject. Finally, an effective supplement to the efforts of CSE was provided by a young

Fulbright and, later, Ford Foundation scholar – Armin Rosencranz – who compiled and commented on the entire gamut of environmental legislation in the early '90s, which has since become the leading text in environmental laws in India.

Corollary

It is evident that environmental concerns have seen a major transformation during the past three decades. Emerging from an almost esoteric concern for wildlife, driven largely by aesthetics, it moved on to a passionate appeal for the protection of traditional peoples and their livelihood. Leading frugal lives modelled on traditional lifestyles was touted as the true solution, to be implemented by appealing to the collective guilt feeling of the human race for its perverse domination over other living creatures of the planet Earth. Only, it doesn't seem to work. The latest is "sustainable development", an attempt to integrate environmental concerns in regular economic activity at all levels.

Ironically, notwithstanding their inherent disparate nature, all three approaches are mainstream environmental approaches in India today. Although the advocates of a pristine, human-free environment are shrinking (much like the last bastions of wildlife territories), they continue to play an important role in influencing decisions on the conservation of rare and endangered organisms.

The second approach of "traditional" development (emphasising social, rather than economic basis of decision-making) is perhaps the dominant view, since it gels well with political expediency, judicial activism and bureaucratic controls. It remains the preferred approach for government and civil society organisations in policy and investment decisions pertaining to the majority of rural India. However, it is increasingly getting apparent that both these approaches work against the stated interests of poverty alleviation either by creating conflicts ("man and wildlife") or by providing incentives to misuse ("the tragedy of commons").

There is, therefore, a growing realisation of the importance of secure property rights, commercial delivery of public services and economic instruments in environment management. The draft environment policy of the Government of India is a clear indication of this trend. Not only does it seek to use "market-based instruments" in pollution control, it actually proposes improved property rights for the marginalised poor dependent on forest and grassland resources. This approach to environment management actually emphasises economic development and views environmental resources as important inputs to the process which need to be owned and managed by those who use them. The tenets of this new paradigm will be based on

(i) use, rather than non-use (or future use), of environmental resources;

(ii) economic, rather than social (and cultural), valuation of environmental goods and services; and

(iii) individual, rather than societal (or governmental), custody of environmental resources.

Appocalypse Now!

An assortment of quotes typifying doomsday environmentalism, showing that even the greatest of minds can underestimate the 'ultimate resource'...

Pollution

The air is polluted, the rivers and lakes are dying, and the ozone layer has holes in it.

> — Ann Landers, Washington Post. April 16, 1989

Hundreds of millions of people will soon perish in smog disasters in New York and Los Angeles...the oceans will die of DDT poisoning by 1979...the U.S life expectancy will drop to 42 years by 1980 due to cancer epidemics.

> — Paul Ehrlich, 1969 in Ramparts magazine. (This one can be seen as a warning - not a flat prediction. Note the precision of the "42 years")

It seems unlikely that much improvement can be expected in this aspect of air pollution until a major shift in our economy takes place. As long as we have an automobile industry centred on the internal combustion engine and a social system which values large, overpowered cars as status symbols, we are likely to be in trouble.

> — Paul Ehrlich, The Population Bomb, Buccaneer Books, Cutchogue, New York, 1968, 1971. p. 103.

Climate change: Blow hot, blow cold....

Global Cooling...

The cooling has already killed hundreds of thousands of people in poor nations. It has already made food and fuel more precious, thus increasing the price of everything we buy. If it continues, and no strong measures are taken to deal with it, the cooling will cause world famine, world chaos, and probably world war, and this could all come by the year 2000.

> — Lowell Ponte, The Cooling, 1976.

The facts have emerged, in recent years and months, from research into past ice ages. They imply that the threat of a new ice age must now stand alongside nuclear war as a likely source of wholesale death and misery for mankind.

> — Nigel Calder, former editor of New Scientist and producer of scientific television documentaries, "In the Grip of a New Ice Age," International Wildlife, July 1975.

At this point, the world's climatologists are agreed...Once the freeze starts, it will be too late.

> — Douglas Colligan, "Brace Yourself for Another Ice Age," Science Digest, February 1973.

Global Warming...

We are not, of course, optimistic about our chances of success. Some form of ecocatastrophe, if not thermonuclear war, seems almost certain to overtake us before the end of the century (The inability to forecast exactly which one – whether plague, famine, the poisoning of the oceans, drastic climatic change, or some disaster entirely unforeseen – is hardly grounds for complacency).

> — John Holdren and Paul Ehrlich in Global Ecology, Harcourt Brace Jovanovich, New York, 1971 p. 279.

As University of California physicist John Holdren has said, it is possible that carbon-dioxide climate-induced famines could kill as many as a billion people before the year 2020.

> — Paul Ehrlich, The Machinery of Nature, Simon & Schuster, New York, 1986, p. 274.

Whether a global compact to reduce CO_2 emissions can be drafted in the near future may depend largely on the severity, over the next year or two, of weather anomalies plausibly attributable to global climate change.

> — John Holdren, 'The Transition to Costlier Energy', in Lee Schipper, Stephen Meyers *et al*, Energy Efficiency and Human Activity: Past, Trends, Future Prospects. Cambridge University Press, Cambridge. 1992. p. 42.

In the long run CO_2-induced warming would melt the polar ice caps, thus flooding many areas, putting New York City, Washington

D.C, and Sacramento, California, under water. Much of eastern England and the low countries would be inundated, and millions of people would be forced by rising waters out of the Tokyo-Yokohama area of Japan. Bangladesh would become submerged, and Rangoon, Calcutta, and Buenos Aires would be gone. Florida would largely disappear, and much of the lower Mississippi River Valley would become an inland sea.

> — Robert Ornstein and Paul Ehrlich, New World New Mind: Moving Toward Conscious Evolution, Doubleday, New York,1989. pp.77-78.

Power and Energy

There is not the slightest indication that (nuclear) energy will ever be obtainable. It would mean that the atom would have to be shattered at will.

> — Albert Einstein, 1932.

The energy produced by the atom is a very poor kind of thing. Anyone who expects a source of power from the transformation of these atoms is talking moonshine.

> — Ernest Rutherford, 1933; after being the first person to split the atom.

Giving society cheap and abundant energy at this point would be equivalent to giving an idiot child a machine gun.

> — Paul Ehrlich, "An Ecologist's Perspective on Nuclear Power", May/June 1978 issue of Federation of American Scientists Public Issue Report.

Many of the conservation measures temporarily undertaken when the mini-crisis was in its acute stage- lowered speed limits, car-pools, reset thermostats, etc... should be instituted on a permanent basis...In the long run, energy should be made expensive, especially for large users, as an incentive to conservation.

> — Paul and Anne Ehrlich, The End of Affluence, Rivercity Press, Riverside, MA, 1974. p.48.

Except in special circumstances, all construction of power generating facilities should cease immediately, and power companies should be forbidden to encourage people to use more power. *Power is much too cheap.* It should certainly be made more expensive and perhaps rationed, in order to reduce its frivolous use.

> — Richard Harriman and Paul Ehrlich, How to Be a Survivor, Rivercity Press, Rivercity, MA, 1975. p.72.

Conservation: Policy and Predictions

Most environmental, economic and social problems of local, regional, and global scale arise from this driving force: too many people using too many resources at too fast a rate.

> — "Blue Planet: Now or Never— a statement" Blue Planet Group, September 1991, Ottawa, Canada.

Only one rational path is open to us – simultaneous 'dedevelopment' of the overdeveloped countries and 'semidevelopment' of the underdeveloped countries, in order to approach a decent and ecologically sustainable standard of living for all in between. By 'dedevelopment' we mean lower per-capita energy consumption, fewer gadgets, and the abolition of planned obsolescence.

> — John Holdren and Paul Ehrlich, in Global Ecology, Harcourt Brace Jovanovich, New York, 1971, p. 3.

It seems clear that the first major penalty man will have to pay for his rapid consumption of the earth's non-renewable resources will be that of having to live in a world where his thoughts and actions are ever more strongly limited, where social organisation has become all pervasive, complex, and inflexible, and where the state completely dominates the actions of the individual.

> — Harrison Brown (1954), quoted in Paul Ehrlich, Anne Ehrlich and John Holdren, Ecoscience: Population, Resources, and Environment. W.H. Freeman, San Francisco. 1977. p. 388.

We must reclaim the roads and the ploughed land, halt dam construction, tear down existing dams, free shackled rivers, and return to wilderness tens of millions of [acres of] presently settled land.

> — David Foreman, Founder of 'Earth First!'

Scientists who work for nuclear power or nuclear energy have sold their soul to the devil. They are either dumb, stupid, or highly compromised.... Free enterprise really means rich people get richer. And they have the freedom to exploit and psychologically rape their fellow human beings in the process.... Capitalism is destroying the earth. Cuba is a wonderful country. What Castro's done is superb.

> — Helen Caldicott, Australian pediatrician, speaking for the Union of Concerned Scientists (as quoted by Elizabeth Whelan in her book Toxic Terror).

The planet is about to break out with fever, indeed it may already have, and we [human beings] are the disease. We should be at war with ourselves and our lifestyles.

> — Thomas Lovejoy, tropical biologist and assistant secretary to the Smithsonian Institution (quoted by David Brooks in The Wall Street Journal article, "Journalists and Others for Saving the Planet", 1989).

Economists as a group have been guiltier than most in perpetuating the most dangerous myths of this troubled age . . . Mineral economists rely on the cornucopian dream, in which advancing technology conjures up ever cheaper minerals while consuming ever increasing amounts of energy and the earth's crust to do it.

> — John Holdren and Paul Ehrlich in Global Ecology, Harcourt Brace Jovanovich, New York, 1971. p. 177.

Technology and the Quality of Life

We've already had too much economic growth in the US. Economic growth in rich countries like ours is the disease, not the cure.

> — Paul Ehrlich

The abolishment of pain in surgery is a chimera. It is absurd to go on seeking it. . . . Knife and pain are two words in surgery that must forever be associated in the consciousness of the patient.

> — Dr. Alfred Velpeau (1839) French surgeon.

When the Paris Exhibition closes electric light will close with it and no more be heard of.

> — Erasmus Wilson (1878) Professor at Oxford University.

The battle to feed all of humanity is over. In the 1970's and 1980's hundreds of millions of people will starve to death in spite of any crash programmes embarked upon now.

> — Paul Ehrlich in the beginning of his 1968 book The Population Bomb.

That the automobile has practically reached the limit of its development is suggested by the fact that during the past year no improvements of a radical nature have been introduced.

> — Scientific American, Jan. 2, 1909.

Radio has no future.

> — Lord Kelvin, ca. 1897.

But what... is it good for?

> — Engineer at the Advanced Computing Systems
> Division of IBM, 1968, commenting on the microchip.

Sources

All chapters are excerpts or edited versions of the following:

Chapter 1

Goklany, Indur M. 2002. *The Globalisation of Human Well-Being.* Policy Analysis No. 447, August 22. Cato Institute. Washington D.C.

Chapter 2

Simon, Julian. 1996. Population Growth, Natural Resources, and Future Generations. Chapter 28 in: Simon, Julian. 1996. *The Ultimate Resource 2.* Princeton University Press. New Jersey.

Chapter 3

Smith, Jr., Fred L. 1993. *The Market and Nature.* The Freeman. September. The Foundation for Economic Education. New York.

Chapter 4

Meiners, Roger E, and Bruce Yandle. 1998. *The Common Law: How it Protects the Environment.* PERC Policy Series Issue Number PS-13. Property and Environment Research Centre. Montana.

Chapter 5

Simmons, Randy T. and John Baden. 1984. *The Theory of New Resource Economics.* Journal of Contemporary Studies. Spring. pp 45-52.

Chapter 6

Morris, Julian. 2002. Reconceptualising Sustainable Development. Chapter 1 in:

Morris, Julian (Ed). 2002. *Sustainable Development: Promoting Progress or Perpetuating Poverty?* Profile Books. London

Chapter 7

Ostrom, Elinor. 1999. *Self-Governance And Forest Resources.* CIFOR Occasional Paper Number 20. February. Bogor, Indonesia.

Chapter 8

Sengupta, Nirmal. 1996. *Common Pool Resources & Indian Legal System.* Revised and edited version of paper originally presented at the Conference on Law and Economics, Indian Statistical Institute, New Delhi, Januray 11-13, 1996.

Chapter 9

Parekh, Trupti and Parth J. Shah. 2003. *Keepers of Forests: Foresters or Forest Dwellers?* CCS and ARCH briefing paper presented at: Keepers of Forests:

Foresters or Forest Dwellers? Conference held in Delhi, 28-29 March 2003. Centre for Civil Society and ARCH. Delhi.

Chapter 10

Powell, Ian; Andy White and Natasha Landell-Mills. 2002. *Developing Markets For The Ecosystem Services of Forests.* Forest Trends. Washington D.C.

Chapter 11

Smith, Robert J. 1982. *Resolving The Tragedy of The Commons By Creating Private Property Rights In Wildlife.* Cato Journal Vol. 1, No.2. Cato Institute. Washington D.C.

Chapter 12

Shikwati, James. 2003. *How To Protect People and Wildlife in Kenya: Right Now, Locals Are Out of The Loop.* PERC Report. March. Property and Environment Research Centre. Montana.

Chapter 13

Shah, Parth J. and Ambrish Mehta. 2003. *Managing Water Resources: Communities & Markets.* CCS briefing paper presented at: Conference on Water under Community Management, held in Delhi, March 24 2003. Centre for Civil Society. Delhi.

Chapter 14

Bate, Roger. 2002. Water – Can Property Rights and Markets Replace Conflict? Chapter 15 in: Morris, Julian (Ed). 2002. *Sustainable Development: Promoting Progress or Perpetuating Poverty?* Profile Books. London.

Chapter 15

De Alessi, Michael. 2002. Sustainable Development and Marine Fisheries. Chapter 13 in: Morris, Julian (Ed). 2002. *Sustainable Development: Promoting Progress or Perpetuating Poverty?* Profile Books. London.

Chapter 16

Gissurarson, Hannes H. 2000. *Overfishing: The Icelandic Solution.* Institute of Economic Affairs. London

Chapter 17

Goklany, Indur M. 2001. Escaping Goblins, Only To Be Captured By Wolves? Chapter 1 in: Goklany, Indur M. 2001. *The Precautionary Principle: A Critical Appraisal of Environmental Risk Assessment.* Cato Institute. Washington D.C.

Chapter 18

Balling, Jr., Robert C. 2002. A Climate of Uncertainty in the Greenhouse Century. Chapter 9 in: Morris, Julian (Ed). 2002. *Sustainable Development: Promoting Progress or Perpetuating Poverty?* Profile Books. London.

Chapter 19

Prakash, C.S. and Gregory Conko. 2002. The Attack on Plant Biotechnology. Chapter 7 in: Ronald Bailey (Ed). 2002. *Global Warming and Other Eco-Myths.* Prima Publishing-Random House.

Chapter 20

Simon, Julian. 1996. Can the Supply of Natural Resources - Especially Energy - Really Be Infinite? Chapter 3 in: Simon, Julian. 1996. *The Ultimate Resource 2*. Princeton University Press. New Jersey.

Chapter 21

Morris, Julian. 1998. *Law, Markets and Waste.* Paper presented to the European Waste Club/World Resource Foundation Seminar, 'Economic Instruments and Waste Management', 15th July 1998. London.

Chapter 22

Simon, Julian. 1996. Coercive Recycling, Forced Conservation, and Free Market Alternatives. Chapter 21 in: Simon, Julian. 1996. *The Ultimate Resource 2*. Princeton University Press. New Jersey.

Chapter 23

Bhagwati, Jagdish. 2000. On Thinking Clearly About the Linkage Between Trade and Environment. In: Bhagwati, Jagdish. 2000. *The Wind of the Hundred Days: How Washington Mismanaged Globalisation*. MIT Press. Cambridge, MA.

Chapter 24

Okonski, Kendra. 2003. *Freedom to Trade Protects the Environment*. Environment and Trade Briefing Paper. August. International Policy Network. London.

Chapter 25

Lal, Deepak. 2000. *The New Cultural Imperialism: The Greens and Economic Development*. The Inaugural Julian L. Simon Memorial Lecture. 9 December 2000, Liberty Institute. Delhi.